Gum Arabic

Gum Arabic

Structure, Properties, Application and Economics

Edited by

Abdalbasit Adam Mariod

ELSEVIER

ACADEMIC PRESS

An imprint of Elsevier

Academic Press is an imprint of Elsevier
125 London Wall, London EC2Y 5AS, United Kingdom
525 B Street, Suite 1800, San Diego, CA 92101-4495, United States
50 Hampshire Street, 5th Floor, Cambridge, MA 02139, United States
The Boulevard, Langford Lane, Kidlington, Oxford OX5 1GB, United Kingdom

Library of Congress Cataloging-in-Publication Data
A catalog record for this book is available from the Library of Congress

British Library Cataloguing-in-Publication Data
A catalogue record for this book is available from the British Library

ISBN: 978-0-12-812002-6

For information on all Academic Press publications visit our website at https://www.elsevier.com/books-and-journals

Publisher: Andre Gerhard Wolff
Acquisition Editor: Nancy Maragioglio
Editorial Project Manager: Amy M. Clark
Production Project Manager: Maria Bernard
Designer: Vicky Pearson Esser

Typeset by Thomson Digital

Contents

Part III—Medical and Health Uses

List of Contributors

Nawal Abdel-Gayoum Abdel-Rahman National Food Research Centre, Khartoum North, Sudan

Hassan H. Abuelhassan Al Neelain University, Khartoum, Sudan

Sara A. Ahmed University of Khartoum, Khartoum North, Sudan

Abdelkareem A. Ahmed University of Nyala, Nyala, Sudan

Rasheeda Hamid Abdallah Ahmed Prince Sattam Bin Abdulaziz University, Al-Kharj, Kingdom of Saudi Arabia

Insaf Abdel Karim Elhag Ali King Faisal University, Al-Ahsa, Saudi Arabia

Yousef A. Alsabah Al Neelain University, Khartoum, Sudan; Hajjah University, Hajjah, Yemen

Abdelaziz E. Atabani Erciyes University, Kayseri, Turkey

Mahiran Basri University Putra Malaysia, Serdang, Malaysia

B.I. Beelly Osman Blue Nile University, Sudan

Taisser H.H. Deafalla Institute of Photogrammetry and Remote Sensing, TU Dresden, Germany

Hatim M.A. Elamin Institute of Gum Arabic Research & Desertification Studies, University of Kordofan, Sudan

Abdelrahman A. Elbadawi Al Neelain University, Khartoum, Sudan; Future University, Khartoum, Sudan

Abdel Moneim El-Hadi Sulieman University of Hail, Kingdom of Saudi Arabia

Khalid H. El-Hag Africa City of Technology, Khartoum, Sudan

Elhadi M.I. Elzain Taif University, Taif, Kingdom of Saudi Arabia; Kordofan University, Elobeid, Sudan

Abdelaziz A. Fadlelmoula University of Jeddah, Alkamel, Kingdom of Saudi Arabia; University of Khartoum, Shambat, Sudan

Hanan S. Fagury Africa City of Technology, Khartoum, Sudan

Abdul Rahman FarajAllah Batterjee Medical College of Science & Technology, Jeddah, Saudi Arabia

Imad Fdl-Elmula Assafa Academy, Khartoum, Sudan

Mai Ahmed Hamed Africa City of Technology, Khartoum, Sudan

Zeinab Mohamed Hammad Institute of Gum Arabic Research and Desertification Studies, University of Kordofan, Sudan; University of Kordofan, Sudan

Elfatih A. Hassan University of Khartoum; Sudan University of Science and Technology, Khartoum, Sudan

I.A. Mofadel Hassan Kenana Sugar Company Limited, Khartoum, Sudan

M.A. Hussein International University of Africa, Khartoum, Sudan

Inaam Awad Ismail National Food Research Centre, Khartoum North, Sudan

Lamis Kaddam AlNeelain University, Khartoum, Sudan

Manal A. Kheiry University of Khartoum, Khartoum North, Sudan

E.H. Konozy Africa City of Technology, Khartoum, Sudan

T.E. Mahmoud University of Kordofan, Elobeid, Sudan

Abdalbasit A. Mariod University of Jeddah, Alkamil, Kingdom of Saudi Arabia; Sudan University of Science & Technology, Shambat, Sudan; Ghibaish College of Science & Technology, Ghibaish, Sudan

Mohamed A. Salih Africa City of Technology, Khartoum, Sudan

Noura Karamalla Mohamed Salih University of Bahri, Khartoum Bahri, Sudan; Infrastructure University Kuala Lumpur, Kuala Lumpur, Malaysia

Hagir Mahagoub El Nasri Mohammed University of Khartoum, Khartoum, Sudan

Mohammed Hamed Mohammed University of Kordofan, Sudan

Muna Mohamed Mustafa National Food Research Centre, Khartoum North, Sudan

Olatunji Ololade University of Lagos, Akoka, Nigeria

Mohamed E. Osman The Gum Arabic Company Ltd, Khartoum, Sudan

Jürgen Pretzsch TU Dresden, Germany

A.A. Rabah University of Khartoum, Khartoum, Sudan

Osama A. Rayis Africa City of Technology, Khartoum, Sudan

Amal Saeed University of Khartoum, Khartoum, Sudan

Suzy Munir Salama University of Malaya, Kuala Lumpur, Malaysia

Sulieman Mohamed El Sanousi University of Khartoum, Khartoum, Sudan

Awad, S. S. Industrial Research and Consultancy Center, Khartoum, Sudan

Muneer E. Siddig Institute of Gum Arabic Research & Desertification Studies, University of Kordofan, Sudan

Mohamed El Nour Taha Kordofan University, Sudan

Mohamed A. Talib Africa City of Technology, Khartoum, Sudan

Isam Abu Zeid King Abdulaziz University, Jeddah, Saudi Arabia

Preface

Gum Arabic (GA), or acacia gum, is a tree gum exudate and has been an important part of commerce since ancient times. Traditionally, the gum has been obtained mainly from the *Acacia senegal* and *Acacia seyal* species. The trees grow widely across the Sahel belt of Africa situated in the north of the equator up to the Sahara Desert and from Senegal in the west to Somalia in the east.

The backbone of GA is made out of 1,3-linked β-D-galactopyranosyl units. Moisture content of GA falls within the range (13%–15%). The protein content of GA ranges from 1.5% to 3.0% from main principal amino acids. The mineral content of the GA are Ca, Mg, Na, K, Fe, P, and nondetectable traces of Pb, Co, Cu, Zn, Ni, Cd, Cr, and Mn. The variation in chemical composition and properties of GA is mainly due to genotypic differences, age, climatic conditions, soil, and plant pathogens. GA is used as an emulsifier and a thickening agent in foods. It has been applied to relieve the effects of chronic renal failure, and it is a good antioxidant, which protects against cardiac and hepatic toxicities. Although there have been articles in journals and chapters in books that have described some chemistry, structure, and health issues of GA, a book devoted entirely to its structure, chemistry, food and feed applications, and medical and health uses has not previously been available.

This book presents three parts divided into 26 chapters on GA. Part 1 presents an introduction to agriculture and economics of GA and is focused mainly on ecology and economic contribution, management practices, and certification and assessment of marketing opportunities of GA-producing trees. This part is covered by five chapters. Part 2 consists of 10 chapters investigating the chemical, physical, and electronic properties of GA; new trends in GA methods of analysis; thermal characteristics of acacia gum; processing of a highly purified grade of GA from gum dust; processing and modification of GA in specific applications; enhancement of color stability in foods by GA; GA as a thickener and stabilizing agent in dairy products; effects of GA (prebiotic) on physicochemical and organoleptic properties of yogurt (probiotic); and extending cloud stability of juice using GA during storage in the refrigerator. Part 3 describes its health benefits, for example, GA as dietary animal feed, treatment for skin lesions, and dietary fiber in addition to its medical uses, such as treatment against kidney failure and the clinical applications and effect of GA on the fecal bacterial mass on healthy human volunteers. This part contains 11 chapters.

This book has been written to ensure that it will be beneficial to industry, medicine, and food scientists. In addition, the book should be useful to academic scientists who require a good source of applications and references. As gums have diverse uses, it is hoped that the use of the book will not be limited to the food and medicine industry, but will extend to related industries, such as the pharmaceutical industry.

Part I

Agricultural and Economics

Acacia Seyal Gums in Sudan: Ecology and Economic Contribution

1

S.S. Awad, A.A. Rabah**, H.I. Ali†, T.E. Mahmoud‡*
*Industrial Research and Consultancy Center, Khartoum, Sudan; **University of Khartoum, Khartoum, Sudan; †Kenana Sugar Company Limited, Khartoum, Sudan; ‡University of Kordofan, Elobeid, Sudan

Introduction

Acacia gum is a natural agricultural resource from the gum belt region of Africa, i.e., countries geographically ranging from east to west: from Sudan, Somalia, Eritrea, and Ethiopia to Chad, Central African Republic, Mali, Niger, and farther west up to Nigeria, Senegal, and even Mauritania (Mohammed, 2011). Economically speaking, acacia gum mainly comes from Sudan, Chad, and Nigeria.

In Sudan, the most important of nonwood forest products is gum Arabic, which exudates from *Acacia senegal* and locally known as hashab, or hard gum, and *Acacia seyal*, which is known as talha, or friable gum. Both species spread naturally in the central belt of the low-rainfall savannah, where they exist in pure or mixed stands, in the clay plains in the East, and the sandy soils in the West (ElKhawad, 2008).

Despite the significant contribution of *A. seyal* to the exports market in the Sudan, farmers have a slightly poor knowledge of postharvest cultural practices of *A. seyal*, which include proper methods of storage conditions and gum stacking. Nevertheless, this chapter is trying to review the existing research in the areas of pre- and postharvest cultural practices, including tree husbandry, tapping, gum collection, handling, and storage. Physical and chemical properties of *A. seyal* establish a clear protocol of the *A. seyal* tree in the Sudan.

Acacia seyal tree

A. seyal trees are up to 17 m tall in Sudan, with a flat top crown. They have a distinctive smooth, powdery bark, which varies from white to greenish-yellow or orange-red, with a green layer beneath. In some populations, both red- and yellow-barked trees can be found. There are two varieties, differing primarily in whether pseudo-galls (ant galls) develop and in bark color. In *A. seyal* var. *seyal*, there are no pseudo-galls, and a reddish bark color prevails, although periodic bark exfoliation exposes a pale, powdery surface, which darkens slowly. In *A. seyal* var. *fistula*, pseudo-galls are present, and the powdery bark typically remains whitish or greenish-yellow.

Gum Arabic. http://dx.doi.org/10.1016/B978-0-12-812002-6.00001-4

Acacia seyal distribution

In general, there are two main varieties of *A. seyal*: variety *seyal* and variety *fistula*. Variety *seyal* is found in both western and eastern Africa and also on the Arabian Peninsula, while variety *fistula* is found in the eastern parts of Africa; this indicates that variety *seyal* is native to northern tropical Africa and Egypt. The two varieties can be easily distinguished; variety *seyal* has a greenish-yellow to reddish-brown bark, while variety *fistula* has white to greenish-yellow bark. Fig. 1.1 shows the distribution of *A. seyal* varieties with respect to rainfall. However, *seyal* trees are naturally grown in the Sudan as long as there has been no reforestation done by humans (Mohammed, 2011).

Acacia seyal ecology

Location

In Sudan, the clay plain is a vast stretch of flat land extending over the eastern part of the country. The White Nile forms a natural boundary to the west, but south of 13°N, the clay plain protrudes farther westward to enclose South Kordofan. Its northern limit corresponds closely to the upper edge of the Sahel, while the southern limit lies within the Guinean zones (Mustafa, 2006). The central clay plain is situated between latitudes 10 and 16°N and longitudes 32 and 37°E.

Soil

The parent materials of the clay plain soils belong to two broad groups: alluvial, deltaic, and paludal sediments from rivers belonging to the Nile system and colluvial–alluvial

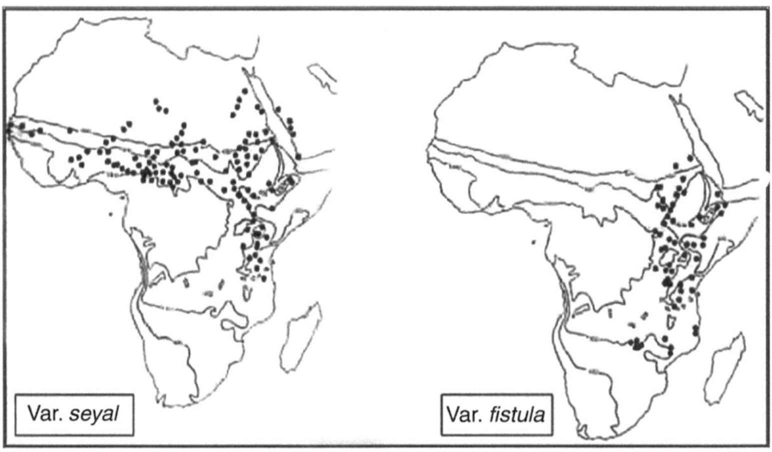

Figure 1.1 Distribution of *Acacia seyal* varieties in Africa with respect to rainfall.

deposits derived from local rock weathering (Balal et al., 2014). Soils dry out and crack during the dry months to restore their permeability; they are poor in organic matter and low in alkaline, nitrogen, and phosphorus percentages. Both traditional and mechanized rain-fed agriculture is practiced in this region to grow sorghum (*Sorghum bicolor*), sesame (*Sesamum orientale*), peanuts (*Arachis hypogaea*), and cotton (*Bombax costatum*). The southern part of the clay plain receives heavy rains during the rainy season and hence is inundated for 4 to 6 months. In general, this area is a region of grasses rather than agricultural crops.

Rainfall

The climate is subtropical with arid characteristics. The annual rainfall increases southeastward from 400 mm to 900 mm, and the rainy season lasts from May to October. However, the rainfall is very erratic in amount and space, and frequencies of cessation could prevail for more than 15 days, which has a paramount impact on the plant establishment. This, in view of the low permeability of the clay soil, shortens the period of sufficient soil moisture for the plant to germinate and grow; only those with high efficiency in absorbing moisture would survive (Balal et al., 2014).

Temperature

Sudan's inherent vulnerability may best be captured by the fact that food security is mainly determined by rainfall, particularly in the rural areas where more than 65% of the population lives. Mean annual temperature lies between 26°C and 32°C, but in some places, it reaches 47°C, causing a lot of stresses and heat-related diseases. Mean annual temperature varies from 18 to 28°C (Orwa et al., 2009).

Acacia seyal husbandry

Most of the farmers in the sand plains areas have adequate awareness of pre-harvest cultural practices of *A. senegal*. This awareness is due to their vast indigenous knowledge of the traditional farming system in addition to the extensive knowledge held by the Forest National Cooperation as a result of a comprehensive research done in the last five decades, which provides a clear protocol for tree husbandry, including tapping of the tree as well as the proper way of collecting gums (Abdel Nour, 1989). However, farmers have a slightly poor knowledge of postharvest cultural practices of *A. seyal*, which include proper methods of storage conditions and gum stacking. Nevertheless, sufficient research has been done in these fields (Abdel Nour, 1989).

Acacia seyal *pre- and postharvest practices*

A protocol for a tree of *A. seyal* is not yet been established, and this is coupled with local farmers' poor indigenous knowledge in the clay plains regarding tree husbandry, tapping, and gum collection. The results of that are inferior quality and poor nodule physical characteristics (GAB, 2007).

Collectors harvest the partially dried gum, and multiple collections up to three times at 3-week intervals from the same tree are possible. A yield of 300–700 g is obtained per tree annually. On the other hand, *A. seyal* is generally collected from natural exudation without tapping. In certain gum sources, the natural exudates were noted as being darker in color compared to gums obtained by tapping (Anderson and Bridgeman, 1985). The method of gum production might be one of the factors causing the difference in color between *A. senegal* and *A. seyal*. Recently, Fadl and Gebauer (2010) reported that the middle stem tapping caused highest *A. seyal* gum exudation. *A. seyal* gum production by tapping might be increased in the future if problems such as serious stress to the tree are avoided (Fadl and Gebauer, 2010). Tapping is often done by using traditional tools, such as the axe, sonki, and makmak, shown in Fig. 1.2. Using the makmak is highly recommended. Fadl reported that Makmak was the best tool for tapping *A. seyal*. The reason can be seen on the wide edge on the top of the tool, which allows better removal of the bark.

Identity and purity of Acacia seyal *gum*

Chemical and physical properties of gum are considered important characteristics for commerce in order to ensure the identity and purity of gum and to avoid mixing and adulteration. Chikamai et al. (1995) illustrated that they were able to distinguish gums from different *Acacia* species as well as suspected contaminants. The analytical method enabled them to differentiate between gums on the basis of optical rotation,

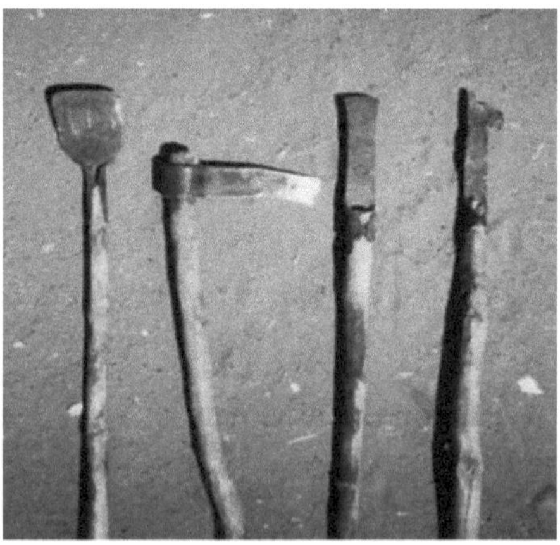

Figure 1.2 Makmak, axe, mohar, and sonkey (from left to right).
Source: Fadl, K.E.M., Gebauer, J., 2010. Effect of different tapping tools and different tapping positions on "talha gum" yield of *Acacia seyal* var. *seyal* in South Kordofan Sudan. In: Kennedy, J.F., Phillips, G.O., Williams, P.A. (Eds.), Gum Arabic. RSC Publishing.

protein content, and viscosity, which were noted as basic parameters for comparison. The method was able to distinguish gums from the other *Acacia* species purely on the basis of optical rotation, though information on nitrogen and sugar composition was also invaluable. Other schools of thought believe that the two parameters emphasized, optical rotation and nitrogen content, are too restrictive and can sometimes result in rejecting otherwise authentic gum Arabic if applied and more so when the analytical method is the only one relied upon. Moisture content gives an indicator for both purity and real weight.

Table 1.1 shows *A. seyal* specifications, whereas Tables 1.2 and 1.3 show the physicochemical properties and cationic composition of *A. seyal*, respectively. These parameters are clearly specified by JECFA (1990).

Table 1.1 **Talha specifications**

Parameter		Range
Viscosity		12 cm/g
Specific optical rotation		+45 to +61
Equivlent weight		1300–1403
pH		3.6–4
Ash content		3.6–4
Nitrogen content		0.147–0.175
Protein content		0.94–1.15
Sugar composition	Galactose %	38
	Arabinose %	46
	Rhamnose %	4
	Glucuronic acid %	6.5

Source: IGADRS, 2015. Production and Value Chain Manual, Institute of Gum Arabic Researches and Desertification Studies (IGARDS).

Table 1.2 **The physiochemical properties of *Acacia seyal* species**

No.	Moisture content	Ash content	pH	S.O.R.	Protein content
1	10.04	3.03	4.77	+51.4	0.96
2	10.08	2.12	4.27	+60.5	0.69
3	9.63	1.92	4.31	+62.8	0.83
4	10.04	1.88	4.43	+57.8	0.85
5	9.63	2.1	4.12	53.9	0.88
6	9.49	2.5	4.31	+59	0.96
7	9.26	2.35	4.02	+56	0.77
8	10.88	2.17	4.46	+62	0.91

S.O.R., Specific Optical Rotation

Source: Obeid, I.S., 2012. Evaluation of Physicochemical, Microbiological and Toxicological Aspects of *Acacia seyal* Gum (Ph.D. thesis in Chemistry). Sudan University of Sciences and Technology.

Table 1.3 Cationic composition of *Acacia seyal*

Elements	*A. seyal*[a] (ppm)	*A. seyal*[b] (μg/g)	*A. seyal*[c] (w/w%)	*A. seyal*[c] (μg/g)
Na	6.54–49.55	111.054	00048–0.03986	4.8125–398.6
K	2400–3558	2802.803	0.1808–0.4605	1807.93–4605.01
Ca	9453–10145	9417.1973	0.60994–6702.5	6093.81–9165.63
Mg	1224–1295	1229.0424	0.136415–0.2159	1364.15–2159
Pb	<0.84	7.5757	0.000228–0.008261	2.275–82.6125
Cr	0.30–0.32	—	0.006449–0.032218	64.4875–378.15
Cu	0.21–2.49	—	0.00008–0.000415	0.8–3.9875
Fe	6.51–17.06	43.9815	—	—

[a] Buffo, R.A., Reineccius, G.A., Oehlert, G.W., 2001. Factors affecting the emulsifying and rheological properties of gum Acacia in beverage emulsions. Food Hydrocoll. 15, 53–66.
[b] Abdelrahman, M. A. (2008). Characterization and thermodynamic properties of gum of *Acacia* origin. PhD thesis. Sudan University of Science and Technology, pp. 32, 46–47.
[c] Obeid, I.S. 2012. Evaluation of Physicochemical, Microbiological and Toxicological Aspects of *Acacia seyal* Gum (Ph. D. thesis in Chemistry). Sudan University of Sciences and Technology.

Table 1.4 Strengths, weaknesses, opportunities, and threats points for *Acacia* gums in Sudan

	Strengths	Weaknesses
Internal characteristics	Very high awareness and indigenous knowledge on traditional farming systems for *A. senegal.*	Poor to moderate awareness and indigenous knowledge on traditional farming system for *A.seyal.*
	Opportunities	Threats
External factors	Good comprehensive research for *A. senegal.* Good Market information of *A. senegal*	Very little comprehensive researches for *A. seyal.* Very little Market information of *A. seyal.*

Strengths, weaknesses, opportunities, and threats analysis for acacia gum in Sudan

Strengths, weaknesses, opportunities, and threats (SWOT) analysis was realized for the identification of the internal (strengths and weaknesses) and external factors (opportunities, threats), as shown in Table 1.4.

Acacia seyal contribution to production and export of gum Arabic

The production of gum Arabic from both *Acacia*s (*senegal* and *seyal*) in the Sudan from 1980 until 2014 is shown in Fig. 1.3. Gum Arabic (both *senegal* and *seyal*) exported from Sudan contributed significantly to the international trade market, as indicated in Fig. 1.3. Besides the significance of gum Arabic's economic role for the

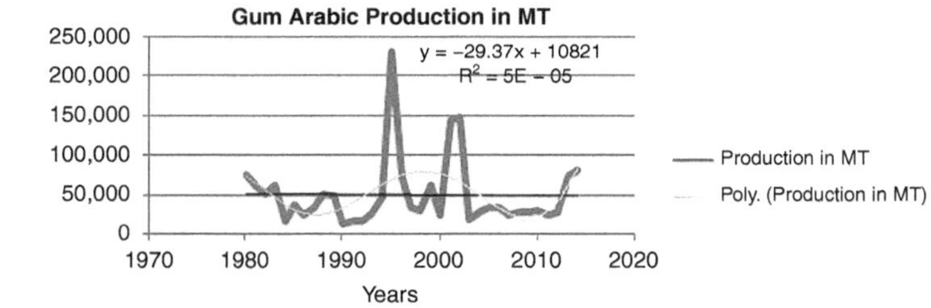

Figure 1.3 Gum Arabic production in MT.
Source: Modified from Gum Arabic Co. (GAC); Forest National Corporation (FNC); Gum Arabic Board (GAB) reports.

country, it plays an important part in rural life, providing a steady income to rural families, especially in dry years when crops fail.

The estimated trend of gum Arabic production (quantities in metric ton) is given over 34 years, and accordingly, the series of these measurements is treated as time-series data (more than 25 years for the period extended from 1980 to 2014). The trend estimation is used to make and justify tendency in the data by relating the quantities of gum Arabic to the time at which they occurred. The model shows that the quantities exhibit a zigzagging trend, where an increasing and decreasing manner appears several times, which is statistically distinguished from random behavior. The least square fitting (R^2) shows the residuals of the data obtained for gum Arabic quantities after fit. The small value of R^2 indicates that the data response around the mean explains very little variability, as shown in Fig. 1.4.

The estimated trend of gum Arabic export (quantities in metric ton) is also given for the 34 years from 1980 to 2014. The model shows that the quantities exhibit an increasing trend over time. The R^2 indicates that the model explains the 27% variability of the response data around its mean (Table 1.5).

General discussion and recommendations

Improving quality in fields increases the likelihood of obtaining chemically good gum. Knowledge of *seyal* quality supplied helps in the process of targeting market niches or reducing the risk of rejection by the buyer.

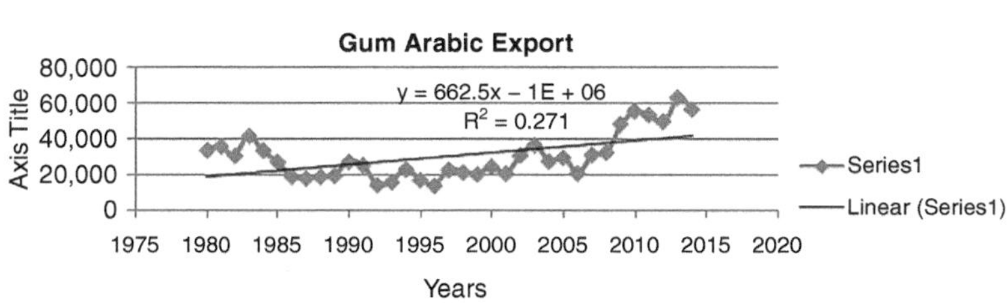

Figure 1.4 Gum Arabic export.
Source: Modified from Gum Arabic Co. (GAC); Central Bank of Sudan (CBoS); Gum Arabic Board (GAB) reports.

Table 1.5 Gum *A. seyal* export contribution from Sudan (2012–2016)

Year	Gum *A. senegal* (hashab)		Gum *A. seyal* (talha)		Total		Gum *A. seyal* export contribution
	Quantities (MT)	Amount ($)	Quantities (MT)	Amount ($)	Quantities (MT)	Amount ($)	Quantities (%)
2012	25,084	—	24,270	—	49,318	82,693,322	49.2
2013	28,122	92,104,000	34,890	42,051,000	63,012	134,155,200	55.4
2014	21,905	62,175,000	37,793	43,799,000	59,698	96,974,000	63.3
2015	24,194	72,437,080	39,286	83,775,580	63,480	156,212,660	61.9
2016[a]	5,523	16,609,738	9,211	12,634,070	14,734	29,243,808	43.2

[a] Three months.
Source: GAB (2016).

The good-quality gum on the field increases the likelihood of obtaining good quality on the basis of invisible attributes. Hence, collectors should be sensitized to put effort in the respect of the good harvest and post-harvest practices so that they supply good-quality gum. Trainings for gum collectors should be regularly conducted, as knowledge and awareness creation are probably the main building blocks for quality-oriented production.

The sensitization of collectors should be accompanied by price incentives by which the high quality is rewarded by a quality premium.

The current study has strengthened the need to understand the role of forest (tree and/or land) management on the quality of the gum. Clear rules of management are needed to counteract the influence of market forces (price) on competition in forests. Forest management is also a pertinent issue, as recurrent competitions may be detrimental to the resource and lead to degradation.

There is need for concerted effort towards strengthening the producer associations, construction of storage facilities in strategic locations, capacity building, provision information on prices and markets, and development of marketing systems.

References

Abdel Nour, H.O., 1989. Gum arabic in Sudan: production and socio-economic aspects. FAO Report, http://www.fao.org/docrep/x5402e/x5402e12.htm.

Anderson, D.M.W., Bridgeman, M.M.E., 1985. The composition of the proteinaceous polysaccharides exuded by *Astragalus microcephalus*, *A. Gummifer* and *A. Kurdicus*—the sources of turkish gum tragacanth. Phytochemistry 24 (10), 2301–2304.

Balal, M.E., Salih, N.K.E., Abdel Magid, T.D., 2014. Ethno-botany of natural forests of Nuba Mountains. J. Forest Prod. Ind. 3 (1), 13–19, South Kordofan State, Sudan. ISSN: 2325–4513. ISSN 2325-453X (online).

Chikamai, B.N., Osman, M.E., Menzies, A.R., Banks, W.B., 1995. An evaluation of methods for characterizing and monitoring gum Arabic of commerce and related Acacia gums. Wood Sci. Technol. 30, 49–61.

ElKhawad, H.E., 2008. Gum Arabic processing and marketing in the Sudan. M.Sc. thesis in Chemical Engineering. University of Khartoum.

Fadl, K.E.M., Gebauer, J., 2010. Effect of different tapping tools and different tapping positions on "talha gum" yield of *Acacia seyal* var. *seyal* in South Kordofan Sudan. In: Kennedy, J.F., Phillips, G.O., Williams, P.A. (Eds.), Gum Arabic. RSC Publishing.

GAB, 2007. Gum Arabic Board reports, Khartoum, Sudan.

JECFA, 1990. Evaluation of certain food additives and contaminants (Thirty-fifth report of the Joint FAO/WHO Expert Committee on Food Additives). WHO Technical Report Series, No. 789 and corrigenda, http://www.who.int/foodsafety/publications/jecfa-reports/en/.

Mohammed, M.H., 2011. Management of Natural Stands of *Acacia seyal* Del. variety *seyal* (Brenan) for Production of Gum Talha, South Kordofan Sudan. Ph.D. thesis in Forest Growth and Computer Sciences. Technical University of Dresden.

Mustafa, R., 2006. Risk management in the rain-fed sector of Sudan: a case Study, Gedarif Area Eastern Sudan. Ph.D. Thesis. Faculty of Agriculture and Natural Sciences. University of Justus Liebig Qiessen.

Orwa, C., Mutua, A., Kindt, R., Jamnadass, R., Anthony, S., 2009. Agroforestree Database: A Tree Reference and Selection Guide Version 4.0. World Agroforestry Centre, Kenya.

Further Readings

Awad, S.S., Rabah, A.A., Ali, H.I., Mahmoud, T.E., 2016. *Acacia seyal* gums in Sudan: a review. In: Proceedings of the 7th Conference of Scientific Research on the Basic and Engineering Sciences. University of Khartoum, Sudan, February 2016. .

Elsiddig, T.A., 2009. Comparative water uptake at germination and early stage of seedling in six provenances of talha (*Acacia seyal* var. *seyal*) from Sudan clay plain. M.Sc. Thesis in Environmental Science and Natural Resources. University of Gezira.

GAB 2015. Gum Arabic Board reports, Khartoum, Sudan.

Hassan, E.A., Al-Assaf, S., Phillips, G.O., Williams, P.A., 2005. Studies on Acacia gums: Part III molecular weight characteristics of *Acacia seyal* var. *seyal* and *Acacia seyal* var. *fistula*. Food Hydrocoll. 19, 669–677.

Mujawamariya, G., Burger, K., 2012. Quality of gum Arabic in Senegal: linking the laboratory research to the field assessment. Quart. J. Int. Agric. 51 (4), 357–383.

Management Practices of Gum Arabic–Producing Trees

2

*Zeinab Mohamed Hammad**,**, *Mohammed Hamed Mohammed***
*Institute of Gum Arabic Research and Desertification Studies, University of Kordofan, Sudan; **University of Kordofan, Elobeid, Sudan

Introduction

Gum Arabic, which has countless applications, is a nontimber forest product of *Acacia* trees, namely *Acacia senegal* and *Acacia seyal*, of the Sudano–Sahelian region of sub-Saharan Africa (JECFA, 1999; Karama, 2002). More than 500 hundred species exist of the genus *Acacia* that produce natural gums (Beyene, 1993). However, gum from *A. senegal* and *A. seyal* is considered to be of high commercial importance. As indicated by Chikamai (1997), gum is an exudate that is obtained naturally from *Acacia* stands. It is worth mentioning that some species, such as *A. senegal*, *Acacia laeta*, *Acacia ehrenbergiana*, and *Acacia karroo*, produce plenty of gum through stimulation by tapping. In the arid rural areas of western Sudan, that is, Kordofan and Darfur, production of gum Arabic is behind economic stability. It is a source of income in the agricultural off-season, and most household members are involved in different gum Arabic production activities.

In Sudan, *A. senegal*, locally known as the hashab tree, is found either in natural stands or in plantations; the gum is attained by tapping, whereas gum from *A. seyal*, locally known as talha, is obtained directly without tapping (Mohammed, 2011). The general trend of utilization of *A. seyal* is production of firewood and charcoal (Elsiddig, 2003; Fadl and Gebauer, 2004; McAllan, 1993). However, a significant amount of gum is produced from the tree, as stated by Ali (2006) and Mohammed (2011). *A. seyal* prefers clay soils and is found naturally either in association with other species or in pure stands (Mustafa, 1997). The species produce natural gum, which is considered inferior to that of hashab; despite that, the gum from this species is cited to have many local uses and industrial applications (Anderson et al., 1984; FAO, 1995; Hall and McAllan, 1993; McAllan, 1993). It represents about 10% of total gum production of the Sudan (GAC, 2008; Mohammed, 2011).

Farming systems in the gum belt of Sudan are mainly *Acacia*-based farming in which *A. senegal* is intercropped with agricultural crops in addition to livestock raising (Fig. 2.1). It is a sustainable farming system that guarantees a stable, healthy environment and proves social and economic advantages to inhabitants of the gum belt of Sudan (Ballal, 2002; Eisa et al., 2008; Hammad, 2014; NEF, 2010).

Gum Arabic represents about 12% of Sudan's GDP; the country produces more than 80% of the total world gum Arabic (Abdulgadir, 2013; Forman, 2012). Gum Arabic plays a vital role in the country's economy. Added to that, it is considered a

Gum Arabic. http://dx.doi.org/10.1016/B978-0-12-812002-6.00002-6

Figure 2.1 Gum Arabic trees . Left: *Acacia senegal*, *Acacia*-based farming in dry lands of North Kordofan. Right: *Acacia seyal* stand in South Kordofan.

stable source of income for rural inhabitants, especially during seasons of crop failure (Abdelnour, 1999). Being the main component of traditional rain-fed farming, *A. senegal* is acknowledged as momentous due to its contribution to household income and the country's foreign exchange earnings (Pretzsch et al., 2014).

In ideal settings of the hashab farming system, the farm is divided into four parts: a sole mature *A. senegal* stand, premature stand, young stand intercopped with agricultural crops, and newly felled trees for sole crops (Abdel Magid et al., 2014). At farm level, the economic and environmental importance of *Acacia*-based production is a complicated system; therefore, efforts are needed to understand and diagnosis this complexity to ensure the attractiveness of the system (Abdelnour, 1999; Barbier, 1992).

The importance of gum Arabic–producing trees

Gum Arabic–producing trees are found in a tropical area known as the gum belt that extends throughout Africa between latitudes of 10–14°N. In Sudan, the belt is described as a zone of low-rainfall woodland savannah in which *A. senegal* and *A. seyal* are naturally and widely distributed (Badi et al., 1989; Ahmed, 2006). Gum arabic–producing trees have many services and benefits; these are environmental, social, and economic. The environmental role in contributing to carbon sequestration gives their importance national, regional, and global dimensions (Oba et al., 2002).

An *A. senegal* stand is a desert control agent, as it considered a shelterbelt and windbreak for the creeping desert (Larson and Bromley, 1991; Olsson, 1984; Pearce et al., 1990). This characteristic is reason for nominating this species a candidate for environmental services, in addition to its advantage in improving soil productivity and reclamation of adverse effects of climate variability and irrational land use (Abdel Magid et al., 2014). At the same time, in the northern part of the clay plain of the gum belt in Sudan, *A. seyal* plays a vital ecological role, as it acts as a defense line for desert creeping (Mohammed, 2011).

A. senegal is characterized by active root nodules having the capacity of nitrogen fixation; hence, it is a valuable source for maintenance and restoration of soil fertility. It therefore improves nutrient cycling and increases crop yields and biodiversity recovery (Elmqvist et al., 2005; Montagnini et al. 2005).

Furthermore, *A. senegal* has many benefits and traditional uses; it is a source of energy and building materials and a favored source of fodder. Additionally, it secures foreign exchange for Sudan and income generation for rural communities, as it is an employment activity in agricultural off-season (Larson and Bromley, 1991; Olsson, 1984; Pearce et al., 1990). Another important advantage of the *Acacia*-based farming system is risk avoidance through diversification of farm products. As indicated by Pretzsch et al. (2014), the importance of the *A. senegal* tree is well recognized by Sudanese at both communal and institutional levels, as it provides a sustained gum for international markets in addition to its socioeconomic benefits; therefore, considerable efforts have been exerted to develop the gum Arabic sector.

Gum Arabic production

Gum Arabic is a natural exudate obtained from the stem and branches of *A. senegal* and closely related species such as *A. seyal* (JECFA, 1999). The introduction of "closely related species" to the definition of gum Arabic is to bound gum Arabic only to those gums that have specific optical rotations ranging from 26 to 34 degrees and have nitrogen content between 0.27% and 0.39% (JECFA, 1999) with the code term E414. However, in Sudan, hashab gum and talha gum are separated in both national statistics and trade (FAO, 1995). The highest quality and quantity gum is obtained from *A. senegal*, which constitutes the bulk of world trade (Beyene, 1993). This gum is produced by proper synchronized activities at individual managed farms known as gum gardens. Gum production activities include tree planting, tending, regular tapping, and gum collection; the amount of gum collected depends on the success of production of gum and combined agricultural crops (Elmqvist et al., 2005). Tapping of *A. senegal* commences in October or November, immediately after rains. Hashab tapping starts 5–7 years after establishment and continues for up to 15–20 years, while first collection begins 45 days after tapping, and the subsequent collections take place every 15 days, lasting for up to seven collections per season. Gum yield depends on the climatic conditions, and the annual average yield per tree varies from 100 to 1000 g, with an average of 250 g per tree. The recommended tapping tool, *sonki*, increases the gum yield (Abdel Magid et al., 2014). Among the crucial factors that effect gum Arabic yield are the amount and distribution of rainfall and hot weather. Although the relationship between the two factors is complicated, it is cited that a season with optimum and well-distributed rain followed by hot weather results in plentiful gum production; however, late rains reduce gum exudation, as the tree keeps green leaves (NAS, 1979; Ross, 1968).

Despite the considerable amount of talha gum production, tapping is not common practice, and only natural exudates are collected by the local population (Beyene, 1993;

Fadl and Gebauer, 2004), although the production potential in Sudan is large as compared to hashab hard gum. Added to that, talha gum production is not promoted by the Sudanese government, although annual exports are estimated to be 3000–5000 tons (Iqbal, 1993; Macrae and Merlin, 2002). *A. seyal* is an important, promising tree species in Sudan with high gum production potential that has never been fully explored (Abdel Magid, 2014). Although talha gum has a significant share of total gum Arabic production in Sudan, scant research has been conducted with regard to productivity and managerial as well as silvicultural aspects that affect gum production. Nevertheless, gum yield ranges from 12–21 g per tree per season (Ali, 2006; Fadl and Gebauer, 2004; Hineit, 2007; Mohammed, 2011).

Land tenure and management practices

Traditional shifting cultivation, agroforestry, which is the dominant crop production system practiced in the tropics (Oba et al., 2002), is the typical *Acacia*-based farming system in the gum belt of Sudan (Hammad, 2014). Management of gum production falls into one of two systems; hashab owner or hashab renter (Elkhidir et al., 2010; ILO, 1985). Hashab owner has either a small-size farm (7–18 ha) or a large-size farm (37–73 ha) (Hammad, 2014). Hashab owners constitute the majority of the producers in the gum belt. Their own small holding gum orchards, which are part of the *A. senegal* rotation system, enable them to practice gum production in one of three ways: tap trees by themselves, hire labor to carry out all production operations, and share crop production with gum workers, whereas large holder farmers include traditional leaders and rich people who depend on hiring labor and sharecropping for production (Elkhidir et al., 2010).

Hammad (2014) cited that tree planting and husbandry has existed in the Sudan for several millennia, during which time gum Arabic has been a prized export commodity. Local socioeconomic conditions determine land tenure practices. Noticeable variations in land tenure with regard to different parts of the country necessitate common regulations: allocating landholding to households by the village leader (sheikh), regulating the *Acacia*-based farming system according to environmentally sound principles, and assessing agricultural and gum Arabic yields. This system has negatively been affected by adverse environmental conditions and irrational resource use as well as political and institutional conditions. Therefore, cultivation in the Gum Arabic Belt and gum tapping have not been stable, and the sustainability is not assured (SWEFDP, 1986). The continuous decrease in production, as shown in Fig. 2.2, has been attributed to low, erratic rainfall and reduction in hashab population by cutting either for wood/charcoal production or the expansion of agricultural field crops areas at the expense of forest lands (Eldukheri, 1997). *Acacia*-based farming systems, such as intercropping, can reduce or offset the initial cost of reforestation, hence providing incentives for *Acacia* planting and contributing to the gum Arabic production and sustainability (Schlonvoigt and Beer, 2001).

Gum Arabic productivity per unit area is influenced by some management practices, including number of trees per unit area and tapping technology adopted. Stocking

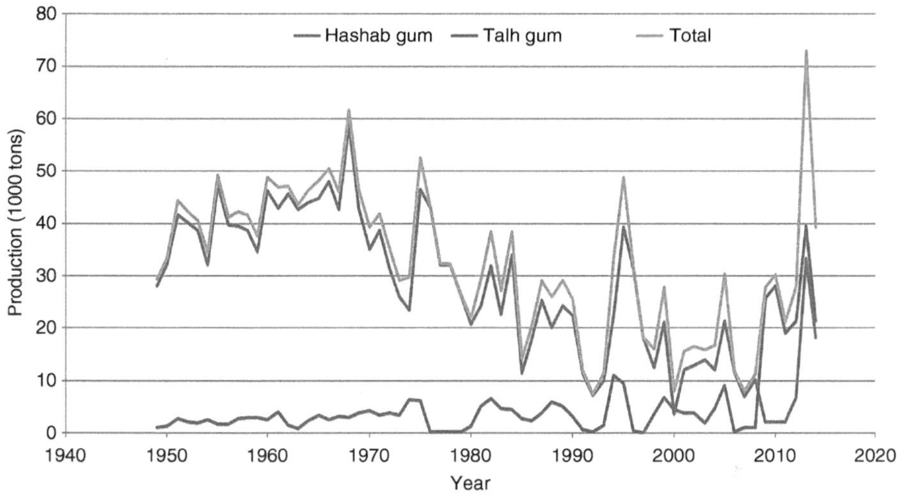

Figure 2.2 Production trend of gum Arabic production in Sudan from 1950 to 2014. Source: Derived from Central Bank of Sudan, Annual reports (2002–2014). Available from: http://www.cbos.gov.sd. and Mohammed, M.H., 2011. Management of natural stands of *Acacia seyal* Del. variety *seyal* (Brenan) for production of gum talha, South Kordofan, Sudan. TUDpress, Verlag der Wissenschaften GmbH, Dresden.

density of 300-400 stems of *A. senegal* per hectare under good management and protection improve the productivity. The density may be as low as 125 stems per hectare in poorly managed gum gardens, and as high as 625 trees per hectare under good management. The spacing accordingly varies from 4 × 4 m to 9 × 9 m. The tree grows rapidly to reach 2–3 m in height in 5 years as a result of improved tending and management practices (Elsiddig et al., 2005). The production in the reserve forests is known to be higher than that of the individual farmer's garden for the reason that the stocking density in the reserve area is in the range of optimum density (Elsiddig and Abdel Magid, 2007).

Limitations to management practices

Natural defoliation is the typical sign of tree maturity and hence, gum production. When defoliation is stimulated by drought, insects, livestock, and other herbivorous wild animals, *A. senegal* ceases to produce gum during that season, but it quickly puts on fresh leaves during the ensuing season unless, of course, the causes of defoliation persist. Four or five years of consecutive drought will, however, arrest almost all exudation and may even damage the tree permanently (Beyene, 1993).

A. senegal trees suffer from damage caused by fires, which reduce the stock density and hence negatively affect the yield of gum Arabic, which may last several years on the stand (Elsiddig and Abdel Magid, 2007). However, Adam (2013) reported a positive effect of slight grass burning on gum yield of *A. seyal*. Gum Arabic tree seeds are susceptible to rodents and insect attacks, such as the buffalo tree hopper *Sticto-cephala bubalus*, and larvae of Coleoptera (bruchids), Lepidoptera, and Hymenoptera, which damage the seeds and seed crops (Dorthe, 2000; Jamal, 1994).

Due to frequent drought spells and climate change in addition to adverse socio-economic conditions of the gum belt inhabitants, the *Acacia*-based farming system has been gradually changed and transformed from the very old shifting cultivation, passing by short rotation to a newly emerging farming system, integrating trees with agricultural crops in the same piece of land (Hammad, 2014). This newly emerging system was also described by Ballal (2002) as a system that integrates sparse *A. senegal* trees with field crops on farmer' fields in North Kordofan. Abdelnour (1999) draws attention to the different constraints and opportunities of improving management practices of the resource. He further noted that the reduction of forest cover in the gum belt of Sudan is caused by drought spells and human-influenced constraints due to extensive expansion in agriculture for crop production, particularly mechanized, rain-fed farming. The same author emphasizes that gum production is influenced by the declining prices of gum as compared to those of such crops as sesame and groundnuts.

Conclusion

Inhabitants of the gum belt appreciate the important role of the gum-producing trees and their integration in farming systems and practices. The appreciation arises from combined services and direct benefits provided to farmers and the environment. The most important advantages of *Acacia*-based farming system are diversification of farm products, provision of income throughout the off-season, and risk minimization. Gum Arabic continues to be a commodity of great importance, as poverty and environmental degradation have led to the difficult socioeconomic situation. In this context, new research strategies are needed, taking into account sustainable agriculture through the *Acacia*-based farming. The environmental role of gum Arabic–producing trees has to be advertised equally with its economic importance.

References

Abdel Magid, A. A., 2014. Book review "Gum Arabic in the Sudan Sudan Vision" (unpublished).

Abdel Magid, T.D., Eltayb, M.T.A., Dirar, A.M.A., 2014. Equalizing gum codal term (a Code E414), of *Acacia senegal* (L) with *Acacia seyal* (Del.), and its impact on gum production and exportation. J. App. Ind. Sci. 2 (3), 144–151.

Abdelnour, H.O., 1999. Gum Arabic in Sudan: production and socio-economic aspects: medicinal culinary and aromatic plants in the near east. In: Proceedings of the International Expert Meeting organized by the Forest Products Division. FAO Forestry Department and the FAO Regional Office for the Near East, Cairo, Egypt.

Abdulgadir, A., 2013. Secretary General of the Gum Arabic Council. Available from: http://news.sudanvisiondaily.com/details.html.

Adam, A.A., 2013. Effects of grass burning, tree size and tapping on gum talha (*Acacia seyal* Del. var. seyal) yield in South Kordofan State. MSc. Thesis, University of Kordofan.

Ahmed, M.E., 2006. The socio-economic role of *Acacia senegal* in sustainable development of rural area in the gum belt of the Sudan. Institut für Internationale Forst- und Holzwirtschft, Technische Universität Dresden, Tharandt, Germany.

Ali, E. Y., 2006. Yield Potential of Acacia seyal Del. variety seyal in Relation to Growth Parameters, Methods and Position of Tapping. MSc. Thesis, University of Kordofan.

Anderson, D.M., Bridgeman, M.M., De Pinto, G., 1984. Acacia gum exudates from species of the series gummiferae. Phytochemistry, 575–577.

Badi, K.H.M., Ahmed, A.E., Bayoumi, A.M.S., 1989. The Forests of the Sudan. Khartoum Agricultural Research Council, Sudan.

Ballal, M. E., 2002. Yield Trends of Gum Arabic from Acacia senegal as Relation to Some Environmental and Managerial Factors (PhD Thesis), Faculty of Forestry, University of Khartoum.

Barbier, E.B., 1992. Rehabilitating gum Arabic systems in Sudan: economic and environmental implications. Environ. Resour. Econ. 2 (4), 341–358.

Beyene, M., 1993. Investing in Acacia senegal lessons from the Sudanese experience to Eritrea. African Arid Lands Working Paper Series No. 4/93.

Central Bank of Sudan, Annual reports (2002–2014). Available from: http://www.cbos.gov.sd.

Chikamai, B., 1997. Production, markets and quality control of gum Arabic in Africa: findings and recommendations from an FAO project. In: Mugah, J.O., Chikamai, B.N., Mbiru, S.S., Casad, E. (Eds.), Conservation, Management and Utilization of Plant Gums, Resins and Essential Oils. Forestry Department, FAO, Proceedings of a regional conference for Africa, October 6–10, 1997. FAO, Rome, Nairobi, Kenya.

Dorthe, J., 2000. *Acacia senegal* (L) Wild Seed Leaflet. Danida Forest Seed Center, Denmark, No. 5, 1–2.

Eisa, M.A., Roth, M., Sama, G., 2008. *Acacia Senegal* (Gum Arabic Tree): Present role and need for future conservation/Sudan. Paper presented on Deutscher Tropentag, Germany.

Eldukheri, I. A. 1997. Past Changes and Future Prospects of Traditional Rain-fed Farming in North Kordofan State. Ph.D Dissertation, Technical University of München, Germany.

Elkhidir, E.E., Zubaidi, B.A.S., Shzee, Y.T., 2010. Estimation of technical efficiency for share contract of producing gum Arabic: Kordofan gum Arabic belt, Sudan. Res. J. Forest. 4, 185–193.

Elmqvist, B., Olsson, L., Elamin, E.M., Warren, A., 2005. A traditional agroforestry system under threat: an analysis of the gum Arabic market and cultivation in the Sudan. Agrofor-est. Syst. 64, 211–218.

Elsiddig, E.A., 2003. Aboveground wood weight and volume relationships for *Acacia seyal* in Eastern Sudan. Sudan Silva 9 (1), 25–39.

Elsiddig, E.A., Abdel Magid, D.A., 2007. Sudan forestry sector review. In: Abdelnour, H.O. (Ed.), Forests National Corporation. National Forest Programme, Khartoum, Sudan.

Elsiddig, E.A., Elballal, M.E., Abdel Magid, T.D., 2005. The *Acacia senegal* agroforestry sys-tem in the Sudan. In: Proceedings of the Planary Garden workshop. Brazil, September 26–28.

Fadl, K.M., Gebauer, J., 2004. Effect of different tapping tools and different tapping posi-tions on "talh gum" yield of Acacia seyal var. seyal in South Kordofan, Sudan. Confer-ence on International Agricultural Research for Development. Berlin, October 5–7, 2004. Deutscher Tropentag.

FAO, 1995. Gums, resins and latexes of plant origin. Non-wood Forest Products 6. FAO, Rome.

Forman, S., 2012. Revitalizing the Sudan Gum Arabic Production and Marketing: Report on Implementation Status and Results, No ISR8088.

GAC, Gum Arabic Company, 2008. Annual reports 1970–2008. Gum Arabic Company Ltd, GAC, Khartoum, Sudan.

Hall, J.B., McAllan, A., 1993. *Acacia seyal*: A Monograph. School of Agricultural and Forest Sciences, University of Wales, Bangor, UK.

Hammad, Z.M., 2014. Transformations in Agroforestry Systems in the Gum Belt of Kordofan. LAP LAMBERT Academic Publishing, Germany.

Hineit, B.M., 2007. Effect of Type, Time of Tapping and Tree Size of *Acacia seyal* var. seyal on the Production of Gum Talha in Rawashda forest (Master Thesis). Faculty of Forestry, University of Khartoum, Gedarif State, Sudan.

ILO, Intensive Labour Organization, 1985. Project Identification mission 1984, Ministry of Agriculture, Sudan.

Iqbal, M., 1993. International Trade in Non-wood Forest Products: An Overview. FAO, Rome, FO: Misc/93/11 Working Paper.

Jamal, A., 1994. Major insect pests of gum Arabic Acacia Vild and *Acacia seyal* (L) in Western Sudan. J. Appl. Entomol. 117 (1), 10–20.

JECFA, (FAO/WHO) Expert Committee on Food Additives, 1999. Compendium of Food Additive Specifications. Appendum 7, Joint JECFA 53rd session June 1–10, Rome.

Karama, M., 2002. Gum Arabic Marketing Issues: Marketing Model, Buffer-Stocking, Processing and Finance. Unpublished Paper Presented by the International Workshop on Promotion of Gum Arabic Production, Processing and Marketing, Khartoum, Sudan.

Larson, B.A., Bromley, D.W., 1991. Natural resource prices, export policies, and deforestation: the case of Sudan. World Develop. 19 (10), 1289–1297.

Macrae, J., Merlin, G., 2002. The prospects and constraints of development of gum Arabic in Sub-Saharan Africa. World Bank, Washington, D.C.

Mohammed, M.H., 2011. Management of natural stands of *Acacia seyal* Del. variety *seyal* (Brenan) for production of gum *talha*, South Kordofan, Sudan. TUDpress, Verlag der Wissenschaften GmbH, Dresden.

McAllan, A., 1993. Acacia Seyal: a Handbook for Extension Workers. School of Agricultural and Forest Sciences, University of Wales, Bangor, UK.

Montagnini, F., Cusack, D., Petit, B., Kanninen, M., 2005. Environmental Services of Native Tree Plantations and Agroforestry Systems in Central America. The Haworth Press, Inc., doi:10.1300/J091v21n01_03; http://www.haworthpress.com/web/JSF.

Mustafa, A.F., 1997. Regeneration of *Acacia seyal* Forests on the Dryland of the Sudan Clay Plain: Tropical Forestry Report No. 15. Department of Forest Ecology, Helsinki University Printing House, Helsinki.

NAS, National Academy of Sciences, 1979. Tropical Legumes: Resources for the Future. NAS, Washington, D.C.

NEF, Near East Foundation, 2010. NEF Launches Gum Arabic Project in Central Sudan.

Oba, G., Stenseth, N.C., Weladji, R.B., 2002. Impacts of shifting agriculture on a floodplain woodland regeneration in dry-land, Kenya. Agric. Eco. Environ. 90, 211–216.

Olsson, K., 1984. Long-term Changes in Woody Vegetation in the North Kordofan—the Sudan: A Study with Special Emphasis on *Acacia senegal*. Lund University, Lund, Sweden.

Pearce, D., Barbier, E., Avil, M., 1990. Sustainable Development: Economics and Environment in the Third World. Edward Elgar Publ. Ltd, Aldershot, England.

Pretzsch, J., Taha, M.E., Siddig, M.E., Dafalla, T.H.H., Elamin, H.M.A., 2014. Valuation of environmental role of *Acacia senegal* Tree in the gum belt of Kordofan and the Blue Nile

Sectors, Sudan. Conference on International Research on Food Security, Natural Resource Management and Rural Development. Tropentag September 17–19, 2014. University of Life Sciences Prague, Czech Republic.

Ross, J.H., 1968. *Acacia senegal* (L) wild in Africa with particular Reference to Natal.

Schlonvoigt, A., Beer, J., 2001. Initial growth of pioneer timber tree species in a taungia system in the humid lowland of Costa Rica. Agroforestry Systems 51.

SWEFDP, 1986. Sudan Wood Energy/Forestry Development Project.

Further Reading

Abdel Magid, T.D., Badi, K.H., 2013. Manual of silviculture for selected tree species in Sudan. LAP LAMBERT Academic Publishing, Germany.

Valuation of the Environmental Role of *Acacia senegal* Tree in the Gum Belt of Kordofan and the Blue Nile Sectors, Sudan

3

*Mohamed El Nour Taha**, Jürgen Pretzsch**, Muneer E. Siddig†, Taisser H.H. Deafalla‡, Hatim M.A. Elamin†*

*Kordofan University, El-Obeid, Sudan; **TU Dresden, Germany; †Institute of Gum Arabic Research & Desertification Studies, University of Kordofan, El-Obeid, Sudan; ‡Institute of Photogrammetry and Remote Sensing, TU Dresden, Germany

Introduction

Having been recognized as the world predominant pioneer in gum Arabic production, Sudan adds to around 79% of the total world gum Arabic production. Sudan adequately controls practically more than 80% of the world market (Abdulgadir, 2013; Anderson, 1993; Forman, 2012; GAC, 1996; Yasseen et al., 2015). Gum Arabic gives an average of 14.8% of the GDP of the nation (Elkhatim, 2016). Gum production represents around 15.3 and 10% of the family income of gum producers and different agriculturists in the gum belt in Sudan, respectively (Elamin and Ballal, 1989; Mahmoud, 2004; Taha, 2006). The total genuine money-related advantages to agriculturists from developing gum Arabic stands were discovered to be positive in numerous various frameworks over the gum belt, even when current market costs were utilized to figure the monetary benefit of such stands (Taha, 2006). Taha (2006) demonstrated additionally that gum production was monetarily more productive and had less escalated labor compared with all other options of cash crops. Mahmoud (2004) also showed that gum Arabic, as a product of a small household farm, has a decent, similar favorable position and moderately higher global competitiveness in connection to other competing cash crops. Gum production is described by an amazing adaptability, and it is a means for fighting poverty; it gives insurance against risks and uncertainties (Karama, 2001; Khalid, 2005). According to Taha (2006), gum Arabic offers two broad-based benefits to the people of the Sudan: (1) commercial benefits in the form of cash for local economies and foreign exchange for the economy as a whole, and (2) environmental benefits in the form of products that are directly useable to the farming communities and in the form of wider antidesertification insurance.

Other than pleasing one-fifth of the number of inhabitants in Sudan and 66% of its domesticated animals' populace, the gum belt is of tremendous agrarian significance

Gum Arabic. http://dx.doi.org/10.1016/B978-0-12-812002-6.00003-8

to the nation (IIES/IES, 1990). The gum belt area is additionally imperative due to its geographical position; it lies just south of thickly populated states whose wood prerequisites are generally gotten from the gum belt range. It is the first and last barrier line securing the rural and domesticated animals production zones toward the south against unavoidable infringement by the Sahara Desert (IIES/IES, 1990). The gum tree has a critical part in providing the family unit with wood energy and feed requests in addition to improving the soil fruitfulness (Barbier, 1992). The tree is additionally utilized as a part of the conventional *Acacia senegal*–based agro-ranger service system, which is perceived and considered as a standout among the best types of common backwoods administration in the tropical drylands (Fries, 1992). Moreover, the gum tree is viewed as economical as far as its natural, social, and biological advantages to a substantial zone of the Sudan (Ballal, 2002; Eisa et al., 2008; NEF, 2010). Also, the *A. senegal* stands offer protection to soils against desertification, which is a marvel of a ceaseless declining soil productivity hazard coming about because of atmosphere variety and human utilization of the land (Fries, 1992). Furthermore, the gum tree is considered as sustainable from the point of its environmental, social, and ecological benefits to a large area of the Sudan (Ballal, 2002; Eisa et al., 2008; NEF, 2010). Conclusions made by Hussein (1983) emphasized the appreciable significant changes, which have been affected by *A. senegal* on the surface soil properties. Gerakis and Tsangarakis (1970) reported similar findings on sand (*qoz*) soils under *Acacia senegal* and traditional agriculture.

People of the Sudan know about the *A. senegal* tree as a national wealth, which should be saved, enhanced, and created, not exclusively to give a maintained yield of gum, but additionally for other vital financial advantages of protection and production. The general target of the review was to recognize natural advantages of *A. senegal* trees inside the most essential delivering territories of the gum belt in the Kordofan and the Blue Nile Regions. The review pointed particularly to discovering values for nonmarketable merchandise and ventures given by gum trees regarding soil security and reclamation of soil fertility.

The study area description

The review covered some chosen regions in the awesome Kordofan sector of western Sudan and in the Blue Nile sector of southeastern Sudan. Officially, the current Kordofan sector includes the North Kordofan, South Kordofan, and West Kordofan states, while the Blue Nile sector is now comprised of both the Sennar and Damazin states. The entire territory of the study exists in the gum belt. It covers distinctive climatic and biological zones, as indicated by the variety of precipitation, relative humidity, temperature, and edaphic elements. The review region epitomizes the possibilities and issues of gum Arabic creation under both sand (*qoz*) and clay soils in the two most essential gum-delivering areas of the Sudan. More than half of gum Arabic created in Sudan originates from the Kordofan sector, and a significant part originates from the Blue Nile sector (Abdulgadir, 2013). The primary financial action of the tenants in the review territory depends on the joining of agribusiness, animal production, and ranger service, and gum Arabic is a fundamental segment of the family cultivated framework (IFAD, 2002; Mahmoud, 2004; Ramly, 2002; Taha, 2006).

Methodology of the study

Due to the large size of the study area, a total census was not possible. For this reason, along with the homogeneity exhibited in the socioeconomic characteristics of rural populations in each sector under study, random sampling of 5% is therefore decided as a method of survey. Hinton (1995) has advocated the advantage of the random sampling. Lists of households were compiled from records of relevant governmental institutions. Based on the estimates of the total population involved in gum production, a random sample of villages and household members was selected from those who own hashab (gum Arabic) stands. In the survey, 173 respondents, representing a total population of 2081 households, were sampled. The actual sampling percentage ranges from 6.7% in the Blue Nile to 10.1% in Kordofan, with an overall average percentage of 8.3%.

Data necessary for achieving study objectives were collected through distribution of purposively structured questionnaires, informal discussions, and interviews with the local people (gum producers) as well as visiting relevant markets. The pretested questionnaire administered to selected gum farmers provided information on local respondents' awareness of the indirect benefits of gum (hashab) stands.

The valuation procedure made to assess the ecological impacts of *A. senegal* trees on land quality depended on two contemplations: (1) the capacity to settle climatic nitrogen and keep up soil fertility, in this manner, enhancing farming production; and (2) arrangement of a lot of nitrogen evaluated as 7.7 kg for each hectare every year (Hussein, 1983).

The measures of nitrogen given by hashab trees were then valued utilizing current market prices of the sensible substitute; the most imperative is the urea (ammonia nitrate fertilizer). The substitution cost approach (supported by Saastamoinen, 1992) that utilizes market-oriented prices was chosen to value the ecological advantages of *A. senegal* trees; the on-location expenses of soil erosion were esteemed by computing the expenses of physically supplanting the lost soil fertility. As a broadly utilized pointer/measure of money related and financial productivity, the inside rate of return of hashab development was figured, considering the natural impacts of such stands on soil quality.

Results and discussion

The attempt made to judge farmers' belief in indirect benefits (mainly the environmental effects) of hashab stands indicated that 46 and 57.5% of interviewed farmers in Kordofan and the Blue Nile sectors, respectively, are considered to believe in indirect benefits of gum tree. Also, results indicate that the majority of respondents (82.1%) all over the study area expressed their willingness to retain hashab stands even if the financial returns from any other competing cash crop are high. The investigated persons stated many factors that justify their unwillingness to exchange land under hashab for any other cash crop and to keep on the preservation of hashab stands. These are ranked in priority order to include the following:

1. It is risky to rely on other farm crops as the only source of income.
2. Returns from gum Arabic compensate for income deficit in the dead season.
3. Timing of gum production activities is convenient and doesn't affect other crops negatively.
4. Hashab stands should be retained for their other indirect protective benefits.
5. Hashab stands provide an important source of fuelwood.
6. Hashab stands can grow in the marginal lands where no other farm crop does.
7. Hashab stands have other advantages that render them much more valuable than other cash crops; for example, they involve low inputs and low costs of production; also, they are easy to work, and the income from their gum crop is sustainable.

Conclusions made by Gerakis and Tsangarakis (1970) and Hussein (1983) emphasized the appreciable significant changes, which have been affected by *A. senegal* on the surface soil properties and traditional agriculture. According to Hussein (1983), *A. senegal* trees provide an amount of nitrogen estimated annually as 770 part per million; this amount is equivalent to 7.7 kg per hectare per year. The urea (ammonia nitrate) fertilizer, the most important fertilizer used to compensate for the loss of nitrogen in the irrigated sector in Sudan, was considered as the reasonable substitute, and its current market prices were then used for valuing the amounts of nitrogen provided by *A. senegal* trees. The ammonia nitrate fertilizer contains, in addition to other elements, 46% nitrogen. Therefore, the amount of nitrogen provided by *A. senegal* trees (7.7 kg/hectare) could be available if 16.74 kg of ammonia nitrate were supplied. The total cost of ammonia nitrate equivalent to the nitrogen provided annually by *A. senegal* trees per 1 hectare was computed as 86,638.37 Sudanese pounds. In other words, the presence of *A. senegal* trees can save a cost of supplying fertilizer equal to 86,638.37 Sudanese pounds per hectare, which is equivalent to approximately $78 USD per hectare annually.

The absolute IRR for a 16-year-rotation *A. senegal* stand was calculated to be approximately 15.2%; it is relatively small when compared to that figure of IRR estimated by Pearce (1988). He calculated the private (financial) internal rate of return from *A. senegal* stands, taking into account production of gum and fodder as well as fuelwood, based on a 16-year rotation, to be around 36%.

When the ecological benefits and social values associated with cultivation of *A. senegal* stands were considered, the recalculated IRR for 16-year-rotation gum stands jumped to approximately 61%.

The general conclusion that could be drawn was that the total financial benefits from gum cultivation were positive. In other words, there were real financial benefits to farmers from growing *A. senegal* stands in many diverse systems across the gum belt, even when current market prices are used to calculate the financial profitability of such stands.

References

Abdulgadir, A., 2013. Secretary General of the Gum Arabic Council. Google Earth., http://news.sudanvisiondaily.com/details.html?rsnpid=219702.

Anderson, D.M.W., 1993. Some factors influencing the demand for gum Arabic (*Acacia senegal* (L.) Willd.) and other water-soluble tree exudates. Forest Ecology and Management, vol. 58, pp. 1–18.

Ballal, M.E., 2002. Yield Trends of Gum Arabic From *Acacia senegal* as Related to Some Environmental and Managerial Factors (Ph.D. thesis). University of Khartoum, Khartoum, Sudan, pp. 105.

Badi, K.H., 1989. The natural forests. The Forests of the Sudan, Khartoum, p. 147. books.google.com.

Barbier, E.B., 1992. Rehabilitating gum Arabic systems in Sudan: economic and environmental implications. Environ. Resour. Econ. 2 (4), 341–358.

Eisa, M.A., Roth, M., Sama, G., 2008. *Acacia Senegal* (gum Arabic tree): present role and need for future conservation/Sudan. Deutscher Tropentag, October 7–9, Hohenheim, Germany, http://www.tropentag.de/.

Elamin, E.M., Ballal, M.E., 1989. Paper on Land Tenure and Usufructuary Rights in the Traditional Gum Belt Agroforestry System in Western Sudan. Elobeid Research Station, Elobeid, Sudan, pp 12.

Elkhatim, A.K., 2016. Economics of non-oil exports of Sudan. Russ. J. Agric. Socio-Econ. Sci. (RJOAS) 10 (58), 45–54.

Forman, S., 2012. Revitalizing the Sudan Gum Arabic Production and Marketing: P110588-Report on Implementation Status and Results., http://documents.worldbank.org/.

Fries, J., 1992. Management of natural forests in the semi-arid areas of Africa, present knowledge and research needs, seminar in June 1990 (English). Fries, J., Swedish University of Agricultural Sciences. International Rural Development Centre, Uppsala, 1992, Earth and Environmental Sciencespp. 47–68, ISBN: 91-576-4610-4.

GAC, Gum Arabic Company, 1996. Paper Presented on Problems and Constraints of Marketing Gum Arabic. Gum Arabic Conference, Friendship Hall, Khartoum, Sudan.

Gerakis, P.A., Tsangarakis, C.Z., 1970. The influence of *Acacia senegal* on the fertility of a sand sheet ('goz') soil in the central Sudan. Plant Soil 33 (1–3), 81–86.

Hinton, P.R., 1995. Statistics Explained. A Guide for Social Science Students, London, New York.

Hussein, S.G., 1983. A preliminary study of soils under *Acacia senegal* (L.) Wild. Sudan Silva, No. 25 vol. V, , Khartoum, Sudan.

IFAD, 2002. Evaluating cooperative societies of gum Arabic producers. Seminar Paper, Records of Forests National Corporation, Om Rowaba Locality, North Kordofan State, Sudan, pp. 8.

IIES & IES, 1990. Gum Arabic Rehabilitation in the Republic of the Sudan (1990). Stage 1 Report. Volume 2. Main Report and Recommendations for Action., International Institute for Environment and Development, London, England and Institute of Environmental Studies, University of Khartoum, Sudan.

Karama, M.M., 2001. Gum Arabic: Present Situation and Future Prospects. Department of Research and Statistics. Gum Arabic Company, Paper No. 3/01, Khartoum, Sudan., 13.

Khalid, A.M., 2005. Annual Reports and Archive Unit, Note No. 28 – 2005. Ministry of Finance, North Kordofan State, Sudan, pp. 17.

Mahmoud, T.E., 2004. The Adequacy of Price Incentives on Production, Processing and Marketing of Gum Arabic in Sudan. A Case of North and West Kordofan (Ph.D. dissertation). TU Dresden, Germany, ISBN: 3-9809816-4-9.

NEF, Near East Foundation, 2010. NEF Launches Gum Arabic Project in Central Sudan., http://www.neareast.org/.

Pearce, D.W., 1988. Natural resource management and anti-desertification policy in the Sahel-Sudan zone: a case study of gum Arabic. Geo J. 17 (3), 349–355.

Ramly, F.M.A., 2002. Evaluating cooperative societies of gum Arabic producers. Seminar Paper No. 1, Records of Forests National Corporation, Om Rowaba Locality, North Kordofan State, Sudanp. 8.

Saastamoinen, O., 1992. Economic evaluation of soil erosion: a survey of methods and approaches. Scand. Forest Econ. 33, 486–509.

Taha, M.E., 2006. The Socio-Economic Role of *Acacia senegal* in Sustainable Development of Rural Areas in the Gum Belt of the Sudan (Ph.D. thesis). TU Dresden, Germany, ISBN: 3-9809816-6-6.

Yasseen, G.A.M., Eljack, A.Y., Ahmed, M.E.D., 2015. Sudan gum Arabic export performance during (2000–2014). Int. J. Chem. Environ. Biol. Sci. (IJCEBS) 3 (5), 360–362.

Impact of Vegetation Cover Changes on Gum Arabic Production Using Remote Sensing Applications in Gedarif State, Sudan

4

Sara A. Ahmed, Manal A. Kheiry*, I.A. Mofadel Hassan***

*University of Khartoum, Khartoum North, Sudan; **Kenana Sugar Company Limited, Khartoum, Sudan

Applications of remote sensing on vegetation cover changes in arid regions

Arid and semiarid areas are described by patterns of fluctuating annual rainfall. Thus, such areas are subjected to drought cycles during the precipitation deficit period that result in deterioration of the vegetation cover, which may recover following a season with good precipitation (Ahlcrona, 1988; Hellden, 1988; Herrmann et al., 2005). However, they provide important an land resource for adopting agricultural production and grazing systems. Monitoring and evaluation the status of vegetation cover in these areas are treated as very important tasks in the context of global climatic change and worldwide desertification dynamics. Studying the topographic features will enable us to determine the form of the heterogeneous distribution of the vegetation cover, taking into account that some topographic features are comparatively stable, whereas others are dynamic. Such assets therefore make up an important contribution to livelihoods.

Ecosystem services provide by forests include soil settlement, climate change mitigation, water conservation, and erosion and desertification control (FAO, 2010; Malagnoux and Atzmon, 2007; Rotenberg and Yakir, 2010). According to Stocking and Murnaghan (2001), land degradation due to natural causes is accepted to happen at a rate that is almost equal to the rate of natural rehabilitation. Thus, human-related elements represent quickened forms of land degradation.

Agriculture plays an important role in Sudan's economy; it contributed a yearly average of 35% to total GDP during 2000–14, and in addition, it provided about 80% of the total employment (Yasseen et al., 2015).

Forestry in Sudan incorporates wood items in the form of firewood, charcoal, timber, and nonwood products such as wild fruits, baobab, and gum products, particularly gum Arabic. Forests also provide watershed protection and fodder for domestic animals and wildlife. Gum Arabic is an essential forest product that is produced in the Gum Arabic Belt, which extends from the western border with Chad toward the eastern outskirt next to Ethiopia. The Gum Arabic Belt covers a territory of around 500,000 km^2; likewise, the belt is home to about one-fifth of the inhabitants of Sudan (Yasseen et al., 2015).

Gum Arabic. http://dx.doi.org/10.1016/B978-0-12-812002-6.00004-X

In semidry regions, agrarian harvests are probably going to be unpredictable, despite the fact that grazing is acceptable (Goodin and Northington, 1985). Dry grounds are, much of the time, subjected to dry spell, which is the fundamental constraining element on biomass generation and product yields. Human-induced factors, such as overcultivation, overgrazing, and other forms of inappropriate land use, when practiced under the situation prevailing in the dry lands, may result in significant degradation of vegetation, soil leaching, and in many cases, desertification (Keller and Goldstein, 1998), where desertification has been defined as land degradation in arid, semiarid, and dry, subhumid areas resulting from different factors, including climatic variations and human activities (IPCC, 2001).

Ideally, remote sensing plays a main role in improving global operational capability for monitoring land degradation and desertification in dry lands. Remote sensing has long been suggested as a time- and cost-efficient method for observing change to desert environments. In this capacity, it can serve both to improve checking endeavors and additionally give profitable data on dry land debasement in particular areas. The invention of remote sensing and geographical information system (GIS) techniques for land use/cover mapping has given a valuable and definite approach to enhance the choice of zones outlined as agricultural, urban, and/or industrial areas of a region (Selcuk et al., 2003). In association with GIS that gives reasonable stage to information, update, and retrieval (Chilar, 2000). The advent of high spatial resolution satellite imagery and more advanced image processing and GIS technologies has brought about a change to more normal and predictable observing and modeling of land use/land cover patterns. Remote sensing has been generally utilized as a part of updating land utilization/cover mapping, and it has become one of the most important applications of remote sensing of the desertification process in arid regions (Lo and Choi, 2004). The role of remote sensing as a tool for detecting degradation is essential in natural resource management and is still in a testing phase. Remote sensing data can support inventory approaches by informing about degradation patterns in forests in combination with ground-based monitoring of forest degradation. Desertification is a continuous threat that cannot be mapped and monitored by occasional searches. Spaceborne Earth observation is the only dependable system to collect systematic specific and quantitative information at frequent rates on the variations of geo- and biophysical parameters over large areas (Sarmap, 2003). However, Khiry (2007) and Adam (2011) manifested that spaceborne remote sensing is capable of monitoring vegetation cover [Acacia trees (hashab and talha)] through time-series analysis and is one of the most reliable devices with a high ability for information and data collection concerning different fields.

The remote sensing–based acreage estimations are aimed to be made available at least 1 month before the harvest of the crop in traditional rain-fed areas in order to cement administrators and planners to make strategic decisions on import–export policy matters and commerce negotiations. An extended project, Forecasting Agricultural produce utilizing Satellite, Agro-meteorological and Land Observations (FASAL), is in progress to supply multiple forecasts at district, state, and national levels (Parihar and Oza, 2006).

Remote sensing has a significant role in evaluating the degradation and desertification in Sudan, especially in the Gum Arabic Belt in eastern Sudan, where it gives good indications for each of the instrumental factors of man-made overcutting, overgrazing, climate factors, or the entry of refugees from neighboring countries (Ahmed, 2013).

According to FAO Global Forest Resources Assessment (FRA, 2000), the total forest cover of Sudan is estimated as 61,630,000 ha and comprises 26% of the country's land area of 237,600,000 ha. The forest cover area in the Sudan decreased from 71,220,000 ha in 1990 to 61,630,000 ha in 2000 (FAO, 2001), a decrease of 959,000 ha/year; 90% of which was for fuel and charcoal making (ADB/EC/FAO, 2003). The steady deterioration and loss of Sudan's forest is attributed to the population's reliance on wood to meet energy required, drought and desertification in the north, the extension of mechanized agriculture, and the lack of efficient forest governance (UNEP, 2007). It can be concluded that the remote sensing technique is very useful in illustrating the deterioration in forests, and very precise image satellites made it easier for much of the detection of deterioration found in the ground (Figs. 4.1 and 4.2).

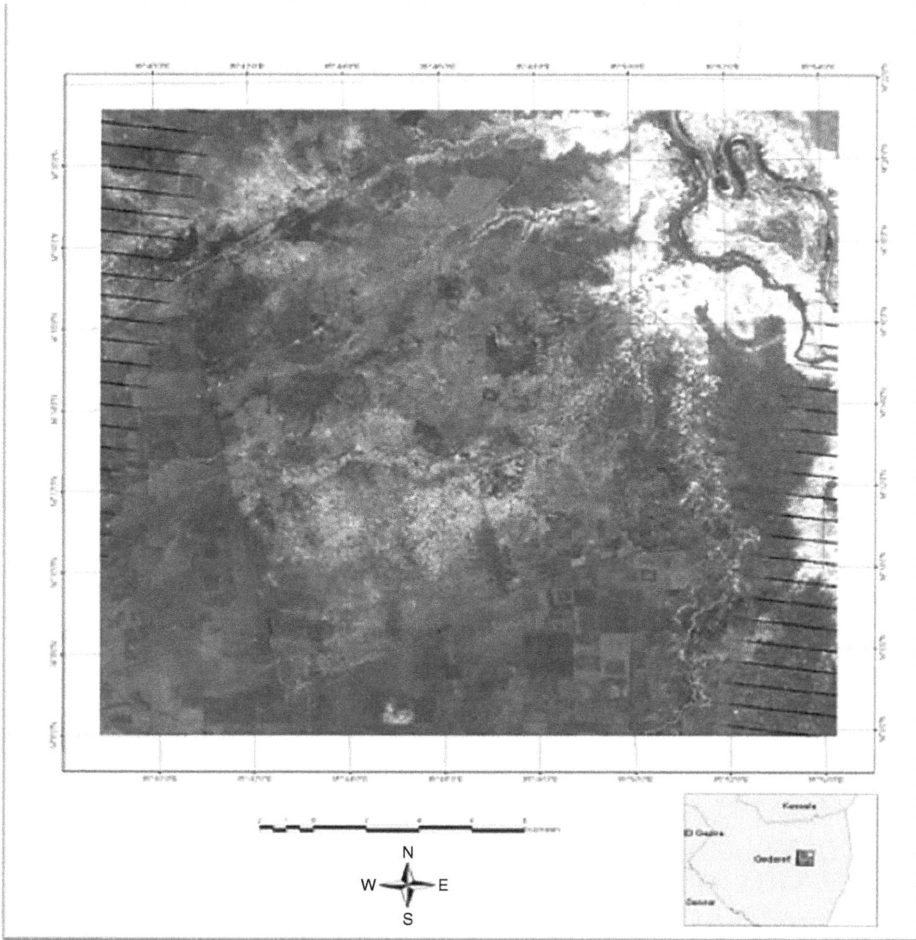

Figure 4.1 This is subset, vegetation cover in 1999.

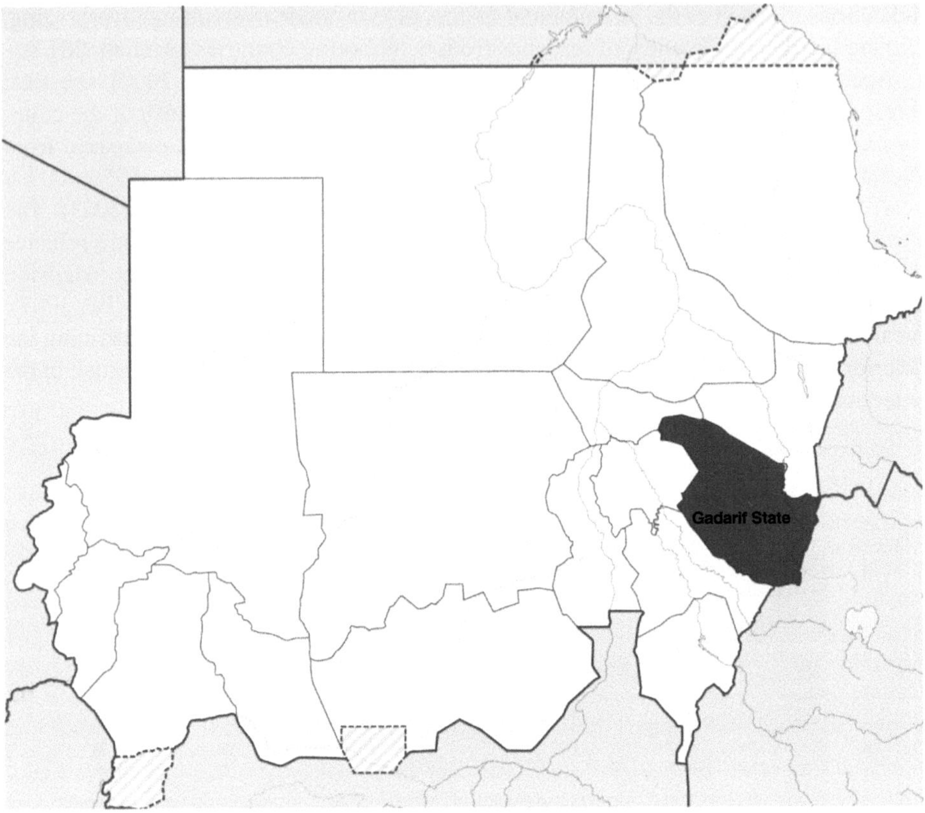

Figure 4.2 Location of AL-Gedarif State area, Sudan.
Source: https://en.wikipedia.org/wiki/Gedarif_(state)

Demographical features of Gadarif state

Gadarif state is one of the eastern region states of the Sudan. It lies between longitudes of 33°30′E and 36°30′E and latitudes of 12°40′N and 15°40′N, with an area of approximately 65,000 km^2. It lies between two frontier tributaries of the Blue Nile: the Atbara River in the east and the Rahad River in the west. The study area of the vegetation cover is located on the western part of the eastern clay plain, which descends down from the basaltic Gadarif–Gallabat edge at 650 m a.s.l. to the Atbara River. The study area lies in the semiarid zone, which has summer rains and warm winters and is characterized by a unimodal rainfall pattern ranging from 400 to 800 mm, with an annual average of 600 mm (Atta El Moula, 1985). In Gadarif state, clay soil prevails as a result of cultivation, which decreases soil fertility and thus impacts vegetation cover. The length of the stormy season is around the four months, from June to September, achieving its peak in August. The greater part of the downpours is from June/July to October/November. Precipitation frequently comes as substantial deluges amid electrical storms, bringing on an overwhelming overflow of water, causing disintegration to inclining, unprotected land because of clay soils. Generally, the vegetation of the area is largely dependent on rainfall and soil types. The area lies in a low-rainfall, woodland savannah belt on clay. Despite the high deforestation trend in the study area, the natural reserve forest vegetation cover is still composed of a great diversity of trees, shrubs, and grasses (FNC, 2013). The main dominant trees are:

- *Acacia seyal* (talha), the dominate tree species, covering about 70% the total area of the forest;
- *Acacia senegal* (hashab), mainly planted but some naturally growing trees are mixed in from the talha stand;
- *Acacia mellifera* (kitr), large shrubs, incidental throughout, more frequent and dominant on the slopes of the major khors draining towards the Atbara River;
- *Acacia nubica* (laot), incidental; small shrubs, few pockets in the northern piece of the forest;
- *Balanites aegyptiaca* (heglig), small to large, dominant tree, scattered all throughout the reserve; and
- *Cortolaria senegalensis* (Safari), *Ocimumbasilicum* (Rehan), *Sorghum happens* (Adar) (Ahmed, 2013) (Fig. 4.3).

The estimated population of the Gadarif state was 363,945 in 2012. About 90% of the population of Gadarif is farmers. The average population density was estimated to be 10 people per km^2 (Mustafa, 2006). The main causes of the vegetation cover changes in Gadarif state are overcutting by nomads and the entry of refugees in the state from the neighboring countries of Ethiopia and Eritrea. Because of the forest being used for farming and sadly also overgrazing, forest agriculture peaks did not

Figure 4.3 Change of vegetation cover in 2013.

develop. Timber cutting and shifting cultivation mechanisms lead to a decline in soil fertility. Gadarif state in eastern Sudan lost its status as one of the real wellsprings of nourishment in the nation because of the expansive scale of corruption of its rich soil and in addition, unsuccessful land utilization strategies and practices (Elsiddig, 2003). Regardless of the accessible records, arranged forest administration in the Sudan has, since 1932, been founded on an administrative approach that limits the entrance of nearby groups to the forests aside from those with unique licenses (Elsiddig, 2003); the result is still a lot of forest deterioration.

Plantations are a positive advantage and ought to help with diminishing the rate of deforestation. Estates expel the timber weight on normal forest does not make an interpretation of in the end into less, but instead into more deforestation. In reality, it is expected that farming extension, which is the primary driver of deforestation in the tropics, may supplant ranger service in the staying regular the way that (Anon., 2001; Cossalter and Pye-Smith, 2003). Whether they are supported by the governmental programmers, these settlers have often colonized the forest by using logging trails or new roads to access the forest for subsistence land (Wilkie et al., 2000). The advancement of these framework ventures are cause for overall worry, since tropical forest clearing represents about 20 for each penny of anthropogenic carbon emanations, obliterating internationally huge carbon sinks (Anon., 2001). Around 21% of tropical forests have been lost worldwide since 1980 (Bawa et al., 2004).

Gum Arabic Belt in Sudan and its economic importance

The Gum Arabic Belt covers about one-fifth of the country; the belt's gross area is estimated to cover 520,000 km^2. It spans over 11 states: Western Darfur, North Darfur, South Darfur, East Darfur, North Kordofan, Western Kordofan, South Kordofan, White Nile, Sennar, Blue Nile, and Gadarif. Sudan was contributing 77% of the world's production of gum Arabic. The percentage from Gadarif state has been about 5%. Gadarif state is comprised of about 41 reserve forests, with *Acacia* trees accounting for most of these forests, including gum Arabic trees from both *A. seyal* and *A. senegal* (Mohamed, 2010). Gum Arabic, which is defined as *A. seyal* and *A. senegal*, enters in many areas in medicine to treat many diseases and is used in many industries, such as yogurt, soda water, and cosmetics (Mohamed, 2010).

In terms of exports, the gum Arabic market its still controlled by Sudan in spite of many years of disruption caused by the civil war, frequent policy intervention, and market failure. Recently in 2013, there was a fluctuation in the exports of gum Arabic, where the exports rose sharply to 63,012 tons as the highest export extent registered for the previous 20 years and fell again to 59,978 tons and 48,700 tons in 2014 (Mahmoud et al., 2014). The general movement for gum Arabic exports for the last 20 years revealed an upward zigzag trend, as presented in Table 4.1.

As it has been previously stated, Sudan is one of the biggest gum Arabic producers in the world, and gum contributes to the household income ratio in the gum belt as cash income in the off-season.

Table 4.1 **Gum Arabic export in Sudan for the period 1999–2015 (quantity in 1000 metric tons and value in USD $100,000)**

Year	Quantity (in 1000 tons)	Value in USD $100, 000
1999	25.67	26.40
2000	24.20	23.10
2001	22.42	24.30
2002	36.19	31.90
2003	36.16	35.40
2004	27.27	60.60
2005	29.21	107.60
2006	20.62	50.20
2007	30.88	51.90
2008	32.22	60.90
2009	47.85	73.20
2010	55.35	77.9 0
2011	52.928	77.80
2012	49.318	82.20
2013	63.012	134.20
2014	59.978	97.99
2015	48.70	84.15
	سبتمبر حتى	سبتمبر حتى

Source: Tarig, G., et al., 2016. Crop Commercialisation and Adoption of Gum-Arabic Agro Forestry.

Impact of vegetation cover changes in the Gum Arabic Belt in Gadarif state

In Gadarif state, there are many reasons to increase the degradation in vegetation cover. Research has been conducted to highlight the causes of degradation and map the changes in vegetation cover, namely in the production of gum Arabic in the study areas. The reasons, including climate change, temperature change, insects, and man-made causes, which include cutting for charcoal, building of homes in proportion to the nature of the area, and the weather conditions that encourage the use of timber and firewood in home building, as well as the entry of refugees from the border, have a significant impact on change in the region. Accordingly, agribusiness is the principle financial movement, trailed by livestock animals ascending in the customary regular transhumance design, town livestock animals rising and, as a current component, animals ascending by substantial scale automated shipper agriculturists putting surplus riches in steers (Ahmed, 2013).

Gum tapping, gathering, exchanging backwoods items, and charcoal consuming are other customary types of financial movement. Also of the most important reasons tapping gum Arabic to the large proportion of farmer's age and the lack of young people to practice this profession. Thus, members of the general population get their pay from different blends of the three fundamental types of land utilization: tapping trees for Gum Arabic, touching, and forest misuse. Automated rain encourages yield creation, but has extensively decreased the land accessible for little holder cultivating and for grazing (Glover, 2005).

Agriculture has caused a decline in crop yields and an increase in infestation of cropland by the hemiparasitic plant *Striga hermonthica*. Mechanized farming has taken over nomads' traditional grazing lands and blocked traditional migration routes (Glover, 2005). This created a fundamental shift in focus from livestock production to crop farming. The land tenure system has created a situation of a few large-scale and many smallholder farmers as well as restricting livestock herders to the marginal areas of the state (Glover, 2005). The gum agroforestry remains an integral part of the farming system in the country. In this type of farming system, in the gum belt in Sudan, the gum Arabic tree (*A. senegal*) is integrated into agricultural cropping through a system called the "gum cultivation cycle" or the "bush fallow system." As noted by Barbier (2000), *A. senegal* is native to northern Sudan and has been cultivated by farmers there for hundreds of years. In such farming systems, the farmers cultivate the land for a period of 4–8 years (in the west) and 5–10 years (in the east) with millet (*Pennisetum americanum*) and sorghum (*Sorghum bicolor*) as the staple food crops and sesame (*Sesamum orientale*), groundnut (*Arachis hypogaea*), watermelon [*Citrullus lanatus* (Thunb)], and roselle (*Hibiscus sabdariffa* L.) as the most important cash crops (Tarig et al., 2016). All these reasons have affected the ratio of gum Arabic in the region in recent years, and that was evidently clear through the change from the years 1999–2013.

Remote sensing tools used

The processing technique was divided into two steps: gathering and analyzing remotely sensed data as well as climatic and field data. Random points were taken irregularly using GPS (Global Positioning System); also, photos have been taken by digital cameras to clarify the deterioration in the study area and analyze the remote sensing data as well as field observation. Data in the form of satellite images were analyzed qualitatively (visual in perpetration) and qualitatively (information is extracted by the aid of computer). The combination of both postclassification methods and spectral change detection techniques were used (Ahmed, 2013). Two cloud free land sat Enhanced Thematic Mapper (ETM) scenes covering (November 22, 1999; December 6, 2013); the study areas were selected for the analysis. The ETM data have been acquired simultaneously in eight spectral bands. In order to assess the vegetation cover change in the study area, three procedures of image preparations are generally employed. These are: digital image processing, where a digital image is sampled and quantized numeric representation of the scenes. The following tasks have performed on both images (1999 and 2013): image classification, where there are many well-developed techniques for land cover change detection using the digital remotely-sensed imagery, and finally the postclassification comparison, which discriminates changes between the two dates of imagery (Ahmed, 2013). It is a common application in land cover change analysis using bitemporal imageries of the same region of interest.

The field survey has been carried out in different localities in Gadarif state study areas to have general descriptions of vegetation covers. Random points were taken irregularly using GPS; also, photos have been taken by digital camera to clarify the deterioration in the study area. Moreover, interviews and discussions were conducted.

Results and discussion

The classification of the image of 2013 compared to the classification of that from 1999 show clear evidence of quantitative and qualitative changes in the study area. The analysis proved that the vegetation of the study area (low and high vegetation) in 1999 was better than low and high vegetation in 2013. Water also increased significantly in 2013, up to 19%. These results illustrated the change in the vegetation, which is considered as clear evidence of degradation in study area over 14 years during the period from 1999 to 2013. From the analysis of the two images, there are obvious changes to the vegetation cover in the study area. The changes in the vegetation cover were clearly quantitative and qualitative. This change was very clearly apparent from the satellite images of the years 1999–2013. This change in the vegetation cover is due to deterioration in the tree cover and as a result of the change in human activities. Regarding the results from the analysis, the vegetation cover in the southern part of the study area is denser than the northern part.

As the ground overview distinguished, the explanations for the debasement of the vegetation in the review zone are as follows:

The impact of nomadic

Migration of refugees from neighboring countries and the misuse of forest resources in the area cause damage from overgrazing and the cutting down of trees to provide building materials and fuelwood. In addition to that, the increased population expansion in the neighboring villages to the forest, (such as Mgargor, Al-Fashaga, and al Tomat), as well as the deducted of part from the forest and cutting down the trees in order to create an Alshowak airport has contributed to the reduction of the forest area; these are additions to the agricultural expansion in the forest.

According to the results, which have been obtained from the analysis of the two images from 1999 to 2013, there is clearly a change in the area covered by vegetation, and there is also an increase in the cropland and water area in the 2013 image. Changes are based on the field observations that have been taken during the ground survey of the samples, which have been identified on the map according to the different colors. The results also proved that the movement of pastoralists with their animals from one region to another has a clear impact on the deterioration of the study area. The effect of overgrazing on the vegetation covers is in quantitative and qualitative terms, which affects the production of gum Arabic and also leads to the deterioration and reduction of the perennials and shrubs. There have been no limitations on the number of animals a pastoralist keeps on the communal land;

each increase their herds at will, but the damage caused by competitive grazing is felt collectively in the limited space, and this could be one of the difficulties encountered by nomads. *A. seyal* (talha), on the other hand, is more brittle than the hard tears of *senegal* gum (wageningenuniversiteit.nl/UK/news agenda/archive/agenda/2006).

Through nomadic system a village based activity support respondents income by 32.9%. Animal rearing could be considered as a complementary activity to agriculture in the study area, where the agricultural residues provide fodder to the livestock, while the hashab trees provide fodder during the dry period. Also, some farmers have shifted to the production of charcoal, which has an attractive price compared to gum Arabic. About one-fifth of the respondents look at the hashab tree as a source for charcoal production. Some respondents (12.7%) look to the hashab tree as a source of building materials. Hashab trees provide the necessary components of the local thatches (locally: *shiba*, *rassas*, *korki*, and *matarig*), as the majority of the houses in the study area are made of local materials obtained from the hashab tree and other natural tree species. Gum Arabic production is no longer attractive to all the farmers despite existence of hashab trees in their lands. Farmers who possess lands through inheritance and purchasing are keen to produce gum Arabic, contrary to those who possess lands through the aptitude management system. Sometimes the hashab trees are cut before the completion of the cycle and converted into charcoal or fuelwood. This situation is accelerated by the low prices of the gum Arabic in the study area (Koli et al., 2013) (Figs. 4.4 and 4.5).

The effect of the overgrazing on the vegetation covers is in quantitative and qualitative terms, which affects the production of gum Arabic and also leads to the deterioration and reduction of the perennials and shrubs. There has been no limitation on

Figure 4.4 Overgrazing in study area (2013).

Figure 4.5 Over refugees in study area (2013).

the number of animals a pastoralist keeps on the communal land; each increase their herds at will, but the damage caused by competitive grazing are felt collectively in the limited space, and this could be one of the difficulties encountered by nomad. *A. seyal* (talha), on the other hand is more brittle than the hard tears of *senegal* gum (wageningenuniversiteit.nl/UK/news agenda/archive/agenda/2006). Other products of acacia tree locally are sources of fuelwood and an important source of browse for livestock.

Shifting cultivation in forests

The clashing objectives between expanded generation and maintainable asset administration ended up noticeably apparent in rain-encouraged horticulture, especially in the vast scale of motorized farming. The extensive agricultural using rain-fed, mechanized and the shifting cultivation that practiced by refugees has contributed on the decline of forest area as a result of cutting trees to use the land to plant crops. The present pattern of trimming, development, and practicing traditional intercropping systems with *A. senegal* was a result of the increasing of agricultural commercialization programs, which led natively to forest degradation in the study area. This situation prompted an uncommon decrease in gum Arabic generation, an increase in soil erosion, and loss of soil as well as decreasing acacia trees and growth of acacias. A definitive outcome is the decrease in harvest efficiency and reducing ranch pay, which speaks to a genuine danger to the agriculturists by placing them in a neediness trap. In view of that, there could be two paths by which the gum agroforestry system may be improved and soil fertility can be maintained (Tarig et al., 2016).

Population pressure

The population pressures in the study area represent one of the most disturbing problems there. The study area was facing displacement from the countryside for grazing animals, agriculture, and even for residential purposes. In additional to that, the refugees from neighboring borders made the situation worse when they used the natural resources for provide their needs illegally. However, speeding up transformation of smallholders to be commercialized farmers without considering their limited resources (labor and land) may lead to undesirable impacts on sustainability of the agroforestry system in the gum belt (Fig. 4.6).

In conclusion, remote detecting was additionally incorporated with programming methods,which made it less demanding and precise to register rates of both *A. senegal* (hashab) and *A. seyal* var. *seyal* (talha). Based on the models developed, it is clear that there is a significant relationship between Land sat bands components and the resources mapped using high-resolution images for both *A. senegal* (hashab) and *A. seyal* var. *seyal* (talha). Remote detecting innovation was observed to be a utilization full device for mapping assets, particularly tree species, because of its uniform and general picture procurement interims. The study concluded that decreased from 30% in 1999 to 12% in 2013, a rate of 1.28 per year in a linear relationship. The resulting change in the amount of gum Arabic is a result of these reasons. This quick change in the cultivating framework from intercropping framework with *A. senegal* to sole trimming framework is prompting unfavorable effects on the maintainability of the cultivating framework, which prompts a radical decrease in the gum Arabic creation and causes more soil disintegration and loss of soil ripeness because of diminishing development of acacias. This has caused the farmers to expand the area allocated to these cash crops in place of *A. senegal* trees. Overcutting and overgrazing are considered as the main reasons behind the degradation of the vegetation

Figure 4.6 Overcutting.

cover in the study area. Based on results obtained, the decrease of vegetation cover will be very significant unless appropriate and updated management solutions for tree cover are adopted.

References

Atta El Moula, M.A., 1985. The problem of resource management in the Sudan. Environ. Monogr. Ser. 4.

Adam, H.E., 2011. Integration of Remote Sensing and GIS in Studying Vegetation Trends and Conditions in the Gum Arabic Belt in North Kordofan, Sudan (Ph.D. thesis). Faculty of Forest, Geo and Hydro Sciences, Dresden, Germany.

Ahlcrona, E., 1988. The Impact of Climate and Man on Land Transformation in the Central Sudan. Lund University Press, 140 pp.ences, 16: 1407–1410.

Ahmed, S., 2013. Assessing and Mapping the Vegetation Cover Using Remote Sensing for the period (1999–2013) in Western Parts of Gadaref State, Sudan (MSc. thesis). University Khartoum, Khartoum, Sudan.

Anon, 2001. Climate Change, 2001. Synthesis Report. Contribution of Working Groups I, II, III to the 3c Assessment Report of the IPCC. IPCC, Cambridge University Press, Cambridge. Rd. UK.

Barbier, E.B., 2000. The economic linkages between rural poverty and land degradation. Some evidence from Africa. Agric. Ecosyst. Environ. 82, 355–370.

Bawa, K.S., Nadkarni, N., Lele, S., Raven, P., Janzen, A., Lugo, A., Ashton, P., Lovejoy, T., 2004. Tropical ecosystems into the 21st century. Science 306, 227–230.

Chilar, 2000. Land cover mapping of large areas from satellites: status and research priorities. Int. J. Remote Sens. 21, 1093–1114.

Cossalter, C., Pye-Smith, C., 2003. Fast-Wood Forestry, Myths and Realities. CIFOR, Bogor Indonesia.

Elsiddig, E., 2003. Management of dryland forest reserves in Sudan based on participatory approach. In: Alsharan, A.A., Wood, W.W., Goudie, A.S., Fowler, A., Abdellatif, E.A. (Eds.), Desertification in the Third Millennium. Belkan Publishers, Lisse, The Netherlands, pp. 361–364.

FAO, 2001. Global forest resources assessment 2000. Main report. FAO Forestry Paper No. 140. Rome.

FAO, 2010. Food and Agriculture Organization of the United Nations: what wood fuels can do to mitigate climate change? FAO Forestry Paper 162. Food and Agriculture Organization of the United Nations, Rome.

FAO, 2003. Unlocking the water potentialof agriculture. Rome, pp. 62.

FNC, 2013. Report of Forest National Corporation in Gadarif State, and Kordufan State, Sudan.

FRA 2000. Forest Resources of Bangladesh, Country Report. Rome 2000. Working Paper No 15. Rome 1999, pp. 93.

Glover, E.K., 2005. Tropical Dry Land Rehabilitation: Case Study on Participatory Forest Management in Gedarif, Sudan (Ph.D. dissertation). Faculty of Agriculture and Forestry of the University of Helsinki, Finland.

Goodin, J.R., Northington, D.K., 1985. Arid and Semi-Arid Lands: Characteristics and Importance Arid and Semi-Arid or Sub-humid Zones are Characterized by Low Erratic Rainfall.

Hellden, U., 1988. Desertification monitoring: is the desert encroaching? Desertification Control Bull. 17, 8–12.

Herrmann, S.M., Anyamba, A., Tucker, C.J., 2005. Exploring Relationship Between Rainfall and Vegetation Dynamics in the Sahel Using Coarse Resolution Satellite Data. www.isprs.org/publications/related/ISRSE/html/papers/293.pdf.

IPCC, 2001. Intergovernmental Panel on Climatic Change, "Impacts, Adaptation and Vulnerability", 2001.

Keller, A.A., Goldstein, R.A., 1998. Impact of carbon storage through restoration of drylands on the global carbon cycle. Environ. Manage. 22 (5), 757.

Khiry, M.A., 2007. Application of Remote Sensing in Monitoring Drylands, North Kordofan State, Sudan (Ph.D. thesis). Faculty of Forest, Geo and Hydro Sciences, Dresden.

Koli, A.O., Eltayeb, A.M., Sanjak, E.M., Mohammed, M.H., 2013. Socio-economic aspects of gum arabic production in Dalanj Area, South Korodofan. Sudan. Pak. J. Biol. Sci..

Lo, C.P, Choi, J., 2004. A hybrid approach to urban land use/cover mapping using Landsat 7 enhanced thematic mapper plus (ETM+) images. Int. J. Remote Sens. 25 (14), 2687–2700.

Mustafa, R.H., 2006. Risk Management in the Rain-Fed Sector of Sudan: Case Study. Margraf Publishers, Gedarif Area Eastern Sudan,Weikersheim.

Mohamed, B.S.O., 2010. Physicochemical Characterization and Rheological Properties of *Acacia mellifera* Gum Exudates From Sudan (Ph.D.), University of Khartoum, Sudan.

Mahmoud, T.E., Maruod, M., Khiery, M., El Naim, A., Zaied, M., 2014. Competitiveness of gum Arabic marketing system at El Obeid crop market, North Kordofan State, Sudan. World J. Agric. Res. 2 (5), 252–256.

Parihar, J.S., Oza, M.P., 2006. FASAL: an integrated approach for crop assessment and production forecasting. In: Kuligowski, R.J., Parihar, J.S., Saito, G. (Eds.), Proc. of SPIE Vol. 6411 641101 Agriculture and Hydrology Applications of Remote Sensing. doi:10.1117/12.713157.

Sarmap, 2003. Treaty Enforcement Services Using Earth Observation Desertification, University of Valencia, EOS.D2C, Chinese Academy of Forest, Final Report, ESA/ESRIN.

UNEP, 2007. Global Environment Outlook (GEO 4)—Environment for Development. Valetta, Malta. http://www.unep.org/geo/geo4/report/GEO4_Report_Full_en.pdf.

Selcuk, R., Nisanci, R., Uzun, B., Yalcin, A., Inan, H., Yomralioglu, T., 2003. TS18.5 Monitoring land-use changes by GIS and remote sensing techniques: case study of Trabzon. In: 2nd FIG Regional Conference Marrakech, Morocco, December 2–5, 2003.

Stocking, M., Murnaghan, N., 2001. Handbook for Field Assessment of Land Degradation. Earth Scan Publication Limited, London, Stering, UK.

Tarig, G., et al., 2016. Crop Commercialisation and Adoption of Gum-Arabic Agro Forestry.

Yasseen, G.A.M, Eljack, A.Y., Ahmed, M.E.D., 2015. Sudan gum Arabic export performance during (2000–2014). Int. J. Chem. Environ. Biol. Sci. 3 (5), 360–362.

Wilkie, D., Shaw, E., Rotberg, F., Morelli, G., Auzel, P., 2000. Roads, development and conservation in the Congo Basin. Conserv. Biol. 14, 1614–1622.

Further Readings

Al Awad, A.A., 1985. Eco Taxonomical Studies on Vegetation of Red Sea State, Sudan (Ph.D. thesis). University of Khartoum, Khartoum.

Daniel, L.C., James, D.H., Emily, H.W., Mingjun, S., Zhenkui, Z., 2002. A comparison of land use and land cover change detection methods, ASPRS-ACSM Annual Conference and FIG 22nd Congress, pp. 2.

Epaphras, A.M., Gereta, E., Lejora, I.A., Ole Meing'ataki, G.E., Ng'umbi, G., Kiwango, Y., Mwangomo, E., Semanini, F., Vitalis, L., Balozi, J., Mtahiko, M.G.G., 2008. Wildlife water utilization and importance of artificial waterholes during dry season at Ruaha National Park Tanzania. Wetland Ecol. Manage. 16, 183–188.

GIZ, 2013. (Deutsche GesellschaftfürInternationaleZusammenarbeit), Combating DesertificationGermany's Commitment to Implementing the United Nations Convention UNCCD) in the Context of International Cooperation, Bonn.

Helldén, U., 1991. Desertification-time for an assessment. Ambio 20 (8), 372–383.

Pooja, K., Kumar, M., Rawat, J.S., 2012. Application of remote sensing and GIS in land use and land cover change detection: a case study of Gagas Watershed, Kumaun Lesser Himalaya India. Quest 6 (2), 342–345, 70.

Rahim, A.H., 2006. Economic Analysis of Deforestation: The Case of the Gum Arabic Belt in Sudan. Wageningen University, Netherlands.

Redfern, J.V., Grant, C.C., Gaylard, A., Getz, W.M., 2005. Surface water availability and the management of herbivore distributions in an African savanna ecosystem. J. Arid Environ. 63, 406–424.

Reynolds, J.F., 2001. Desertification. Encycl. Biodivers., 66–78.

Reynolds, J.F., Stafford Smith, D.M, 2002. Global Desertification: Do Humans Cause Deserts? Dahlem University Press, Berlin, vol. 88.

Singh, P., Gupta, P., Singh, M., 2014. Hydrological inferences from watershed analysis for water resource management using remote sensing and GIS techniques Egypt. J. Remote Sens. Space Sci. 17, 111–121.

Souza, C.M., Dar, A.R., Mark, A.C., 2005. Combining spectral and spatial information to map canopy damage from selective logging and forest fires. Remote Sens. Environ. 98, 329.

UNFCCC, 2007. Report on the Second Workshop on Réduction Émissions from Déforestation in Développant Countries. FCCC/SBSTA/2007/3.

Gum Arabic: Certification and Assessment of Marketing Opportunities

5

Hagir Mahagoub Elnasri Mohammed
University of Khartoum, Khartoum, Sudan

Introduction

Certification envisages recovery of degraded Sudanese gum Arabic production and the opening of new market opportunities abroad, since it assesses the extent of institutional, socioeconomical, and safe environmental management aspects of trade products from natural resource origins against certain international set quality standards. As has been highlighted by Fall Brook Center. Website: http://www.web.net/fbcja the assessment extends across management plans, monitoring and rating, ecological harvesting activity, biodiversity conservation, soil and water conservation, protection of high conservation value areas, regulation of chemical substances, integrated pest management, tenure and customary use rights, air returns and adequate benefits, safe and healthy working environments, impact on local and indigenous communities, and economic viability.

The gum Arabic commodity has dominated the world markets for generations, as it is a basic ingredient to many food and medicinal items. Sudan dependence on the export earnings of gum Arabic has been threatened in recent years by poor production and marketing policies that discouraged small producers from producing large quantities with good quality that fit with the international standards. The Forest Certification Handbook (1995) provides the principles for forest products certification, while three international organizations carry out implementation of the assessment of nontimber forest products (NTFPs); these include the Forest (timberland) Stewardship Council (FSC), the International Federation of Organic Farming Movements (IFOAM), and the Fairtrade Labelling Organization (FLO). The three are of interest because they each have a set of generic standards in their specific field of expertise, they play a coordinating role for organizations working in their sector, and they each accredit or coordinate certification organizations that do the actual assessments of the product chain.

Gum Arabic is a strategic NTFP of Acacia trees of the Sudano–Sahelian region of sub-Saharan Africa that grows naturally in the Sudan. The inherent characteristics of gum Arabic are exemplified by its emulsification, low viscosity at high assiduousness concentration, acid stability, adhesive and binding properties, and good mouth feel, which make it widely used in beverages, insecticides, pharmaceuticals, ceramics, lithographic plates, candies, and countless other applications (Karama, 2002).

Gum Arabic. http://dx.doi.org/10.1016/B978-0-12-812002-6.00005-1

The marketing policy of gum Arabic in Sudan was unstable and had been partially responsible for the ongoing deterioration of production and foreign trade of this product. The marketing of gum Arabic had a complicated market chain associated with fluctuating demand and supply prices (Karama, 2002). There was no reference to product quality and grade price differential to producers of gum Arabic. The effect of those policies resulted in increased smuggling of gum Arabic, severe deforestation of the gum belt, reduced income, and substitution of the gum trees by other crops such as sesame, groundnuts, and rosella. The application of certification to gum Arabic had been considered as essential, since the small producers of this NTFP have been subjected to increased poverty and their land exposed to continued environmental degradation.

Absence of certification is assumed to be one of the critical factors causing fluctuations and nonsustainability of domestic and international supply and prices of gum Arabic of Sudan. This concern calls for exerting more effort to analyze the basic factors for the failure in market, institutions, and policies that had led to such problems in order to study whether certification and current certification initiatives are of real use in slowing the Gum Arabic Belt loss and deterioration and assuring the objectives of sustainable management practices. The identification of quality forestry as an interim and achievable goal regarding social, economic, and environmental costs and benefits is crucial in this respect. The socioeconomic impact of gum Arabic production and marketing are social impact and migration of gum Arabic producers in search of work, which leads to loss of indigenous knowledge, change of welfare, and change in production credit and marketing relationships at the field level. The environmental impact includes the movement of the gum belt southward, where there is absence of indigenous knowledge and expertise, effects of climate change on gum production even under rehabilitated areas, depletion of soils, and decline in crop production under shifting cultivation of the Acacia crop. All of these affect the quality of tapped gum Arabic (certification) and sustainability of products, so the key challenges are the management information system, effect on production–marketing chain (chain of custody in certification system), and international market demand and prices (fluctuations and development of synthetics substitutes to reduce dependence on natural gum Arabic).

The aim of this chapter is to highlight the factors that may facilitate the certification of gum Arabic to access more assured marketing opportunities worldwide.

Gum Arabic production policies

The policies and strategies adopted by the state since the dawn of the last century and the National Forests Plan were directed toward the protection and development of the tree resources and increasing gum Arabic production to the level that secure the foremost position of the Sudan in the international trade. Section 6 of the 1932 Forests' Policy reads that the role of the provincial authorities is focused on the maintenance and protection of the existing (gum) areas, supervision over the markets, improvement of communications, afforestation, and encouraging the opening up of new areas.

Governors of provinces are entrusted with supervision of the markets; the role of the forestry units in the provinces is restricted to technical aspects concerned with surveys and the provision of field data on all factors that affect gum Arabic production.

The amended forest policies of 1986 prescribed the development of gum forests, increasing their productivity by forestation in forest areas and gum gardens, selection of high-yielding strains, improvement of production methods, and encouragement of the rural dwellers to restock the gum gardens. The Comprehensive National Strategy of 1993 was deeply concerned with gum Arabic and developments of its production, setting a production target of 60,000 tons per year. The government policy encourages local and foreign capitals to invest in gum Arabic production and industries by granting concessions to export raw gum to those who plant 50,000 feddans of hashab or invest five million tons or more in the gum industry. Thus, there are many companies involved in gum Arabic export besides those of the concession company (Elnasri, 2000).

The draft of Sudanese forest policy in 2006 devoted appreciable space to gum Arabic as the most important nonwood forest product in the Sudan. The policy option focused on increasing the benefits to the land owners and producers by encouraging the establishment of local societies, cooperatives, and producers' unions and improving their investment and industrial capabilities, increasing productivity and their competitive abilities through the provision of loans and finance, and increasing investment in infrastructures and services in the production areas. The policy focuses also on encouraging the private sectors and setting out the organizational frameworks that safeguard the gum quality and the upkeep of the Sudanese position in the international markets, securing supplies and market stability through the protection and stocking of the gum belt, expanding the production base, and the establishment of reserve against fluctuations in production. Unfortunately, all of these polices did not mention the certification of gum Arabic as a policy tool for the sustainability of the products or as a market-base instruments for the gum trade, the nonsustainable supply and prices of gum Arabic of the Sudan were the obvious result of adopting arrangements and policies in addition to the way that producers practice the business. Advancement of gum exchange gum—liberalization of the gum foreign trade—and the foundation of direct connection among gum Arabic business stakeholders. That would help to clarify the aspects in supply and demand that would prompt great benefits for both the tree owners and producers (Abdelazim, 2007).

The role of stakeholders involved in gum Arabic business

In the gum Arabic certification process, there is a wide range of players (stakeholders), varying from the local community to the government; each group has its own goals and aims. These players are:

1. *The community*: These are people who live in and around the woodlands (local people). They depend on the woodland for their livelihoods and are usually poor rural dwellers. In practice, they are the owners of the lands, and any decision on forest matters affects them directly. Economic development efforts in the rural areas are aimed at improving the life of this lot. Apart from subsistence farming, there are no industries in rural Sudan, and the infrastructure

is undeveloped. Their goal is daily sustenance. What they have been told during the visit is that certification will bring increased incomes from the *A. senegal* trees that they sell through identified system and organizations. In view of this, the group has welcomed certification, with high expectations. Local communities are organized into villages that are headed by traditional leaders (tribal sheikhs) who come under a chief. In some villages, there is a gum Arabic producer association.

2. *The chiefs and traditional leaders*: These form the traditional authorities (tribal sheikhs) and have a significant influence among their subjects. They administer natural resources in their respective chiefdoms. Their goal is socioeconomic improvement in the lives of their subjects. They welcome certification as long as it does not usurp their powers.
3. *Local government*: The local government system is made up of state governments and municipalities. Their goal is provision of services to the local communities in their areas of jurisdiction. They are only interested in certification because it will bring more commerce to their areas, which will lead to raising their revenues through taxes.
4. *Forest department*: The department is the government arm; its task is to manage and supervise the usage of forest resources. Its goal is the sustainability management of the forest resources. Hence, any measures that promote sustainability are welcome. The concern of the department is the redistribution of powers that will result from the certification process as the certificate holder becomes more involved in the forest management plan. It is therefore the desire of the department to be closely involved in the process of certification so that their role is not sidelined.
5. *Donors and NGOs*: These have been involved in improving the livelihoods in rural areas. In Sudan, they have recognized and are well aware about the important role that NTFPs can play in order to improve rural incomes. To this end, they have to promote processing and marketing of all products nationally with the aim of maximizing income from NTFPs, including from *A. senegal* gum. In doing this, they will understand the fact that the source of NTFPs should be managed sustainably, but this can only be achieved when forest users receive adequate incomes or have access to credit and funds. This group adopts certification in order to access the more lucrative foreign marketplace.
6. *Private companies*: These are in business for profits and include Acacia Company and Dar Savannah. However, due to the size of the internal market, it is not possible to increase production without seeking external port markets, which are inaccessible without gum certification. From the foregoing information, it is evident that the goals of all players have not been a source of conflict.

Trading channels

With the exception of Sudan, which exports small quantities of kibbled gum, most other producing countries export only raw gum Arabic (Elnasri, 2000). Some of this is used by end-use industries without further processing, but in most cases, kibbled or spray- dried gum is used. For many consuming countries, principally where a small amount of gum is involved or where there is no domestic processor, imports are mainly processed gum that come from an intermediate country, or from any other secondary source, rather than the conventional producers. A significant proportion of gum Arabic in international trade is therefore in the form of reexports from countries such as France, the United Kingdom, and the United States, which have large processing capabilities and capacities.

There is much is to be gained, therefore, by examining export statistics for these intermediate destinations, as well as those (where they exist) for primary producers. This situation can be represented schematically as thus: once the gum Arabic reaches its final destination, it is either sold promptly to other, smaller brokers, prepared and processed by the similar organization which imports it, and afterward sold on to the end-use industries or utilized by the importer himself, who additionally happens to be an end user. In some cases, such as the importation of Sudanese gum Arabic into Germany, Italy, and the United States, some agent facilities the transaction between exporter and produced country but does not otherwise physically intercede in the transfer of goods.

Certification is driven by international markets. In Sudan, both the gum Arabic producers and consumers are not aware about certification, and certification itself is not yet developed. The idea of certification is to link trading channels to the sustainable management of forest resources by providing consumers with information on the current status of the production systems of the forests from which timber and other NTFPs products come.

International factors

The initiatives of timber certification began in 1992 following the Earth Summit, the United Nations Conference on Environment and Development in Rio de Janerio. The Rio conference recognized that the issues of need, poverty, nourishment, and food security were connected to cutting of trees and indebtedness of developing nations. Numbers of intergovernmental approaches and protocols provide sets of principles and criteria for developing of certification standards all over the world (Bass, 1998). Groups concerned about the environment, NGOs, and other interested groups began the certification initiatives, leading to the foundation of the FSC in 1992 (Ng'andwe, 2003). Other certification schemes emerged from there. After Rio, global awareness, nature protection, and environment conservation expanded in many countries. Other studies' results also demonstrated that expanded harm and damage to the environment of the Earth in one territory influences the life quality of a person somewhere else. The alert was being raised about the ozone layer, and it warned of the global costs. All of these were acknowledged, realized, and had starting points originating in the uncaring manner in which the woodland assets and natural resources products were being utilized. For encouraging better management for wood and nonwood resources, it was thought about wisely, and the effort was to empower and enforce measures that would support and encourage the approach of sustainable management of these woodland assets. The consumers from Europe and the United States also began to demand more natural, as opposed to synthetic, products. The best way to strike a balance is to adopt and come up with certification. The thought and idea were to confine and control markets for the nations and institutions that administered and managed their resource environments in an unsustainable manner. With this confinement, it has turned out to be troublesome and very difficult for the national institutions that are attempting to export timber and nontimber products to do so. National institutions are interested in either poverty alleviation or industrial development. Since the internal market is small and depressed, high income generations and salaries and increasing production from natural resource industries can only be attained through access to external markets for both timber and NTFPs. Sudan is a signatory of more than 22 local, regional, and global environment-related agreements, conventions, and treaties, some of which have been approved. The conventions and treaties provide for policy arrangement structures, which direct the country's worldwide policy on woodlands. The national forest strategy does not mention wood and Non-Wood Forest Products (nwfps) accreditation; the main topics and issues are sustainability of the management plan, conservation, and utilization (Felix, 2004).

Certification has turned to be the hottest and most debated issue in numerous natural resource sector. The harvest of NTFPs has recently become a topic of great interest for a variety of certification projects and programs.

Definitions of certification

Certification is a procedure by which a woodland owner voluntarily asks for an inspection to examine the forest and determine and decide whether predefined management principles, criteria, and indicators (standards) are being met.

Standards

The term standard is used generally to describe a quality or measure serving as a basis, example, or principle to which others conform or should conform, or by which the accuracy or quality of others is judged, as defined by the *Oxford English Dictionary*. Also, standards are documented agreements containing technical specifications or other precise criteria to be used consistently as rules, guidelines, or definitions of characteristics to ensure that materials, products, process, and services are fit for their purpose.

As defined by the International Organization for Standardization (ISO), there are different types of standards, which can be categorized as being either external or internal to the local forest management unit (LFMU).

External standards are those set by third-party independent bodies. These are usually international or national bodies, and all stakeholders usually participate in the standard-setting process. Internal standards are those developed by LFMUs to describe the level of performance that their forestry activities must reach.

Internal standards are interpreted from external standards and are specific to the LFMU. A vital part of the certification process is an assessment of this interpretation by the certification body.

The process for evaluating if the woodland is managed sustainably, and a way to impart natural data about woodland to consumer.

Definition of nontimber forest products

The FSC working group suggested a definition for NTFPs as "every biotic item other than timber that can be reaped and utilized for subsistence and additionally for exchange." NTFPs may originate from different types of woodlands, such as primary, secondary, and natural forests, as well as forest plantations (FSC, 1997). Sorts of NTFPs incorporate sustenance, drinks, flavors, seasoning, fragrances, solutions, paints, shines, construction materials, and concentrates utilized as part of manufacturing chemicals. Recently, not less than 150 NTFPs have been significant in terms of worldwide exchange, for example, nectar, gum, rattan, bamboo, stopper, nuts, mushrooms, essential oils, and plants or creatures' parts utilized for pharmaceutical industries (FAO, 1997).

Certification systems for NTFPs

NTFPs can be evaluated and labeled according to acknowledged principles, criteria, and indicators (standards) of the best management (certified). NTFPs can be certified according to the principles, criteria, indicators, and in view of the norms of the three frameworks, which are forest stewardship, fair trade, and natural farming.

Certification of NTFPs can create and enhance market access, add to the value paid to the collectors and processors through cost premiums, and educate purchasers. They provide conscientious buyers with clear evidence to choose the products, and they encourage sustainable utilizing approaches (Baharuddin and Simula, 1996) (Newsletter No. 43, November 2000). The FSC in Bonn, Germany, was established by NGOs and the private sector. The FSC is an accredited organization for certification bodies. This organization gives full authority to its individuals to carry out its central goal and vision. The enrollment is to professionals, experts in the fields of financial, environmental, and social affairs, since 1994, nine to ten FSC principles of Sustainable Forest Management (SFM) rules and related criteria. FSC standards are oriented to assess the execution of Forest Management Unit (FMU) performance standards (Table 5.1).

FMU is defined as a clear, demarcated area of land covered predominantly by forests and managed according to a set of explicit objectives and a long-term management plan. Since it began, the FSC has grown rapidly. In 5 continents, it has over 800

Table 5.1 Separation of management and performance dimensions

Management dimension:
a. Area management: area demarcation (compulsory requirements)
b. Production management: performance (core activities)
c. Organizational management: managerial quality (desired requirements)
Performance dimension:
a. Production sustainability
b. Environment sustainability
c. Social sustainability

members in over 80 countries. The members vary from large players in forest sectors to small companies, also from community forestry groups, to local peoples' organizations and associations, and from organizations in which the environment is the specific concern to progressive forestry and wood retail companies. All members are working in a balanced partnership to improve and sustain forest management all over the world.

With certified forests in 80 countries on 5 continents, FSC is the leading standard for responsible certification. Currently, 7% of the world's forest area, around 180 million ha, as of June 14, 2013, are FSC certified (http://www.greenpeace.org/). The certificates enable promotion and rational management of forests. They promote well-managed forests through the application of criteria and indicators that address sustainable management aspects of ecology, social, and economics related to the forest sector. The FSC has relatively recent certification programs that resulted from concern over the fate of the world's forests, particularly in the tropics (http://www.fscoax.org/).

Table 5.2 shows how the management and production dimensions are combined and that each indicator represents a combination of dimensions.

Forest (timberland) Stewardship Council principles

Forest management certification under FSC has the most well-rounded criteria and is perhaps the most natural choice for NTFPs. The principles are:

- Principle 1: Compliance with laws and FSC principles
- Principle 2: Tenure and use rights and responsibilities
- Principle 3: Indigenous people's rights
- Principle 4: Community relations and workers' rights
- Principle 5: Benefits from the forest

Table 5.2 Matrix showing the management and production dimensions

Management dimension (principles)	**Production dimension (principles)**		
Strategies for achieving results	**Production sustainability**	**Environment sustainability**	**Social sustainability**
1. Area management (Compulsory requirements)	Indicator	Indicator	Indicator
2. Forest management 2.1 Production management 2.2 Environmental management 2.3 Social management (core activities)	Indicator	Indicator	Indicator
3. Organizational management (desirable)	Indicator	Indicator	Indicator

Source: Indonesian Eco labeling 5000 (LEI 5000).

- Principle 6: Environmental impact
- Principle 7: Management plan
- Principle 8: Monitoring and assessment
- Principle 9: Maintenance of high conservation value of the forest
- Principle 10: Plantation
- Principle 11: Nontimber forest products

Organic agriculture

Introduction

IFOAM is the equivalent world body for organic agriculture; the origin of IFOAM was of a global farmers' movement; its main focuses are on building soil fertility and the avoidance of chemicals in agriculture. Organic farming is referring to the production methods that do not use synthetic pesticides, chemical fertilizers, or genetically modified organisms. As highlighted by Falls Brooks NTFP, page 10, organic techniques focus on improving soil quality, assuring crop diversity, protecting the health of workers and consumers, and reducing the environmental impacts of production. Organic farming activities contribute to the millennium development goal of improved health and food security, environmental conservation, and economic development. Organic agriculture is characterized by two main features, which are the recycling of nutrients and natural means of pest and disease control according to both traditional and modern scientific knowledge. Nevertheless, organic farming is more than just a system of production that includes or excludes certain inputs, particularly agrochemicals and genetically modified organisms, because it builds on and enhances the ecological management skills of the farmers, fishermen, and pastoralists, and it includes soil standards. The organic principles and basic standards formulated by IFOAM are applied in both certified and noncertified organic agriculture and are adhered to and cherished.

Principles of Organic Agriculture

1. Health
2. Ecology
3. Fairness
4. Care

Fair trade: Fairtrade Labeling Organizations International

Introduction

FLO developed out of the alternative trade movement. They seek to ensure that business contributes to the well-being of workers and communities. Fair trade also considers that producers receive a fair deal and that they strive to reverse the usual tendency of international trade, which all too often increases global and regional inequality.

With the emphasis on the social components of production, fair trade–certified enterprises are often, though not exclusively, worker cooperatives or associations (Falls Brook Center's NTFP Pages, 2008).

Concept of fair trade

Fair trade is a way of trading that seeks to reduce poverty and empower producers who are disadvantaged by the structures of conventional trade. Those involved in fair trade seek to empower producers by adhering to certain principles and standards of behavior in their trading relationships and also campaign to challenge mainstream trade structures. The goals of fair trade are to improve the livelihoods of producers (improve market access, provide fair terms of trade, and maintain long-term, stable trading relationships), promote development opportunities for disadvantaged producers, and protect children from exploitation in the production process in addition to raising awareness and educating and encouraging consumers to buy fair trade goods. They also aim to set an example of partnership in trade through dialogue, transparency and respect, challenging the rules and practice of conventional trade, and promoting social justice, sound environmental practices, and economic security.

Principles of fair trade

1. Principle 1: Creating opportunities for economically disadvantaged producers
2. Principle 2: Transparency and accountability within the activity
3. Principle 3: Assuring fair trading practices
4. Principle 4: Payment of a fair price
5. Principle 5: Ensuring no child labor and forced labor
6. Principal 6: Assuring gender equity
7. Principle 7: Ensuring good working conditions
8. Principle 8: Providing capacity building
9. Principle 9: Promoting fair trade
10. Principle 10: Respect for the environment

Current situation of gum Arabic business in Umm Ruwaba and El Rahad, Sudan production

The survey data (questionnaires and interviews with eight gum Arabic producers unions in Umm Ruwaba and El Rahad) were analyzed using simple statistical methods for estimating frequencies and percentages. The study findings revealed that a majority (86%) of the respondents were not aware of the policies. Despite the existence of policies, regulations, and laws, a very small portion (23.7%) were found to be acquired about laws and regulation regarding the illicit felling and grazing protection law, provision of certificates to safeguard land tenure, local decree for collection of gum Arabic in sacks, laws of restriction of a gum trade monopoly, taxes, customs, and fees.

The Sudan leading position of gum Arabic production was threatened in recent years by poor production and marketing policies and high competition from neighboring countries, and synthetic products posed another threat to the livelihoods of the small producing households in the gum belt of Sudan. The poor management of the production and conservation of the gum Arabic resources diverted the producers from growing gum into growing of other cash crops, and led to severe tree cutting for making charcoal and firewood. The end result was a degrading environment with southward movement of the gum belt. The social factors constituted one of the main elements that had influenced the performance of gum Arabic production and standardization.

The married status is high among the gum Arabic producers. This provides a positive sign of culture for carrying out sustained family responsibilities and searching for adequate income that would supply the basic needs of the household. This could be achieved by improving and maintaining the quantity and quality of production of gum Arabic. The early marriages result in extended families, which provide a good source for family labor that hedge against migration of young people out of the Gum Arabic Belt (Table 5.3).

The existence of young and middle-aged producers reflects willingness and experience in pursuing gum Arabic production system as a form of livelihood. Nevertheless, the decreasing numbers of old producers warns against depletion of indigenous knowledge about gum Arabic production, which needs to be augmented with intensive extension services to provide skills in the area of sustainable natural resources management (Elnasri, 2010). This has been supported by information on farmers' ages, which indicated the existence of young to middle-aged farmers, with ages ranging between 18 years and less than 60 years, which exhibits the presence of farmers with good experience who are willing to follow orders and regulations and are open to possibilities of accepting new innovations (Table 5.4). The data also detected the presence of farmers above 60 years old (28% of the sample).

Table 5.3 Distribution of households in Umm Ruwaba and El Rahad areas by family size

Family members	Frequency	Valid percent	Cumulative percent
1–6	50	36.0	36.0
7–11	66	47.5	83.5
More than 11	23	16.5	100.0
Total	139	100.0	

Table 5.4 Distribution of households in Umm Ruwaba and El Rahad areas by age

Farmer's age	Frequency	Valid percent	Cumulative percent
18–38	28	20.1	20.1
39–59	71	51.1	71.2
60–80	38	27.3	98.6
More than 80 years	2	1.4	100.0
Total	139	100.0	

Table 5.5 Distribution of gum Arabic farmers in Umm Ruwaba and El Rahad areas by level of education

Level of education	Frequency	Valid percent
Illiterate	33	23.7
An informal education (khalwa)	28	20.1
Primary school	63	45.3
Secondary	15	10.8
Total	139	100.0

The responses among the gum Arabic producers ensured the high possibility of accepting new innovation, following rules and regulations for forest conservation and involvement in marketing operations. However, the present education level among gum Arabic producers (Table 5.5) warrants the use of appropriate means of communication for transfer of new knowledge and awareness programs. The use of audio–visual aids might be a recommended means of communication, given the prevailing status quo. This is because a large number of gum Arabic producers do not have access to radio and television facilities.

While involvement in crop production and animal raising forms the major source of income and social prestige, small producers depend on gum Arabic production as a second-best source of livelihood (Taha, 2000). This dependence occurs without realizing the benefits that can accrue from certification of their gum Arabic collections. Certification becomes a prerequisite for enhancing the socioeconomic status of the small households since it expands the opportunities for new markets, assures fair prices offered in the world market, and controls against unfair domestic marketing practices.

There are few laws that regulate gum Arabic production and trade in Sudan that can lay a foundation for implementing certification of the product. Such laws deal mainly with natural resource conservation, including the prohibition of illicit felling of trees and severe grazing. Domestic animals pose a great source of threat to gum Arabic trees, as they eat the crowns of the trees, nab and break the emerging branches, graze on the leaves, and trample on the soil. However, certification requires additional laws that provide for safeguarding land tenure system, issuing local decrees for collection of gum Arabic in sacks, and protecting of forests using fire lines. Long-term land tenure and usufruct rights to forest resources shall be clearly defined, documented, and legally established. Furthermore, there are other laws that restrict gum trade like monopoly of the gum Arabic Company (GAC), the royalties, taxes, and states fees that levied against the movement of gum Arabic across states. The different types of taxes, customs, and values, and the institutions are displayed in Table 5.6.

The Forests National Corporation (FNC) is responsible for the planning and management of forest trees and products, but remains nonfunctional when it comes to gum Arabic production, marketing and export policies, and management. The severe felling of gum Arabic trees for making firewood and charcoal and the construction of shelters, as well as the turnover of producers toward intensive cultivation of cash crops, reveal the noneffective policies followed by the GAC and conflict with the local forests authorities' mandates and functions. All of this severe natural resource

Table 5.6 The types of taxes, customs, values, and sources of fees collection

Taxes or customs	Value SDD	Institution that received the value
Marketing taxes	5000	Locality
Production taxes	7000	State
Transport	2000	Locality
Zakat	10,000/sack	Zakat

Table 5.7 Type of land ownership for gum Arabic producers in Umm Ruwaba and El Rahad areas, Sudan

Description	Frequency	Valid percent	Cumulative percent
Customary inherited land tenure	99	71.2	71.2
Rent	2	1.4	72.7
Purchased	31	22.3	95.0
Inherited and purchased	7	5.0	100.0
Total	139	100.0	

degradation takes place irrespective of the available nonoperational laws forbidding tree felling. The involvement of the Gum Arabic Producers' Associations in trade aspects gives them a leverage of knowledge and experience of the negative outcomes of the GAC and taxes on their income and livelihoods.

The main land use systems practiced in the Gum Arabic Belt are composed of growing crops, forests, and usage of land ranges based on usufruct rights. The community chiefs grant the privilege of growing the crops and having the animals graze on the forests and gum Arabic gardens. The landholdings of gum Arabic vary among users (Table 5.7); less than 50% of the respondents were used to recording and documenting land ownership, and all lands are used under customary rights being inherited or purchased within the community system. It could be concluded that inheritance might lead to land fragmentation (small-size landholdings).

The traditional leaders protect the ownership rights on behalf of their people without receiving any payment in exchange. This role extends into the protection of land against illicit grazing and caring for conflict resolution among people and tribes using the same resources They sometimes act as security guarantees for credit institutions in supplying loans to gum Arabic producers. Almost all of the producers are members of one of the gum Arabic producers' associations. Nevertheless, they suffer from close linkages and common communication grounds that can facilitate their effective organization into unions to protect their production and price rights and enhance their access to better credit and marketing services. It is difficult to find adequate access to credit since the main sources of credit are the GAC (Table 5.8), the Agricultural Bank of Sudan, and money lenders and merchants at the village level. The major alternative source to loan borrowing is the wide sale of animals for obtaining supplemental financing. The timing of repayment of loans varied among lending sources, according to prevailing situations, and the repayment takes place in kind and in cash.

Table 5.8 Credit accessibility for gum Arabic producers in Umm Ruwaba and El Rahad areas, Sudan

Credit accessibility	Frequency	Valid percent	Cumulative percent
No answer	4	2.9	2.9
Yes	33	23.7	26.6
No	102	73.4	100.0
Total	139	100.0	

Table 5.9 Types of services offered by the government in Umm Ruwaba and El Rahad areas, Sudan

Types of services	Frequency	Valid percent	Cumulative percent
No answers	12	8.6	8.6
Water, health, and education	53	38.1	46.8
No services provided by government for community development	71	51.1	97.8
Establishment of cooperative associations	1	7	98.6
Water, health, and education and establishment of cooperative associations	2	1.4	100.0
Total	139	100.0	

There are many institutions that are involved with gum Arabic production and marketing in one way or another. These include the FNC, the GAC, local governments, the small gum Arabic processing private companies, banks, and traders. Given such a diversity of stakeholders, these institutions failed to provide their mandated services except for training in tapping and gum Arabic collection.

The gene bank policy of FNC to conserve the genetic characteristics of the gum Arabic trees (hashab, talha, and others) is not well implemented. The efforts to conserve the genetic characteristics included provision of certificates to safeguard land tenure, reforestation, extension services, provision of seedlings, and establishment of nurseries. Efforts of forest research and FNC used demonstration farms in the past, which are not in existence anymore due to financial and technical limitations.

As shown in Table 5.9, the services provided by the local government are focused on the supply of drinking water and provision of health and education services. Still, the government support for community development is considered minimal.

All producers practiced intercropping to diversify their local economy and avoid dependence on a single forest product. The hashab tree offers several benefits during and after its life cycle period: it provides gum, wood for building, and furniture, firewood, and charcoal. The estimated income generated by the hashab tree ranged between Sudanese Dinnar (SDD) 100 and 57,000. The percentage share of the contribution of the hashab tree out of total income varied among producers.

Table 5.10 The percentage share of revenue out of total income accrued from the hashab tree in Umm Ruwaba and El Rahad areas, Sudan

Category of revenue share	Frequency	Valid percent	Cumulative percent
0%	1	0.7	0.7
Less than 15%	49	35.3	36.0
15%–30%	59	42.4	78.4
31%–45%	14	10.1	88.5
46%–60%	11	7.9	96.4
More than 60%	5	3.6	100.0
Total	139	100.0	

Table 5.11 The percentage distribution of gum Arabic producers in selecting criteria for sorting gum Arabic nodules

Criteria of sorting nodule	Frequency	Valid percent	Cumulative percent
Volume	55	39.6	39.6
Color	2	1.4	41.0
Moisture content	3	2.2	43.2
No grading for gum	54	38.8	82.0
Others (purity)	17	12.2	94.2
All of the above mentioned	8	5.8	100.0
Total	139	100.0	

The highest revenue accounted between less than 15% and 30% of the total income (Table 5.10).

Most of the producers sell their gum Arabic at village level and receive their payment of money on sales promptly, and the others receive their payment in advance or a year later. In all, the producers were not satisfied with the prices they received, as they did not cover the cost of production, especially the high cost of tapping the gum.

The concept of grading is familiar to the gum Arabic producers. The producers sort the gum granules by nodules volume, color, moisture content, and purity (Table 5.11). Most of the grading is done by the household members, relatives, friends, and by nafir.

The gum collected directly from the tree is more clean and pure than the other gum collected that had fallen on the ground. The purity of gum is the affected by the place of collection and mixing of hashab gum with other gum. The degree of purity affects the quality and prices of gum sold in the market. The less-pure gum would result in lower prices, penalizing the high-quality hashab gum. The concept of a management plan was conceived and implemented by the producers. The required criterion indicated in principle 7 was customarily applied, but not written. The gum Arabic producers are acquainted with the role of the tree and its value to the environment, economic, and social aspects. The economical values of the tree include profit out of tapping and collecting the gum and making of firewood and charcoal, and are rising with their standard of living. The average number of planted hashab trees in a typical garden ranged between 100 and 1,000,000 trees, which were used as park area and arranged as windbreaks and hedges. Most of the gum Arabic producers adopted the shifting cultivation system for gum Arabic production, and the others adopted the fallow period system, while few practiced both systems.

The hashab regenerates naturally from seeds or uses direct sowing, gum–bush cultivation cycle, and artificial regeneration of seedlings. The limiting factors that determine the liability of the trees for tapping are temperature of the area, the age of the tree, and announced prices.

Regarding the tools used for tapping, the producers were found to prefer using the traditional axe, followed by both sonki and farar.

The right and the best time for tapping the gum trees is in October, as specified by 97 producers. They can tap in November, and if that fails, it will be redone preferably in December or February for the same production seasons. The gum Arabic producers tap the same tree only once per season. Tapping and picking is done mainly by household members, hired labor, nafir, and by assistance of relatives and friends (Table 5.12).

The producer usually taps between 40 and 10,000 trees per year. The number of people picking differs; it is determined by existing ecological factors, and it ranges between five and eight pickings. The period between tapping and picking differs according to ecological and physiological reasons (Karma, 2000). The minimum production of one tree ranges between 0 and 1.7 kg, while the maximum production may go up to 7.6 kg. Winter was considered to be the best time for collection of gum Arabic, but 15% preferred the summer collection.

On monitoring and assessing the performance of the gum Arabic tree, it was indicated that the number of gum trees had increased due to natural regeneration, use of seeds from mother trees, and reseeding by humans, while another view believed that the number of the trees had decreased due to illicit cutting, low rainfall, and attacks of pests and diseases. Management activities of high-value forest products shall maintain and enhance the attributes of such forests. The majority of the producers asserted the absence of exotic trees species in the area. All of the existing species are indigenous and dominated by *A. senegal*, *Balanites aeygyptiaca*, *Azadirachta indica*, *Compretum hartmunimum*, *Faidherbia albida*, *Zyisphus spina-christi*, and *Acacia nubica*.

Table 5.12 Labor force involved in tapping gum Arabic trees in Umm Ruwaba and El Rahad areas, Sudan

Categories of tappers	Frequency	Valid percent	Cumulative percent
Family labor (household)	77	55.4	55.4
Nafir (communal labors)	3	2.2	57.6
Assistance of relatives and friends	2	1.4	59.0
Hired laborers	19	13.7	72.7
Family members and hired labor	30	21.6	94.2
Family members, nafir and assistance of relatives and friends	3	2.2	96.4
All the above	5	3.6	100.0
Total	139	100.0	

Cost of certification

The FSC is an expensive and time-consuming certification program, and it is easy to apply to many NTFPs since it fits the conventional forest harvest structure of FSC, i.e., large-scale industrial operations. Therefore, FSC is appropriate for large-scale, industrial NTFP operations. The FSC provides a better guarantee of price premiums for certified products and access to a range of specialized markets. It is important to remember that the major cost of getting certified is usually not the cost of the certification process itself, but rather the cost of improving operations to meet certification criteria. Some common examples of areas where improvement is needed are in the development of management plans and maps, better recordkeeping, and in addressing landscape management issues. The certification cost can be classified as direct cost of initial forest assessment plus required annual audit and reassessment, indirect cost of improved forest management practices (i.e., reduced harvest or increased expenditures), and cost of chain of custody audit.

Chain of custody auditing is the second major component of forest management certification, complementing a forest assessment. Chain of custody can be defined as "an unbroken trail of accountability that ensures the physical security of samples, data, and records." Chain of custody is a critical element of any certification system, since it provides the ultimate link between the "consumers" and the "producers." The term "consumers" is used here in a broad sense to indicate individuals, retail companies, or suppliers in countries where wood, wood products, and nonwood are distributed, sold, or used. The producers may be individual landowners; a community organization (such as a cooperative); a company that manages forests, brokers, and products and/or processes wood; or a government.

The desire by consumers to assure themselves that the wood or other forest products they sell or purchase are from well-managed sources is one of the key motivations underlying attempts to develop credible certification systems; if the consumers seeks assurance, the responsible producer provides this assurance through a certificate that the product does indeed derive from a certified, well-managed forest. Ideally, a secure chain of custody system will have the following basic elements (1) physical evidence (such as documents, tags, and labels) that the goods originate from a particular source; and (2) an "auditable" data recording and communication system that runs in parallel with and links to the physical evidence identifying each product Groves et al., 1996.

Challenges of gum Arabic certification

The main challenges facing gum Arabic certification in Sudan are:

- The main challenges facing gum Arabic certification in Sudan are:
- High cost associated with forest certification,
- Poor roads and other infrastructure systems in the study area make forest certification costly to set up and maintained.
- Lack of certifiers with experience in the ecology and regeneration of gum Arabic.
- Lack of standard policies.

- Weak supervision of FNC (damages by animals, Illegal cutting and grazing by nomads)
- Weak forestry institutions especially for the implementing of forestry regulation and enforcement of forest laws
- Financing problems
- The low level of information on Forest Certification in general
- Political instability
- Poor capacity of local civil society organizations, in the rural community and local NGOs to monitor sustainable forest management.

The opportunities

- Rich in ecological, social, and economic knowledge about *A. senegal*;
- Available baseline data at the Sudan forestry sector;
- Assured minimum annual sustained harvest;
- Resiliency of specific ecological niche for the species dominant at the gum belt;
- Collaborative small producers who produced gum Arabic and other crops for cash and subsistence;
- The quality control measures (standards) for gum Arabic tapping, harvesting, and processing that are well developed;
- Based on traditional customary land tenure and access usufruct rights;
- Ability of the certified gum Arabic to command a price premium in the marketplace;
- Well-developed market for certified forest nontimber products that gives a value for standardized product;
- Increasing the numbers of environmentally and socially concerned customers who are asking for forest products that are friends to the environment and emanating from sustainable forest management;
- There are considerable areas of gum Arabic (the belt) that help to motivate efforts toward forest certification as a tool for both sustainable forest management and market-based instruments; and
- Government institutions are becoming increasingly open to the involvement of the civil society in forest management and monitoring.

Conclusions

The findings of the study revealed a rather bleak situation for gum Arabic in the Sudan. Gum Arabic is mismanaged and therefore depleted. The responsible institutions for conservation, development, and maintenance of these forests are weak. In spite of the existence of several policies, laws and regulations on their implementation are noneffective. The land tenure patterns reduce the economic incentives, and the marketing practices are not often conducive for achieving sustainable management and use of gum Arabic forests. National government and donor partners give low priority to forest issues. A result of all this is that the potential contribution of gum Arabic forest resources to poverty reduction, economic development, and ecological and hydrological stability are far from being realized. However, it is possible to turn the present negative trend in the gum belt. One tool toward sustainability management is forest certification.

References

Abdelazim, M.I., 2007. The Gum Arabic Issues: The Resources,The Commodity and the Producers. Workshop on the Future of Gum Arabic in Sudan. Forest National Corporation and Donor Support Fund, July 2007, Khartoum.

Baharuddin, H.G., Simula, M., 1996. Timber certification in transition study on the development in the formulation and implementation of certification schemes for all internationally traded timber and timber products. Report ITCC (XX)/8 Rev. 1 for the 20th ITTC Session, May 15–23, Manila, Philipin.

Bass, S., 1998. Introducing timber certification-A report prepared by the timber certification advisory group (FCAR) for DG VIII of the European Commission. European Forest Institute.

Elnasri, H.M., 2000. Forest Policy of the Sudan: Past, Present and future (M.Sc. thesis). Khartoum, Sudan.

Elnasri, H.M., 2010. Certification of gum Arabic in Sudan: An Approach for Sustainable Management (Ph.D. thesis). Khartoum, Sudan.

Falls Brook Centre, 2008. 125 South Knowlesville Road, Knowlesville, New Brunswick E7L 1B1Canada (Tel: +1 506 375-8143; Fax: +1506 375-4221; Website: http://www.web.net/fbcja.

FAO, 1997. State of the Worlds forests. FAO, Rome, Italy.

Felix, C.N., 2004. Paper Presented at the Symposium of Forest Certification in Developing and Transitioning Societies: Social, Economic and Ecological Effects. Yale School of Forestry and Environmental Studies, New Haven, Connecticut, USA.

FSC, 1997. Forest Stewardship Council. Regional Working Group, Draft Principle 11.

Groves, M., Miller, M., Donovan, R.Z., 1996. Chain of Custody. In: Viana, V.M., Ervin, J., Elliot, C., Gholz, H. (Eds.), Certification of Forest Products: Issues and Perspectives. Island Press, Washington, D.C.

Karma, M.M., Trading: Concession or Liberalization. A Paper Presented to the Regional Workshop on Gum Arabic at the Friendship Hall, May 2000. Khartoum.

Karama, M.M., 2002. Gum Arabic Marketing Issues: Marketing Model, Buffer- Stocking, Processing and Financing. International workshop on promotion of gum Arabic production processing & marketing.

Ng'andwe P., 2003. Timber Certification, Optimization and Value Added Wood Products: A Case Study of Zambia (M.Sc. thesis). University of Wales.

Taha, M.E., 2000. The Socio-Economic Role of *Acacia Senegal* in Sustainable Development of Rural Areas in The Gum Belt (Ph.D. thesis) (1999) Dresden, Germany.

Upton, C., Bass, S., 2004. Forestry. Oxford Center Earthscan Ltd, United Kingdom.

Further Readings

Falls Brook Centre, 2005. 125 South Knowlesville Road, Knowlesville, New Brunswick E7L 1B1, Canada.

FAO, 1996. A review of production, marketing and quality control of gum Arabic in Africa. Technical co-operation programme. Project No. TCP/RAF/4557, Food and Agriculture Organization of the United Nations, Rome.

Forest Stewardship Council Principles and Criteria, 2000. Document1.2.
IIES&IES, 1990. Gum Arabic Rehabilitation in the Republic of the Sudan. Stage 1 Report Volume 2 Main Report and Recommendations for Action. International Institute for Environment and development, London, England and Institute of Environment Studies, University of Khartoum, Sudan.
Mallet, P., 2005. Falls Brook Centre, 125 South Knowlesville Road, Knowlesville, New Brunswick E7L 1B1Canada (Tel: +1 506 375-8143; Fax: +1506375-4221; Email: pmallet@web.net; Website: http://www.web.net/fbcja.

Part II

Chemistry: Food and Feed Applications

Chemical Properties of Gum Arabic

6

*Abdalbasit A. Mariod**, **
*University of Jeddah, Alkamil, Kingdom of Saudi Arabia; **Ghibaish College of Science & Technology, Ghibaish, Sudan

Introduction

A chemical property is any of a material's properties that ends up plainly apparent amid or after a chemical reaction, that is, any quality that can be built up just by changing a substance's chemical identity. Essentially, chemical properties can't be resolved just by survey or touching the substance; the substance's inner structure must be influenced significantly by its chemical properties to be researched. At the point when a substance goes under a compound response, the properties will change definitely, bringing about a chemical change. Be that as it may, a reactant property would likewise be a substance property (William et al., 2009). A number of chemical methods were used to characterize the chemical properties of the gums.

Chemical composition of gum Arabic

The chemical composition of gum Arabic (GA) is mind-boggling, and various investigations have been carried on GA composition. GA is a spread-chain, complex polysaccharide, either neutral or somewhat acidic, found as a blended calcium, magnesium, and potassium salt of a polysaccharide acid (Arabic acid). The skeleton is made out of 1,3-connected β-D-galactopyranosyl units. The side chains are made out of two to five 1,3-connected β-D-galactopyranosyl units, joined to the primary chain by 1,6-linkages. Both the fundamental and the side chains contain units of α-L-arabinofuranosyl, α-L-rhamnopyranosyl, β-D-glucopyranosyl, and 4-O-methyl-β-D-glucopyranosyl, the last two generally as end units. The chemical composition of GA can change with its source, the age of the trees from which it was gotten, climatic condition, and soil condition (Azzaoui et al., 2015).

Osman et al. (1993) fractionated GA by hydrophobic cooperation chromatography to yield four fractions, all of which had a comparable carbohydrate composition; however contrasted in their substance of protein, amino acid composition, and atomic weight distribution. Every one of the four parts responded with a variety of antiarabinogalactan–protein monoclonal antibodies by means of anticarbohydrate epitopes and were precipitated by Yariv reagent, which demonstrated that each of the four fractions belonged to the arabinogalactan-protein fraction (AGP) family (Osman et al., 1993).

Gum Arabic. http://dx.doi.org/10.1016/B978-0-12-812002-6.00006-3

The differences between *Acacia seyal* and *Acacia senegal* gums in terms of sugars and amino acid compositions are mostly known, as well as the major differences in molecular weight (Mw) distribution. However, *A. seyal* gum is richer in minerals and polyphenols, less rich in proteins, more compact, more unstable in solution, less charged, less surface active, and less hydrolysable by enzymes. The structure and conformation of AGP from *A. seyal* are unknown and their functional properties badly known (Sanchez et al., 2017).

Moisture

Kheir (2005) studied the solubility of eight GA formulations (GAF) in water and some other chemical solvents. She reported a highly significant difference between the GAF in water solubility. The high solubility was found in the spray-dried GA (98.8%), and was lowest for the sifting formulation (97.4%). This author found highly significant differences between GAF in the levels of solubility in ethanol, acetone, and chloroform. Concerning moisture contents of the eight GAF ranged between 9.90% in the hand-picked selections and 8.32% in spray-dried (Kheir, 2005). Moisture content determines the hardness of the gum. The studied samples of gum loss on drying of good-quality gum does not exceed 15% for granular and 10% for spray-dried material (FAO, 1999). In a recent study reported by Karamalla (1999) of over 1500 authentic and commercial *A. senegal* var. *senegal*, the results of that study indicate the mean value of the moisture content was 10.75%. The range of moisture content for *A. senegal* was 8.1%–14.7%. Ibrahim et al. (2013) reported an average value of moisture content of *A. senegal* var. *senegal* samples as 13.49%. Lelon et al. (2010) studied the moisture content of GA obtained from four different areas in Kenya; they reported that two areas showed moisture content in the range of 14.9%–15.0%, which fell within international specifications (13%–15%), while the other two areas showed the moisture content of 15.4%–17.5%, which fell outside the international specifications.

Ash

In a study of 800 authentic formulations of gum from *A. senegal* var. *senegal* collected from 32 different localities of the gum-producing belt of the Sudan, it was shown that the type of the soil had no significant effect on the ash content of the gum (Karamalla, 1999). Ibrahim et al. (2013) reported an average value of ash content of *A. senegal* var. *senegal* samples as 3.27%. Anderson et al. (1991) found that the value of the ash content of commercial GA to be 3.6% for Sudanese formulations. FAO (1999) reported that the ash content of GA did not exceed more than 4%. Kheir (2005) reported ash content ranged between 3.91 for the kibbled 121 and 3.05 for the spray-dried gum samples.

Ash content of GA from four Kenyan sites fell within the international specifications. Ash content of GA from Solit and Kapkun (2.94 and 3.16%, respectively) was higher ($p < 0.05$) than those from Kimorok and Maoi (2.88 and 2.72%), respectively (Lelon et al., 2010).

Nitrogen and protein content

GA is a polymer with about 3% protein (Anderson et al., 1991). The protein fraction is responsible for the emulsification properties of the gum. The role of nitrogen and nitrogenous components in the structure, physicochemical properties, and functionality of GA was recently subjected to intensive investigation. An adsorbed layer of protein at oil/water interface gives the essential balancing out structure in numerous food colloids (Dickinson, 1994). The inconstancy in the emulsifying properties of gums from various *Acacia* species is reliant not just on the aggregate protein (on polypeptide content), but additionally on the distribution of the protein–peptide between the low- and high-molecular-weight parts and on the molecular availability of the protein peptide for absorption, as indicated by Randall et al. (1989). The US pharmacopoeia and EU specifications define the minimum standards of the protein content for good-quality GA as 3%.

Investigations of protein in *A. senegal* gum have been carried out by Akiyama et al. (1984); they reported that GA contained 2.0% protein, and they established that amino acid of GA is rich in hydroxyproline and serine, while alanine content is low. GA is a proteinaceous polysaccharide with a protein content ranging from 1.5% to 3.0%. The variation was mainly due to different localities. Osman (1998) reported 0.33%–0.36% nitrogen and 2.14%–2.16% protein for *A. senegal* gum, and Jurasek et al. (1993), in a chemometric study for different *Acacia* species, reported 0.27%–0.38% nitrogen for commercial samples of GA from Sudan. Awad Elkarim (1994) reported that the average nitrogen contents of different commercial grades are around 0.28%. Karamalla (1999) reported that the average nitrogen content of different commercial grades is around 0.33%. Kheir (2005) reported protein content ranged between 2.4 and 2.7% in eight GAF.

Gashua (2016) detailed a higher protein substance of the Nigerian gums than the Sudanese gums. What's more, since the gum is known to be discharged inside a zone between the inward bark and the cambial district as a major aspect of the plants' wound/push reaction, the watched distinction in protein substance could be because of genotypic contrasts between the Nigerian and Sudanese trees from which the gum was reaped, the age of the trees, climatic conditions, the soil in which the trees were planted, plant pathogens, or a mix of components (Al-Assaf et al., 2005; Renard et al., 2006; Siddig et al., 2005).

In connection to the use of the gum in the food business, it has been indicated beforehand that the capacity of Acacia gum exudates to settle oil-in-water emulsions is because of the proteinaceous segments introduced inside the gum (Padala et al., 2009). Thus, the Nigerian samples may have improved emulsification properties. The Nigerian *A. senegal* gum samples appeared to contain around double the measure of protein when contrasted with the gum acquired from *A. seyal*, which is in concurrence with past findings (Idris and Haddad, 2012). The gum gotten from *A. seyal* is not ready to successfully stabilize oil-in-water emulsions.

Amino acid composition

The main amino acids of GA are hydroxyproline, serine, leucine, threonine, histidine, and aspartic acid. *A. senegal* has the highest amount of total amino acids, while the *A. seyal* has the least of all the samples (Gashua, 2016).

Gashua (2016) studied Nigerian and Sudanese gum samples; he found the principal amino acids present in them are hydroxyproline, serine, aspartame, threonine, and proline, which are present in a ratio that approximates to 4:2:1:1. The difference he observed between the three *A. senegal* samples is that a significantly higher value for phenylalanine (0.130 g/100 g) was observed for the Nigerian *A. senegal* sample compared to the Sudanese *A. senegal* gum sample (0.0745 g/100 g).

Naima et al. (2012) investigated four gum sorts (*A. senegal, Acacia polyacantha* and *Acacia laeta* and *A. seyal*); their outcomes demonstrate that *A. laeta* gum has the most elevated aggregate amino acid substance, followed by *A. senegal* gum, then *A. polyacantha* gum, and lastly, *A. seyal* gum. Serine (0.26%) and aspartic acids (0.22%) were the two predominant amino acids present in *A. senegal* gum. Serine (0.22), aspartic acid (0.14%), and glumatic acid (0.13) were the overwhelming amino acids exhibited in *A. seyal* gum. Aspartic acid (0.44%) and serine were the overwhelming amino acids present in *A. seyal*, aspartic acid (0.24%), and serine (0.21%) were the prevailing amino acids in *A. polyacantha* gum, and serine (0.43%) and hydroxyproline (0.31%) were the predominant amino acids introduced in *A. laeta* (Naima et al., 2012).

Acidity and pH

Comparative studies show that gum from *A. senegal* has higher content of rhamnose (12%–14%) and lower arabinose content (24%–29%) compared to rhamnose and arabinose of other *Acacias* (Karamalla, 1999). The main content of GA is arabian (acid substance), and when it was decomposed, it gave arabinose. Thus, GA is therefore called arabic acid, and the gum solution is slightly acidic, with pH 4.66, as reported by Karamalla (1999).

Solubility

GA is one of a kind among the natural hydrocolloids on account, to a great degree, of its high solvency in water, and it can yield solutions of up to 60% concentration. It is really dissolvable in cold water; different gums are either insoluble in cold water or form colloidal suspensions, "not genuine solutions" (G.A.C., 1993).

Acacia gum is quickly soluble in hot and cold water. Most gums cannot be dissolved in water at concentrations higher than 5% due to their high viscosity, but acacia gum can yield solutions of up to 50% concentration. The capacity of acacia gum to shape these concentrated solutions without an over-the-top increase in viscosity is due to the high degree of branching within the gum structure and, therefore, small hydrodynamic volume. Acacia gum is insoluble in oils and in most organic solvents and is soluble in aqueous ethanol up to a limit of about 60% ethanol (Idris, 2017).

Equivalent weight, molecular weight, and uronic acid

Titrable acidity represents the acid equivalent weight of the gum from which the uronic acid content could be determined. Karamalla (1999) assessed the potentials of new parameters such as equivalent weight and total uronic acid content as additional qualifying indices. They found that the mean values for gum of *A. senegal* for the equivalent weight was 1436 and for uronic acid was 13.71%. Gum from *A. senegal* was reported to have high molecular weight average 4 × 105 Mv. The most common method used to determine the Mw was the intrinsic viscosity measurements. However, Eric et al. (1991) reported an average of 2.2 × 105 to 3.1 × 105 for Mw of degraded products of GA. The procedure used by Kheir (2005) of determining apparent equivalent weight is quite different than that followed for obtaining Mw. The former procedure was analytical and the latter computational. Hence, they recorded significant differences in Mw among eight GAF, and that was an expected result, since their calculations depended to a large extent on the intrinsic viscosity.

Reducing sugars

The presence of reducing sugars would give evidence to the reducing power (free reducing groups) of the gum. It is usually calculated as arabinose. The range of 0.16%–0.44% value of reducing sugar was reported for *A. senegal* gum. It was noticed that the *A. seyal* gum had altogether brought down rhamnose and glucuronic acid substance; however, it had a higher arabinose level than the gums gotten from *A. senegal*, as has been accounted previously. Since the core of the arabinogalactan (AG) macromolecules present in gum Arabic (which constitute ~90% of the gum by weight) is believed to consist of a backbone of β 1,3-linked galactose residues, the increased arabinose content recorded for *A. seyal*gum sample suggests that it possesses more and/or longer branches, which may account for its more compact structure (Al-Assaf et al., 2005; Siddig et al., 2005). There are no noteworthy contrasts seen between the sugar structures of the *A. senegal* gums from Nigeria and Sudan (Gashua, 2016).

Mineral composition

Naima et al. (2012) thought about four gum types (*A. seyal*, A. *polyacantha*, *A. laeta*, and *A. seyal*); the qualities for Na were observed to be pretty much the same in all gum types, aside from *A. polyacantha*, which had a low Na value. The estimations of K were higher in *A. seyal*, *A. polyacantha*, and *A. laeta*, and *A. seyal* had the highest values. Mg, Mn, Br, and Zn had about similar values in all gum samples. Values of Fe, Cu, and Zn were similar with the exception of *A. seyal*, which had a more extensive range contrasted with the other three types. *A. laeta* had a more extensive range of Rb, and *A. polyacantha* had a more extensive range of Sr (Naima et al., 2012). Yebeyen et al. (2009) determined the mineral substance of gum from *A. senegal* trees in the

Central Rift Valley of Ethiopia; they announced that the mineral substances of the GA (g/100 g) are Ca 0.7%, Mg 0.2%, Na 0.01%, K 0.95%, Fe 0.001%, P 0.6%, and nondistinguishable traces of Pb, Co, Cu, Zn, Ni, Cd, Cr, and Mn.

Minor components of Acacia gums

Acacia gums contained 3%–6% minor components, which deserve much more attention. Part of variability in gum functional properties must originate from these minor compounds. Among these components, the major part is composed by minerals (3%–5%). The concentration and nature of ions are important since they are supposed to impact the charge density of AGP, which in turn plays an important role in the hydration, solubility, and compactness of AGP as well as in the stabilization of colloidal suspensions. Polyphenol-type molecules are present in colored gum nodules, depending on the gum type (*A. senegal* vs. *A. seyal* gum) and origin. The nature of these polyphenols and their concentration need to be determined; they can result in oxidation phenomena and modify functional properties of AGP. Traces of lipids (the glycophosphatidylinositol anchor) linked to high-Mw AGP has been reported in *A. senegal* gum. It is obvious that the presence of lipids impacts the gum surface properties (Sanchez et al., 2017).

References

Akiyama, Y., Eda, S., Kota, K., 1984. Gum Arabic is a kind of arabinogalactan protein. Agric. Biol. Chem. 48 (1), 235–247.

Al-Assaf, S., Philips, G.O., Williams, P.A., 2005. Studies of acacia exudate gums. Part I. The molecular weight of *Acacia senegal* gum exudate. Food Hydrocoll. 19, 647–660.

Anderson, D.M.W., Millar, J.R.A., Weiping, W., 1991. Gum Arabic (*Acacia senegal*) from Nigeria composition with other sources and potential agrofrestry development. Biochem. Syst. Ecol. 19 (6), 447–452.

Awad Elkarim, M., 1994. Analytical Studies on Some Crude and Processed Gum Arabic Samples With Regard to Quality Aspects (M.Sc. thesis). University of Khartoum, Khartoum, Sudan.

Azzaoui, K., Hammouti, B., Lamhamdi, A., Mejdoubi, E., Berrabah, M., 2015. The gum Arabic in the southern region of Morocco. Morocco J. Chem. 3 (1), 99–107.

Dickinson, E., 1994. Protein-stabilized emulsions. J. Food Eng. 22, 59–74.

Eric, D., Galaz, V.B., Anderson, D.M.W., 1991. Emulsifying behavior of gum Arabic. Part 2: The effect of the gum molecular weight on emulsion droplet-size distribution. Carbohydr. Polym. 14, 385–392.

FAO, 1999. Specification for Identity and Purity of Certain Food Specification for Identity and Purity of Certain Food Additives. Food and Nutrition paper No. 52, Addendum 3. FAO, Rome, pp. 83–85.

G.A.C. (Gum Arabic Company), 1993. Gum Arabic: A Product of Nature. The Gum Arabic Company Ltd., Khartoum.

Gashua, I.B., 2016. An Investigation of the Molecular Structure, Composition and Biophysical Properties of Gum Arabic (Ph.D. thesis). University of Wolverhampton, Wolverhampton, UK.

Ibrahim, O.B., Osman, M.E., Hassan, E.A., 2013. Characterization and simple fractionation of *Acacia senegal*. J. Chem. Acta 2, 11–17.

Idris, O.H.M., 2017. What is gum Arabic? An overview'. Int. J. Sudan Res. 7 (1), 1–14.

Idris, O.H.M., Haddad, G.M., 2012. Gum Arabic's (gum Acacia's) journey from tree to end user. In: Kennedy, J.F., Phillips, G.O., Williams, P.A. (Eds.), Gum Arabic, special ed. RSC Publishing, Cambridge, UK, pp. 3–19.

Jurasek, P., Kosik, M., Phillips, G.O., 1993. Chemometric study of the *Acacia senegal* (gum Arabic) and related natural gums. Food Hydrocoll. 7 (1), 73–85.

Karamalla, K.A., 1999. Gum Arabic: Production, Chemistry and Application, Manager Research and Development Department. Gandil Agricultural Company Ltd., Sudan.

Kheir, M.E.M., 2005. Physicochemical, Functional and Microbial Properties of Crude and Processed Gum Arabic (*Acacia senegal*) (M.Sc. thesis). University of Khartoum, Khartoum, Sudan.

Lelon, J.K., Jumba, I.O., Keter, J.K., Chemuku, W., Oduor, F.D.O., 2010. Assessment of physical properties of gum Arabic from *Acacia senegal* varieties in Baringo District, Kenya. Afr. J. Plant Sci. 4 (4), 95–98.

Naima, E.S.A., Awad Elkarim, A.M., Fageer, A.SH.M., Nour, A.A.M., 2012. Physicochemical characteristics of some *Acacia* gums. Int. J. Agric. Res. 7, 406–413.

Osman, E.M., 1998. Microbiological and Physicochemical Studies on Gum Arabic: Quality and Safety (M.Sc. thesis). University of Khartoum.

Osman, M.E., Menzies, A.R., Williams, P.A., Phillips, G.O., Baldwin, T.C., 1993. The molecular characterization of the polysaccharide gum from *Acacia senegal*. Carbohydr. Res. 246, 303e318.

Padala, S.R., Williams, P.A., Phillips, G.O., 2009. Adsorption of gum Arabic, egg white protein, and their mixtures at the oil-water interface in limonene oil-in-water emulsions. J. Agric. Food Chem. 57, 4964–4973.

Randall, R.C., Phillips, G.O., Williams, P.A., 1989. Fractionation and characterization of gum from *Acacia senegal*. Food Hydrocoll. 3, 65.

Renard, D., Lavenant-Gourgeon, L., Ralet, M.C., Sanchez, C., 2006. *Acacia senegal* gum: continuum of molecular species differing by their protein to sugar ratio, molecular weight, and charges. Biomacromolecules 8, 2637–2649.

Sanchez, C., Nigen, M., Mejia Tamayo, V., Doco, T., Williams, P., Amine, C., Renard, D., 2017. Acacia gum: history of the future. Food Hydrocoll. 2017, 1–21. doi: 10.1016/j.foodhyd.2017.04.008.

Siddig, N.E., Osman, M.E., Al-Assaf, S., Phillips, G.O., Williams, P.A., 2005. Studies on Acacia exudate gums. Part IV: Distribution of molecular components in *Acacia seyal* in relation to *Acacia senegal*. Food Hydrocoll. 19 (4), 679–686.

William, L., Masterton, Cecile, N., Hurley, 2009. Chemistry: Principles and Reactions, sixth ed. Brooks/Cole Cengage Learning, p. 13.

Yebeyen, D., Lemenih, M., Feleke, S., 2009. Characteristics and quality of gum Arabic from naturally grown Acacia senegal (Linne) Willd. trees in the Central Rift Valley of Ethiopia. Food Hydrocoll. 23 (1), 175–180.

Physical and Electrical Properties of Gum Arabic

7

*Abdelrahman A. Elbadawi**,**, *Yousef A. Alsabah**,†, *Hassan H. Abuelhassan**
*Al Neelain University, Khartoum, Sudan; **Future University, Khartoum, Sudan;
†Hajjah University, Hajjah, Yemen

List of abbreviations

C	*Acacia polyacantha* (cacamot)
E_g	Energy band gap
ESR	Electron spin resonance
eV	Electron volt
FTIR	Fourier-transform infrared spectroscopy
FWHM	Full width at half maximum
GA	Gum Arabic
H	*Acacia senegal* (hashab)
n	Refractive index
N	*Azadirachta indica* (neem)
t	Transmissivity
T	*Acacia seyal* (talha)
UV-vis	Ultraviolet-visible
ε	Emissivity
α	Absorptivity
ρ	Reflectivity
λ	Wavelength

Introduction

Gum Arabic (GA) is a natural polymer that signifies an important role in our daily lives. It is the firstborn and the most extreme recognized of all the natural gums in the world. It is described as dried, gummy exudation, and it comes from the stems and branches of acacia trees of the legume. GA is produced when the trees are in an injurious state of major nutrition or are of little wetness, or during hot weather. It harvests in breakdowns or wounds in the tree bay, and it shows in the form of spherical balls and droplet nosae family. Around 1100 classes of acacia extend over tropical and subtropical areas of Africa, Central America, southwest North America, India, and Australia, but only discretely; they rarely are very important commercially. The significant ones are made in the areas of West Africa, Sudan, and several smaller neighboring African countries (Wickens, 1995). GA is a composite polysaccharide consisting of Ca, Mg, K, N, and P. It is, principally, incomprehensible to both humans and animals. It is not destroyed in the small intestine

Gum Arabic. http://dx.doi.org/10.1016/B978-0-12-812002-6.00007-5

only, but also inflamed in the large intestine by microorganisms to short-chain oily acids, mainly propionic acid. Such humiliation products are absorbed in the human colon and subsequently utilized energetically in metabolism. The dietary fibers have no direct role in nutrition, but they play a main role in preventing certain common diseases such as colon cancer, diabetes mellitus, coronary thrombosis, obesity, chronic renal failure, and others (Nishi and Jayakrishnan, 2004). It is one of the major exported goods from Sudan; more than 67% of world production is from Sudan. GA has many uses: in foodstuff and as an adhesive material due to its high viscosity, and it is also used as an additive to build a stable suspension mixture for medical syrups, lithography, textiles, paint, inks, and cosmetics (ElKhawad, 2015; Macrae and Merlin, 2002). GA can be used in electron spin resonance (ESR) technique, radiation dosimetry, and modified magnetic nanoparticle, biological, and medical imaging, for example, toxicological experiments using mammalian cells and conducting biomaterial, redox behavior and sensor (Özer, 1986). GA is produced from many species of *Acacia* from African sources. Chemically, Senegal gum is an Arabian galactose protein collected of arabinose (17%–34%), galactose (32%–50%), rhamnose (n = 16%), glue carbonic acid (3%–50%), and protein (1.8%–16%), with an optical cycle of 28 degrees to 32 degrees (Mokhtar, 2016), where the solubility in water is 98%, 19% in a polar solvent, 15% in a nonpolar solvent, 97% in a protonated solvent, and 15% in protonated solvent. The refractive index of GA is 1.3337 (Elnour, 2015), and the molecular weight about 250,000 (Al-Assaf et al., 2007). The good culture methods of the GA acting as the main role in its properties, so the sandy gums are very pale yellow weightiest to colorless spherical nodules, while the clay soil gums are brown to dark brown irregular nodules. All gum types are soluble in cold water, and there is no significant difference in physical properties, with the exception of a minor increase in viscosity of sandy gum (Fincher et al., 1983; Siew and Williams, 2008). GA is used mainly in the food industry as an additive. It is set and has E number E414; GA is a key ingredient in traditional lithography and is used in lithography, paint creation, glue, cosmetics, and other areas, counting viscosity control in the inks and the textile industries, although less expensive material competes with it in many of these areas. The difference in chemical and physical properties of GA from various *Acacia* types is quite wide. The differences can be quite great; for example, nitrogen content may vary from 0.02 to 1.66% and optical rotation from +66 to 108. The smallest amount of protein presentation the GA molecule is integral (Lelon, 2008). From its properties, GA attracts researchers to do their actual works and research. This review aims to focus on the physical and electrical properties of some types of GA and identifies their new applications.

Gum Arabic structure

The GA hydrocarbon is an isoprene polymer developed in the formula of a continuous series. The succession of isoprene units in the chain is perfectly regular, with every fourth carbon atom in the chain carrying the methyl (CH_3) side group. The presence of the double bond is also very noteworthy, since it is mostly determined by the chemical molecule reactivity and its ability to react off the other process (Dror et al., 2006; Nie et al., 2013).

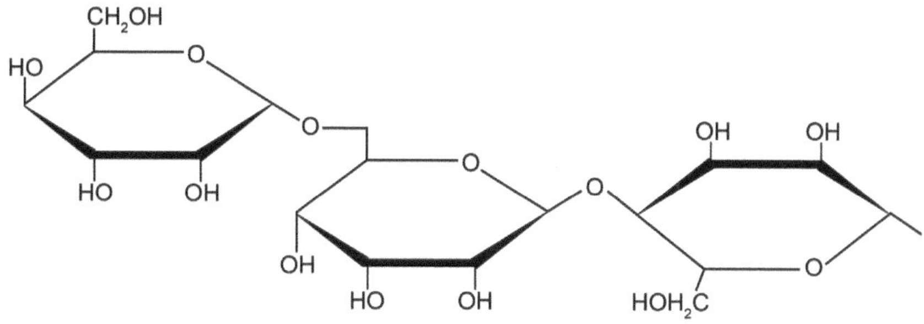

Figure 7.1 Gum Arabic structure.

The double bond is also in charge of the susceptibility of the GA molecule to oxidation or other derivative reaction principal to a decline of physical properties.

One class of ingredient is often explained as texturing capabilities, which is called hydrocolloids, or more commonly, "gum"; these are long-chain, higher molecular weight polymers that dissolve or diffuse in water to give a thickening or gelling effect (Kadajji and Betageri, 2011; Fig. 7.1).

Chemical and physical properties of gum Arabic

Chemical properties

GA is a complex chemical that consists of the macromolecules group categorized by a high percentage of carbohydrates (97%), which mainly possess D-galactose and L-arabinose units and a little part of protein (<3%). The chemical composition of GA may contrast vaguely in its origin, climate, harvest season, tree age, and processing conditions, such as spray drying (Koègel-Knabner, 2002).

Physical properties

GA is a complex creation: a mixture of hydrophilic carbohydrate and hydrophobic protein modules. GA occurs as white to yellowish with white spherical nodules and tears of varying size. It is tasteless and odorless, dissolves readily in hot and cold water, and a gum solution of up to 50% by weight (of gum) can be set. In addition, it is nontoxic, noncaloric, and nonpolluting. It does not revolutionize the flavor, odor, or color of the products to which it is added (Dickinson, 2003, 2008).

Radiation properties

The heat transfer processes can be defined from conduction and convection, and they necessitate the presence of a medium (Bergman and Incropera, 2011). Radiation heat transfer is mainly different from the other types; there is no need for a medium to transfer, and in fact, it reaches maximum efficiency in a vacuum. The main physical quantities in this type of electromagnetic radiation transfer are the frequency and wavelengths. There are two theories describe the radiation; however, neither of them is absolutely acceptable.

The first theory is earlier, which initiated from the insight of a hypothetical medium referred as ether. Ether theoretically fills all evacuated or nonevacuated spaces. The light transmission of radiant heat is tolerable by the electromagnetic waves spread in the ether (Kerker, 2016). All electromagnetic waves are transportable at identical speeds; therefore, shorter wavelengths of the radiation mean high frequencies. Due to the molecules the vibration; everybody or fluid is submerged in the ether, and can potentially commence an electromagnetic wave. All bodies can generate and receive electromagnetic waves at the outlay of their stored energy (Nelson, 1996). The second radiation theory is best known as the quantum theory and was first discovered by Max Planck in 1900 (Klein, 1961). According to this theory, energy is expressed by quanta. As Max Planck maintained, the frequencies of vibration are similar to the wave theory (Rogers, 2005). The energy E can be determined from the equation $E = h\nu$, where h represents Planck's constant, and ν is the frequency. In the quantum theory, higher frequencies are created by high temperatures, which create increasing energy. The electromagnetic wave propagation of all wavelengths is regularly referred as thermal radiation, which is frequently constrained to the visible and infrared regions. For engineering purposes, it may be avowed that thermal radiation is a form of electromagnetic radiation, which varies with surface nature and temperature (Schaefer, 2010). Radiation waves may travel in unusual patterns compared to conduction of heat flow. Radiation heat transfer allows waves traveling from heated body to cold one, absorbing or moderately absorbing medium and accomplish a warmer body again (Fu and Zhang, 2006) for example, the case of the radiation traveling waves from the sun to the Earth.

Emissivity (ε)

The emissivity of a given surface is the measure of capability to emit radiation energy in comparison to a blackbody at the unchanged temperature. The surface emissivity varies from 0 and 1. This is a possession a surface performs as a blackbody. The emissivity of a real surface varies as a surface temperature function of the wavelength and the direction of the emitted radiation (Dewitt and Nutter, 1988; Zhengming and Dozier, 1989). The innovative emissivity of a surface at a given temperature is the spectral directional emissivity, which is defined as the radiation-emitted intensity ratio by the surface at a particular wavelength and direction to that emitted by a blackbody under the same conditions (Gaussorgues and Chomet, 2012). The total directional emissivity is defined in the same fashion: by using the total intensities incorporated over all wavelengths. In practice, a more suitable method is used: hemispherical properties. These properties are spectrally and directionally averaged (Howell et al., 2010). The emissivity of a surface at a specified wavelength may vary as temperature changes, since the emitted radiation spectral distribution changes with temperature. Therefore, the gray and diffuse approximation methods are commonly used to perform radiation calculations (Dewitt and Nutter, 1988).

Absorptivity (α), reflectivity (ρ), and transmissivity (t)

When the electromagnetic radiation strikes a surface, the amount of radiation energy is absorbed, reflected, and transmitted; it will be measured in total energy percentage of the incident electromagnetic beams. Three groups of the total energy would be divided, and they are described by absorptivity (α), reflectivity (ρ), and transmissivity (t) (Pifferi et al., 2004; Singh et al., 2001).

$$\alpha+\rho+\mathrm{t}=1 \tag{7.1}$$

- The fraction of irradiation absorbed by a surface is called absorption.
- The fraction reflected by the surface is known as reflectivity.
- The fraction transmitted by the surface is titled transmissivity.

A body is regarded as transparent if it can transmit some of the falling radiation waves on its surface. In the opaque substances, electromagnetic waves are not transmitted through them, so when radiation waves hit the opaque body surface, some of the waves are reflected back while the other waves are absorbed by a thin layer of the material close to the surface. For engineering purposes, all materials are thick enough that they can be considered opaque, reducing Eq. (7.1) to:

$$\alpha+\rho=1 \tag{7.2}$$

The other properties of radiation are reflectivity, which deviates from its bidirectional nature, whereas it directly depends on the incidental radiation direction as well as the direction of the reflection. Therefore, the reflected rays of a radiation spectrum incident on a real surface in a specified direction form an irregular shape that is not easily predictable. In practice, surfaces are assumed to reflect in a perfectible specular or diffuse manner. In specular reflection, the angles of reflection and occurrence are equal. In diffuse reflection, radiation is reflected equally in all directions. Reflection from smooth and polished surfaces can be assumed to be specular reflection, whereas reflection from rough surfaces approximates diffuse reflection (Alsabah et al., 2017; van Ginneken et al., 1998). In radiation analysis, a surface is defined as smooth if the height of the surface roughness is much smaller relative to the wavelength of the incident radiation.

Therefore, GA is considered good matter to interact with radiation, where the properties and characterizations can be determined and noticed.

Fourier-transformation infrared spectroscopy

The main useful technique for determination of compounds functional groups is infrared spectroscopy. Compounds that have covalent bonds, organic and/or inorganic, absorb various frequencies of electromagnetic radiation in the infrared region of the electromagnetic spectrum. In terms of wave numbers, the infrared vibration extends from about 400 to 4000 cm^{-1} (Alsabah et al., 2016).

In the infrared absorption process, molecules are excited to a higher energy state when the infrared radiation is absorbed. The energy range radiation corresponds to the surrounding range through the stretching and bending vibration frequency bands in most covalent bonds. As for many harmonic oscillators, the vibration energy is continuously and periodically changing from kinetic to potential energy and back again. The total amount of energy is proportional to the vibration frequency (Alsabah et al., 2016).

The four types of GA were collected from the Ministry of Agriculture and Irrigation (plant sector management) and the market, and the traditional manually from many steps to be ready for measurements. Fourier-transform infrared spectroscopy (FTIR) is used for finding the bonds between the atoms, and this becomes a good approach to examine the materials, especially for the GA types that are shown in Fig. 7.2.

Fourier-transform infrared spectroscopy result

From the FTIR results, it can be noticed that Fig. 7.2 explains different pure types of GA in FTIR, and hence, the volume increases linearly (hashab, talha, and neem), but the volume decreases in cacamot, which tends to shift red. Tables 7.1–7.4 show the FTIR results for pure hashab, talha, cacamot, and neem, respectively.

Tables 7.1–7.4 show the wave number, bond types, absorption, and appearance of the GA different types.

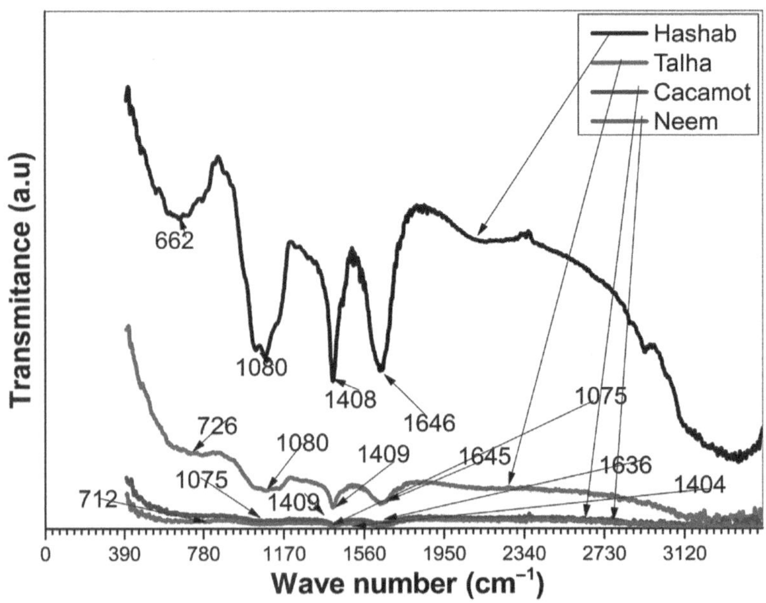

Figure 7.2 The Fourier-transform infrared spectroscopy diagram of the pure gum Arabic types.

Table 7.1 The Fourier-transform infrared spectroscopy result for pure hashab gum

Wave number	Bond type	Absorption	Appearance
664	C—X	Chloroalkanes	Weak to medium
1080	C—X	Aliphatic amines	Often overlapped
1408	O—H	Phenols	Broad and medium
1646	C—C	Acyclic C—C	Medium

Table 7.2 The Fourier-transform infrared spectroscopy result for pure talha gum

Wave number	Bond type	Absorption	Appearance
726	C—X	Chloroalkanes	Weak to medium
1080	C—X	Aliphatic amines	Often overlapped
1409	O—H	Phenols	Broad and medium
1645	C—N	C═N	Similar conjugation effects to C═O

Table 7.3 The Fourier-transform infrared spectroscopy result for pure cacamot gum

Wave number	Bond type	Absorption	Appearance
712	C—X	Chloroalkanes	Weak to medium
1075	C—X	Fluoroalkanes	Often overlapped
1409	O—H	Phenols	Broad and medium
1636	N—H	Primary amines	Strong

Table 7.4 The Fourier-transform infrared spectroscopy result for pure neem gum

Wave number	Bond type	Absorption	Appearance
712	C—X	Chloroalkanes	Weak to medium
1074	C—X	Fluoroalkanes	Often overlapped
1404	O—H	Phenols	Broad and medium
1640	N—H	Primary amines	Strong

UV-visible spectroscopy

Electronic absorption, or UV-visible (UV-vis) spectroscopy, is one of the simplest, most informal, and useful tests; it's the most suitable optical technique for optical and electronic properties of nonmaterials. Most of the organic molecules and functional groups are transparent in the ratio of the electromagnetic spectrum, which is named by the UV and/or visible areas. That is the area where wavelength ranges from

190 to 800 nm. The modest standard of electronic absorption spectroscopy is created on the measurement of light absorption produced by electronic transition in the sample. The absorption spectrum caused by energy adsorption through atoms and molecules passes from a grounded state to an excited state. Then the energy difference can be calculated by both the ground and the excited states. The main law here is Beer's law, which deals with the absorbance or optical density as a function of wavelength. In Beer's law; *A* can be calculated from the incident light intensity I_0 and transmitted light intensity *I*, respectively, concentration of a solution sample *c*, path length of the sample *l*, absorption coefficient α, and molar absorptive (formerly known as molar extinction coefficient, ε) by the following equation (Alsabah et al., 2015; Pace et al., 1995):

$$A = \log I_o / I = \varepsilon lc = \alpha c \tag{7.3}$$

Any colored compounds that have absorption in the visible region are likely to contain a long-chain conjugation system or a polycyclic aromatic chromophore. A few functional groups like nitro, azo, nitroso, polybromo, α-diketo, etc., have compounds that also show color in the visible region.

A clear difference between many compounds is their color. For example, egg is white, grass is green, the sea and sky are blue; and aspirin and ice are colorless (Zollinger, 2003). The absorption of light plays a good role for colored compounds where the electromagnetic spectrum lies between 400 and 800 nm. The electromagnetic spectrum is too clear, whereas the UV-visible range is only a small part of it and is generally defined from wavelengths of 190 nm at the high-energy UV end to about 750 nm at the low-energy red end of the spectrum. Light of other regions of the electromagnetic spectrum give rise to different types of transitions, and from this time, different types of spectroscopy, like IR radiation, are usually not energetic enough to cause electronic transitions, but can excite vibrations of molecules (Hurle and Hertzberg, 1965), and this probability can be very different, depending on the nature of the grounded and excited states of the orbitals.

UV-visible spectroscopy analysis

The diffuse reflection of a sample is measured by shining broadband light on a powder material and the collecting light that reflects off of the sample. The powder sample is placed approximately 3 mm below the double-fiber end of the bifurcated per optic cable. We used a ZFQ-9803 Bifurcated Optical Fiber with a fiber diameter of 455 μ and Ocean Optics for the UV-visible measurement (Russell, 2011). Figs. 7.3 and 7.4 show the powder diffuse reflection and absorption spectrum of the samples. The mean exciton absorption was found in convergent values. By the Kubelka–Munk theory equation (Kubelka, 1931), the absorption was found as a function of the reflectivity:

$$\text{Absorption}\, f(R) = k / s = (1 - R)^2 / 2R \tag{7.4}$$

where *k* is absorption coefficient, and *s* is scattering coefficient.

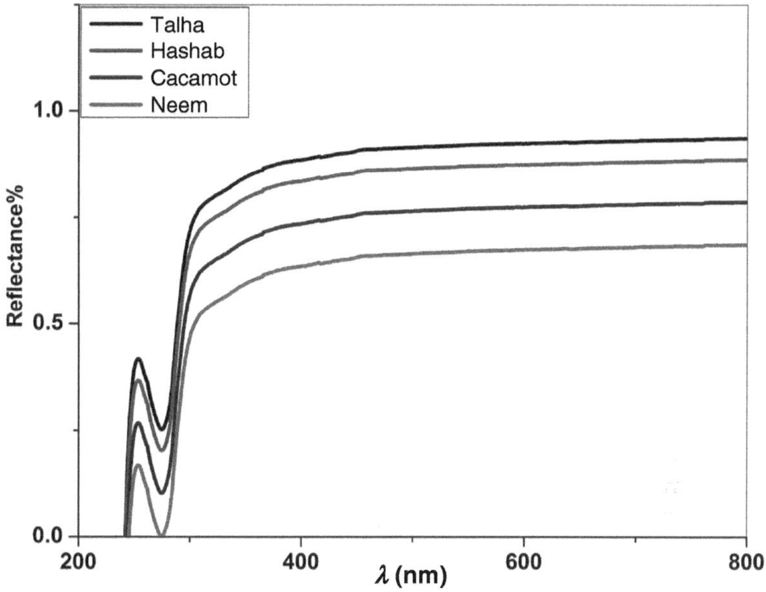

Figure 7.3 Reflectance of all the samples.

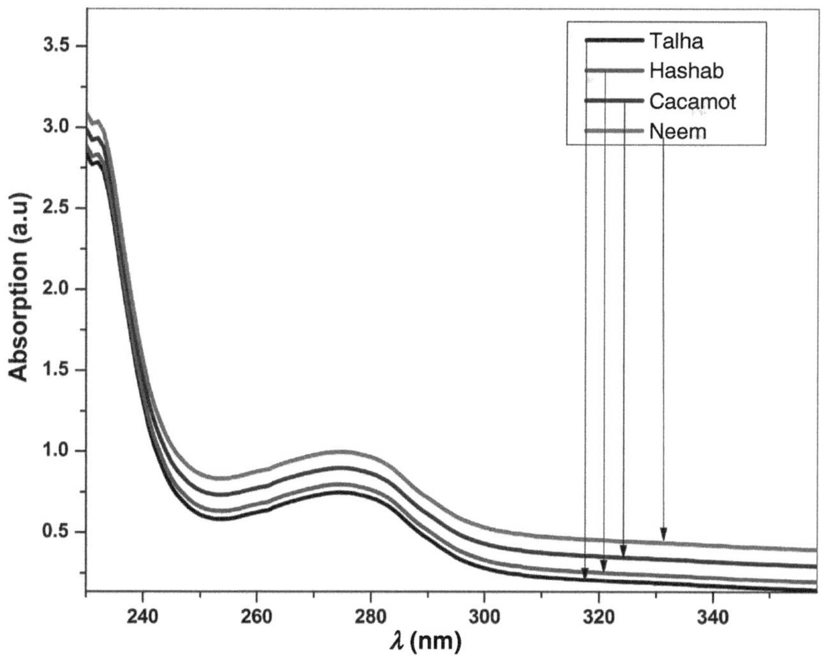

Figure 7.4 The relation between the absorption and wavelength as a function of the reflectivity.

Determination of optical constant

We note that the absorbability occurred in visible light at a wavelength of almost 396 nm, and by using the Tauc plot method, the energy gap was calculated for GA types (Sánchez-Vergara et al., 2012).

$$\alpha h\nu = A(h\nu - E_g)^n \tag{7.5}$$

where E_g is the band gap, A is the constant for different transitions, ($h\nu$) is the energy of a photon, and the n value is, respectively, 1/2 and 2 for direct and indirect transition. Fig. 7.5 refers to the Tauc plot of samples. The band gap energy was found to be E_g = 5.13, 5.10, 5.09, and 5.05 eV for talha, hashab, cacamote, and neem GA, respectively.

The absorption coefficient is given by α = 2.303 A/d (Luque et al., 2015), where d is the thickness of the sample and A is the absorbance. Fig. 7.6 shows the relation between the absorption coefficient and wavelength of all samples.The refraction index of the sample is given by the following relation (Al-Mudhaffer et al., 2012):

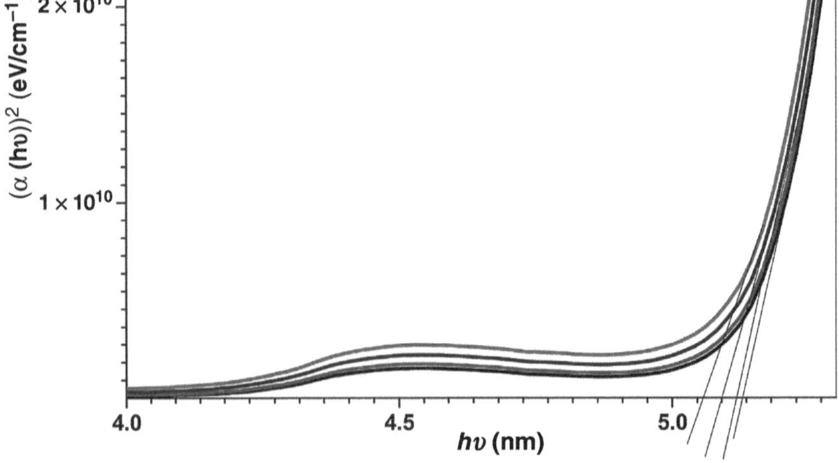

Figure 7.5 The Tauc plot of samples.

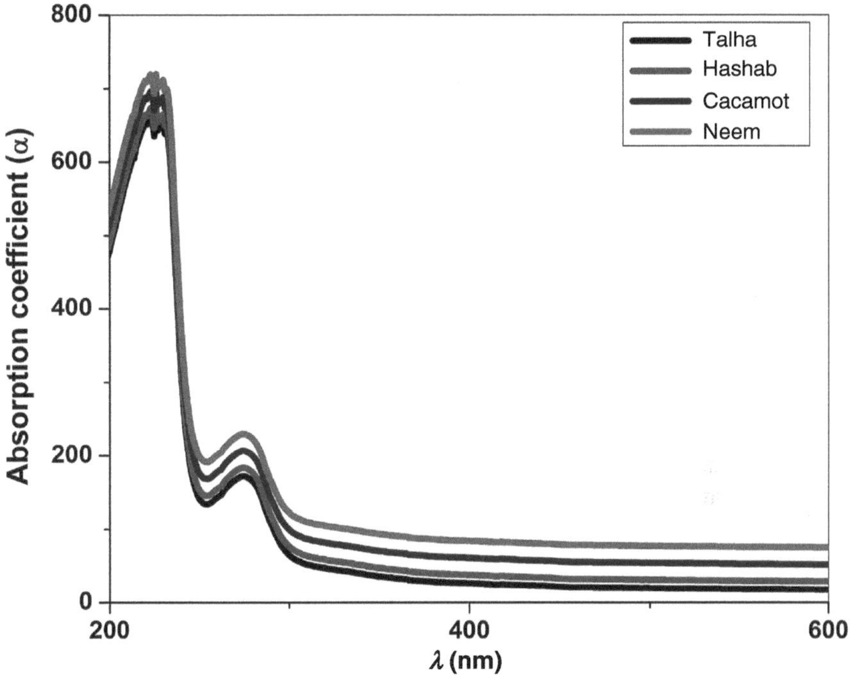

Figure 7.6 The relation between the absorption coefficient and wavelength of all samples.

$$n = \left(\frac{1+R}{1-R}\right) + \sqrt{\left(\frac{4R}{(1-R)^2} - k^2\right)} \tag{7.6}$$

where $k = (\alpha\lambda/4\pi)$ is the extinction coefficient (Narayanan and Peter, 2012), which is shown in Fig. 7.7, and it is noticed that the refractive index (Dmitriev et al., 2013) has a maximum value of 1.28, 1.25, 1.20, and 1.16 for talha, hashab, cacamote, and neem, respectively, at wavelength 230 nm, as shown in Fig. 7.8, where the interactions take place between photons and electrons. The refractive index differs with the variation of the incident light beam wavelength.

The optical losses are caused by absorption and scattering of light, where it decreases the amplitudes of the transmission intensity oscillations at shorter wavelengths. Fig. 7.9 shows the variation of the real dielectric constant (ε_1) with a wavelength of thin film samples, which is calculated from the relation. The imaginary dielectric constant (ε_2) versus (λ) is shown in Fig. 7.10; this value was calculated from the relation where the real dielectric (ε_1) is the normal dielectric constant. From Fig. 7.9, the variation of (ε_2) is following the refractive index, which is decreased in the region of $\lambda > 300$ nm, and hence, the absorption of the film of these wavelengths is small where the polarization was increased.

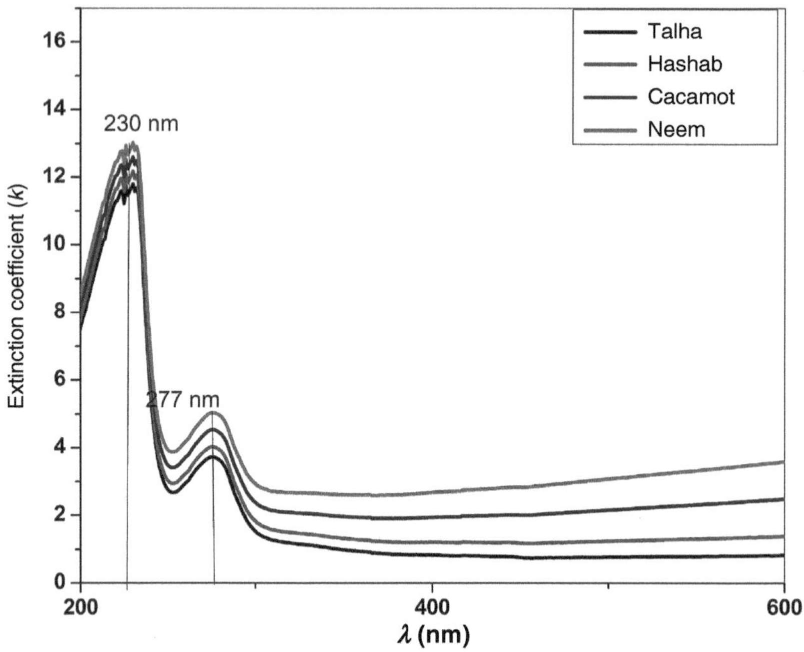

Figure 7.7 The extinction coefficient of samples.

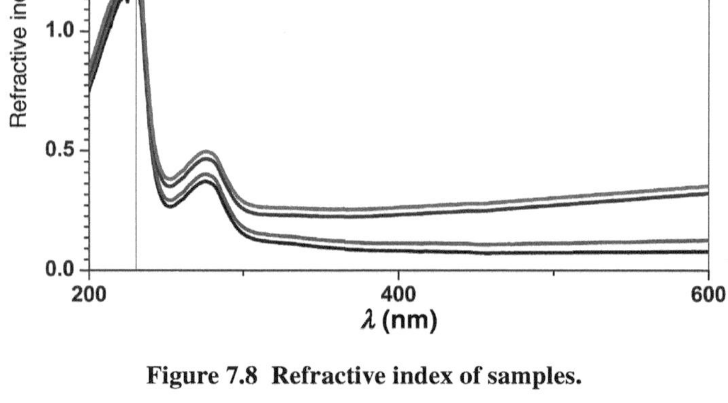

Figure 7.8 Refractive index of samples.

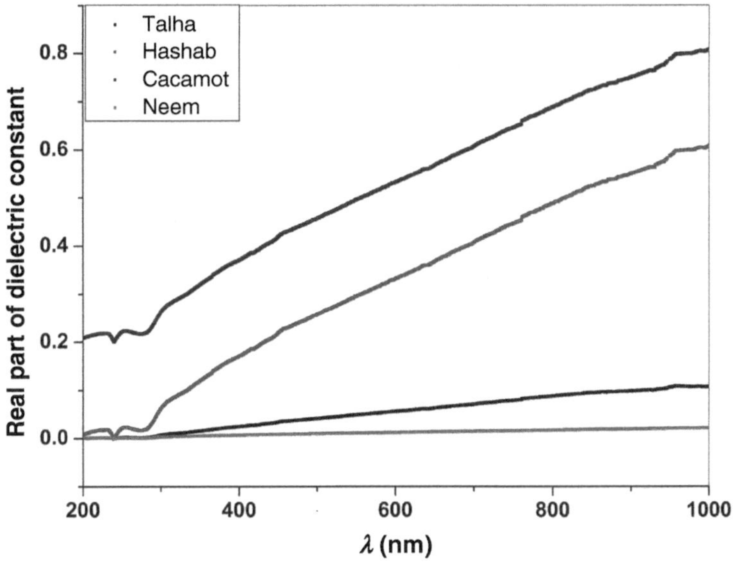

Figure 7.9 Variation of the real part of the dielectric constant versus wavelength of the samples.

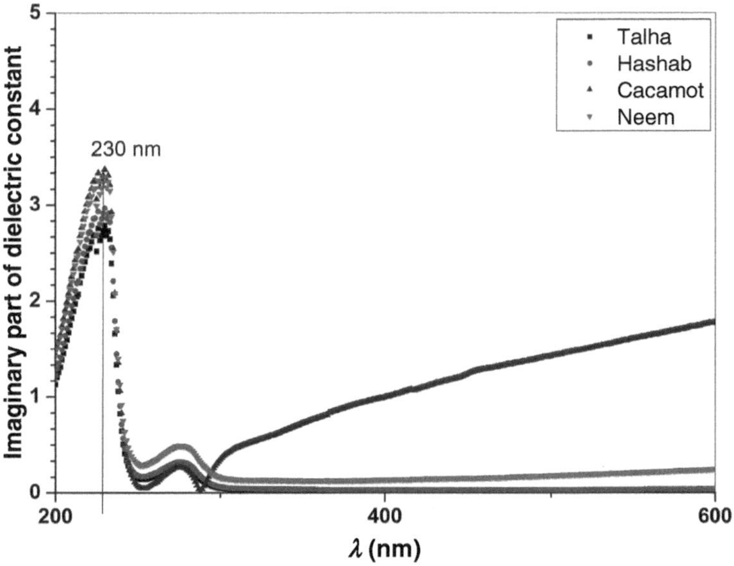

Figure 7.10 Variation of the imaginary part of the dielectric constant versus wavelength of the samples.

The optical conductivity σ_{opt} is a measure of the frequency response of a material with a light, which is determined using the following equation (Kaur et al., 2014; Fig. 7.11):

$$\sigma_{opt} = (\alpha nc / 4\pi) \quad (7.7)$$

where c is the speed of light and σ_e is the electrical conductivity, which can be estimated using the following equation (Fig. 7.12; Al-Mudhaffer et al., 2012):

$$\sigma_e = (2\lambda\sigma_{opt} / \alpha) \quad (7.8)$$

The highest magnitude of optical conductivity was found at 5.5 eV, and the photo energy for all samples confirms the presence of a very high photo response. The increase of optical conductivity at high photon energy is due to the high absorbance of the samples, and it may be also due to electron excitation of photon energy.

In conclusion, the GA color and shape represent good factor. The optical and electrical properties, namely, refractive index, absorption, transmission, and dielectric coefficients, have been determined. Before that, the main introduction clarifying the multifunction of GA that has been noticed, discussed, and determined.

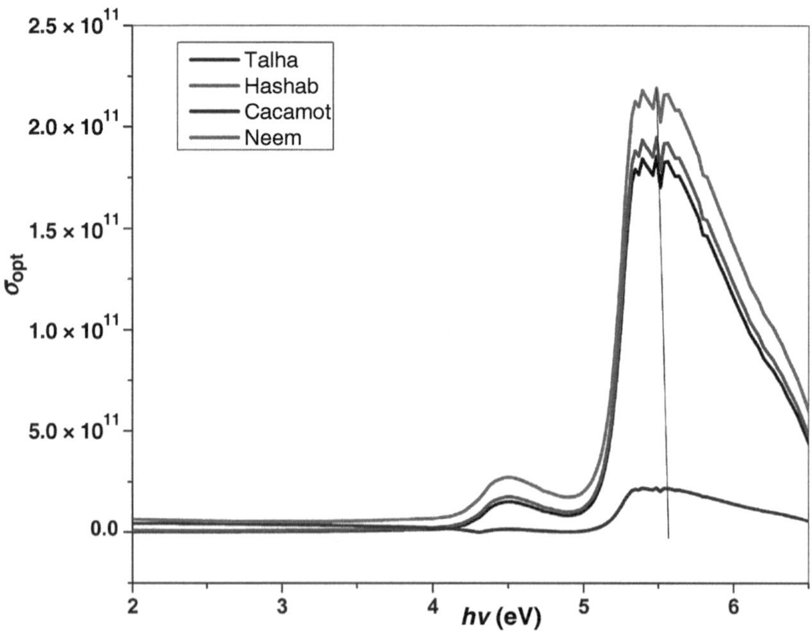

Figure 7.11 Optical conductivity of the samples.

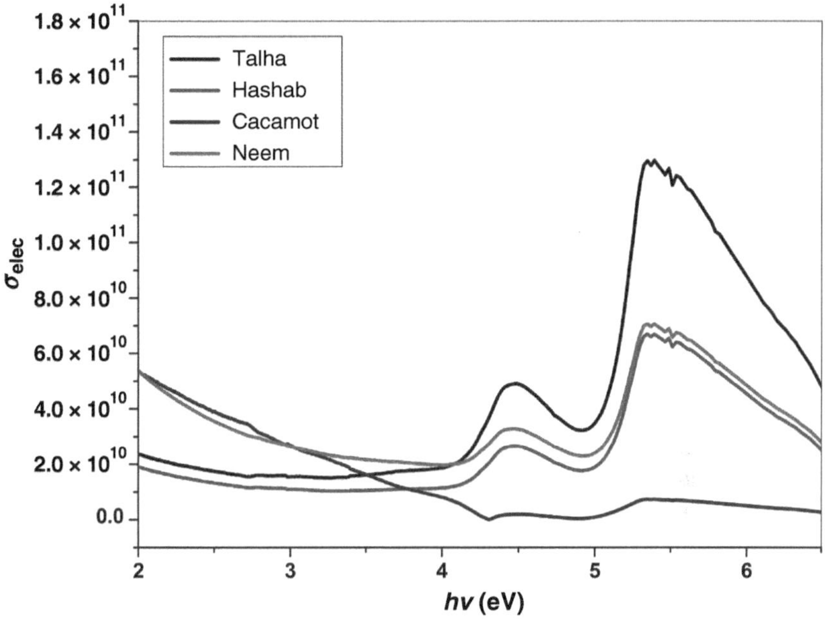

Figure 7.12 Electrical conductivity of the samples.

References

Al-Assaf, S., Phillips, G.O., Aoki, H., Sasaki, Y., 2007. Characterization and properties of *Acacia senegal* (L.) Willd. var. senegal with enhanced properties (Acacia (sen) SUPER GUM™): Part 1—Controlled maturation of *Acacia senegal* var. senegal to increase viscoelasticity, produce a hydrogel form and convert a poor into a good emulsifier. Food Hydrocoll. 21, 319–328.

Al-Mudhaffer, M.F., Al-ahmad, A.Y., Ali, Q.M., Emshary, C.A.-H., 2012. Optical Parameters and Absorption Studies of Benzene-sulfonamide Azo Dye Thin Film Prepared by Spray Pyrolysis Method. Conference Babylon University, Iraq.

Alsabah, Y.A., Alsalhi, M.S., Elbadawi, A.A., Mustafa, E.M., 2017. Influence of Zn^{2+} and Ni^{2+} cations on the structural and optical properties of $Ba_2Zn_{1-x}Ni_xWO_6$ ($0 \leq x \leq 1$) tungsten double perovskites. J. Alloys Compd. 701, 797–805.

Alsabah, Y.A., Elbadawi, A.A., Mustafa, E.M., Siddig, M.A., 2016. The effect of replacement of Zn^{2+} cation with Ni^{2+} cation on the structural properties of $Ba_2Zn_{1-x}Ni_xWO_6$ double perovskite oxides (X = 0, 0.25, 0.50, 0.75, 1). J. Mater. Sci. Chem. Eng. 4 (2), 61–70.

Alsabah, Y.A., Elbadawi, A.A., Siddig, M.A., 2015. Physical properties of X_2AlVO_6 new double perovskite oxide. J. Sci. Nat. 6 (1), 56–62.

Bergman, T.l., Incropera, F.P., 2011. Introduction to Heat Transfer. John Wiley & Sons, New Jersey, USA.

Dewitt, D.P., Nutter, G.D., 1988. Theory and Practice of Radiation Thermometry. John Wiley & Sons, New Jersey, USA.

Dickinson, E., 2003. Hydrocolloids at interfaces and the influence on the properties of dispersed systems. Food Hydrocoll. 17, 25–39.

Dickinson, E., 2008. Interfacial structure and stability of food emulsions as affected by protein–polysaccharide interactions. Soft Matter 4, 932–942.

Dmitriev, V.G., Gurzadyan, G.G., Nikogosyan, D.N., 2013. Handbook of Nonlinear Optical Crystals. Springer, Berlin, Germany.

Dror, Y., Cohen, Y., Yerushalmi-Rozen, R., 2006. Structure of gum arabic in aqueous solution. J. Polym. Sci. B Polym. Phys. 44, 3265–3271.

Elkhawad, H.E.M.A., 2015. Gum Arabic Processing and Marketing in the Sudan (M.Sc. thesis). University of Khartoum, Khartoum, Sudan.

Elnour, A.A., 2015. Fractionation; Physicochemical and Functional Properties of Acacia Polyacantha Gum. Doctoral dissertation. University of Khartoum, Khartoum, Sudan.

Fincher, G.B., Stone, B.A., Clarke, A.E., 1983. Arabinogalactan-proteins: structure, biosynthesis, and function. Annu. Rev. Plant Physiol. 34, 47–70.

Fu, C., Zhang, Z., 2006. Nanoscale radiation heat transfer for silicon at different doping levels. Int. J. Heat Mass Transf. 49, 1703–1718.

Gaussorgues, G., Chomet, S., 2012. Infrared Thermography. Springer Science & Business Media, Berlin, Germany.

Howell, J.R., Menguc, M.P., Siegel, R., 2010. Thermal Radiation Heat Transfer. CRC press, Boca Raton, Florida, United States.

Hurle, I., Hertzberg, A., 1965. Electronic population inversions by fluid-mechanical techniques. Phys. Fluids 8, 1601–1607.

Kadajji, V.G., Betageri, G.V., 2011. Water soluble polymers for pharmaceutical applications. Polymers 3, 1972–2009.

Kaur, J., Rani, M., Tripathi, S., 2014. Determination of optical constants of annealed Sn doped CdSe thin films prepared by thermal evaporation technique. Invertis J. Renew. Energy 74.

Kerker, M., 2016. The Scattering of Light and Other Electromagnetic Radiation. Elsevier, Amsterdam, Netherlands.

Klein, M.J., 1961. Max Planck and the beginnings of the quantum theory. Arch. Hist. Exact Sci. 1, 459–479.

Koègel-Knabner, I., 2002. The macromolecular organic composition of plant and microbial residues as inputs to soil organic matter. Soil Biol. Biochem. 34, 139–162.

Kubelka, P., 1931. The Kubelka-Munk theory of reflectance. Z Tech. Phys. 12, 539.

Lelon, J.K., 2008. Uptake of Micronutrients by *Acacia senegal* Varieties and Its Possible Effect on Gum Arabic Quality. University of Nairobi, Nairobi, Kenya.

Luque, A., Panchak, A., Mellor, A., Vlasov, A., Martí, A., Andreev, A., 2015. Empiric k·p Hamiltonian calculation of the band-to-band photon absorption in semiconductors. Phys. B Condens. Matter 456, 82–86.

Macrae, J., Merlin, G., 2002. The Prospects and Constraints of Development of gum Arabic in Sub-Saharan Africa. A Document Based on the Available Literature and Field Trips to Chad, Mali and Niger for the Africa Tree Crops Initiative. The World Bank, Washington, DC.

Mokhtar, L., 2016. Electric conductivity of gum Arabic from *Acacia senegal*. Int. J. Sci. Res. (IJSR) 5 (2), 583–593.

Narayanan, M., Peter, A.J., 2012. Pressure and temperature induced non-linear optical properties in a narrow band gap quantum dot. Quantum Matter 1, 53–58.

Nelson, S.O., 1996. Review and assessment of radio-frequency and microwave energy for stored-grain insect control. Trans. Am. Soc. Agric. Eng. 39, 1475–1484.

Nie, S.-P., Wang, C., Cui, S., 2013. A further amendment to the classical core structure of gum Arabic (*Acacia senegal*). Food Hydrocoll. 31, 42–48.

Nishi, K., Jayakrishnan, A., 2004. Preparation and in vitro evaluation of primaquine-conjugated gum Arabic microspheres. Biomacromolecules 5, 1489–1495.

Özer, A., 1986. Applications of the electron spin resonance (ESR) technique to flint. PACT, 133–140.
Pace, C.N., Vajdos, F., Fee, L., 1995. How to measure and predict the molar absorption coefficient of a protein. Protein Sci. 4, 2411–2423.
Pifferi, A., Swartling, J., Chikoidze, E., Torricelli, A., Taroni, P., 2004. Spectroscopic time-resolved diffuse reflectance and transmittance measurements of the female breast at different interfiber distances. J. Biomed. Opt. 9, 1143–1151.
Rogers, D., 2005. Einstein's Other theory: The Planck-Bose-Einstein Theory of Heat Capacity. Princeton University Press, 41 William Street, Princeton, New Jersey, USA.
Russell, J.A., 2011. Measurement of Optical Bandgap Energies of Semiconductors.
Sánchez-Vergara, M.E., Alonso-Huitron, J.C., Rodriguez-Gómez, A., Reider-Burstin, J.N., 2012. Determination of the optical GAP in thin films of amorphous dilithium phthalocyanine using the Tauc and Cody models. Molecules 17 (9), 10000–10013.
Schaefer, H.-E., 2010. Nanoscience: The Science of the Small in Physics, Engineering, Chemistry, Biology and Medicine. Springer Science & Business Media, Berlin, Germany.
Siew, C.K., Williams, P.A., 2008. Role of protein and ferulic acid in the emulsification properties of sugar beet pectin. J. Agric. Food Chem. 56, 4164–4171.
Singh, A., Mehra, R., Buthrath, N., Wakahara, A., Yoshida, A., 2001. Highly conductive and transparent aluminum-doped zinc oxide thin films prepared by pulsed laser deposition in oxygen ambient. J. Appl. Phys. 90, 5661–5665.
van Ginneken, B., Stavridi, M., Koenderink, J.J., 1998. Diffuse and specular reflectance from rough surfaces. Appl. Opt. 37, 130–139.
Wickens, G.E., 1995. Role of *Acacia species* in the rural economy of dry Africa and the Near East. Food & Agriculture Organization, Rome, Italy.
Zhengming, W., Dozier, J., 1989. Land-surface temperature measurement from space: physical principles and inverse modeling. IEEE Trans. Geosci. Remote Sensing 27, 268–278.
Zollinger, H., 2003. Color Chemistry: Syntheses, Properties and Applications of Organic Dyes And Pigments. John Wiley & Sons, New Jersey, USA.

Further Reading

Balanis, C.A., 2016. Antenna Theory: Analysis and Design. John Wiley & Sons, New Jersey, USA.

Analytical Techniques for New Trends in Gum Arabic (GA) Research

8

*Elhadi M.I. Elzain**, **, *Abdalbasit A. Mariod*†, ‡
*Taif University, Taif, Kingdom of Saudi Arabia; **Kordofan University, Elobeid, Sudan; †University of Jeddah, Al kamil, Kingdom of Saudi Arabia; ‡The Indigenous Knowledge & Heritage Center, Ghibaish College of Science & Technology, Ghibaish, Sudan

Introduction

The plants gums are comprised of a rather set number of conceivable sugars, including: L-arabinose (Ara), D-galactose (Gal), D-mannose (Man), and L-rhamnose. Onset temperatures associated with the second mass loss were characteristic for the different gum samples, implying different gum structures and different degrees of stability. (Rham), D-xylose (Xyl), L-fucose (Fuc), and D-glucose (Glu). Moreover, there are two sugar acids, D-Galacturonic (GalAc) and D-Glucuronic acid (GluAc). The proportions of the various chemicals in gum Arabic (GA) vary widely, to make it suitable for many important applications, such as food and nonfood products.

The most important criteria for choosing the appropriate technique and tools for qualitative and/or quantitative characterization of any material depend on its nature and the target behind the results. According to the biopolymeric nature of gums, many methods that are applied to biomaterials, food chemistry, polymer physics, or drugs may possibly be tried for gums.

The analytical studies of GA started with purifying the gum by dissolving it in distilled water, followed by electrodialysis, precipitation, ion exchange chromatography, or electrophoresis. Then fractionating the gum components by precipitation using saturated salt solution, and the differential solubility in water, or organic solvent, or by using a suitable complexion agent, or by using ion exchange chromatography, or electrophoresis techniques. There are lots of well-defined classical techniques and tools to investigate the classical uses of GA (AOCS, 2001). Accordingly, some normal qualities for worldwide determinations of value parameters of GA from the *Acacia senegal* species of the gum belt region of Sudan are recorded in Table 8.1 (FAO, 1990; Lelon et al., 2010).

Recent results show that GA has some electric conductivity, magnetic inductivity, and optical properties (Table 8.2).

These intrinsic electric, optical, and magnetic properties could further be improved by changing the structure through some physical stress, such as magnetic induction, thermal, electrical or optical energy, or by changing the chemical composition. Baldwin (2003) produced electrically active, water-soluble intrinsic conducting polymers, biosynthetic macromolecules composed of the GA. Other methods to synthesize

Gum Arabic. http://dx.doi.org/10.1016/B978-0-12-812002-6.00008-7

Table 8.1 **Specifications of quality parameters of gum Arabic**

Property	Range
Moisture content (%)	13–15 (dried at 105°C for 6 h)
Ash content (%)	2–4 (ignited at 550°C for 1 h)
Volatile matter (%)	30–39
Optical rotation (degrees)	51–65 Okalebo et al. (2002)
Nitrogen content (%)	−26 to −34
Rhamnose (%)	0.26–0.39
Arabinose (%)	12–14
Galactose (%)	24–29
Glucuronic acid (%)	36–42
Protein (%)	16–17
Molecular weight	2.16–2.41
Stand density	1.64×10^5
Copper (ppm)	0.5 Lelon (2008)
	52–66
Iron (ppm)	730–2490
Manganese (ppm)	69–117
Zinc (ppm)	45–111

Source: FAO, 1990. JECFA specifications for identity and purity of certain food additives. Food Nutr. Pap. 49, 23, Rome; Lelon, J.K., Jumba, I.O., Keter, J.K., Chemuku, W., Oduor, F.D.O., 2010. Kenya Forestry Research Institute (KEFRI), Nairobi, Kenya. University of Nairobi, Nairobi, Kenya.

Table 8.2 **Average electric, magnetic and optical properties of gum Arabic**

Physical property	Range
Maximum electric conductivity	9.88×10^{-7} ohm^{-1}.cm^{-1}
Maximum electric permittivity	2.8×10^{-4} C/Nm^2
Maximum current passing through the capacitor	19.99 mA
Magnetic susceptibility	7.9×10^{-6} ($0 <$ paramagnetic materials < 1)
Magnetic permeability	2.8×10^{13} N
Energy gap (E_g)	0.5 eV > 0.1–0.25 eV for ordinary semiconductors

Source: Elzain, E.M.I., Mobarak, L.M., Dirar, M., 2012. Investigating the electric conductivity, magnetic inductivity, and optical properties of gum Arabic crystals. J. Basic Appl. Chem. 2(6), 35–49.

polymer-based nanoparticles and composites of electrically active redox biomaterials from GA with induced electric, magnetic, and optic properties. Such polymer compounds, due to their advantage of being both a polymer and filler component, are also subject to recent research to investigate their electric conductivity, relaxation, energy gap, electric permittivity, magnetic susceptibility, magnetic permeability, and optical properties (Tiwari, 2007). Such attempts to synthesize environmentally degradable polymers with redox properties, as to replace the inorganic semiconductors in many practical applications such as: conducting coating, capacitors, semiconductors, rotating machines, biosensors, electrostatic dissipation, optelectronics, electroinactive materials, resisters, solar cells, optic filters, etc. Hence, these induced chemical and physical properties need to be characterized by some new and precise analytical methods.

Chromatographic techniques

The majority of the strategies that are produced for diagnostic purposes in food science are likewise utilized for the recognizable proof of plant gums in imaginative or historic works. In this way, chromatographic methods have demonstrated to be extremely valuable for the investigation of historic and artistic works. In early phases of advancement, paper chromatography was applied, a piece of the inconveniences of a long improvement time, poor determination, and a constrained detectability. Use of thin-layer chromatography gives some fascinating outcomes. Gas chromatography (GC) is presumably the most broadly utilized strategy these days (Mejanelle et al., 2002; Pitthard and Finch, 2001; Schneider and Kenndler, 2001), albeit extra steps are required to acquire the volatile derivatives. Moreover, systems like high-pressure liquid chromatography (Colombini et al., 2002), mass spectrometry (MS), and also blends of the strategies previously mentioned, like GC, in addition to IR spectrometry or GC–MS, have been recently used in this field. GC–MS has been applied for the recognizable proof of plant gums and animal glues, taking after the hydrolysis and gathering particular partition of amino acids and sugars (Schneider and Kenndler, 2001).

Chromatographic techniques are used as both separation and analytical techniques. Thus, the separation and purification of gums could be done by using one or a combination of more than one chromatographic method. The capillary electrophoresis (CE) with amperometric detection, which has been built up for quick and compelling estimation of impurities in pharmaceutical formulations, can also be applied in gum research. The chromatographic separation is affected by charged components, such as: amount of ion concentration, amount of charge on the ions, $M^{3+} > M^{2+} > M^{+}$ and $X^{3-} > X^{3-} > X^{-}$, rate of movement of the ions, which depends on type of solvent, viscosity, applied voltage, temperature, ionic size, $Na^{+} > K^{+} > Cs^{+}$ and $F^{-} > Cl^{-} > Br^{-}$, temperature, pH, and ionic mobility and conductivity.

Sugars are extremely powerless acids that ionize just at firmly alkaline pH (pH > 12). At this pH, the carbohydrates are available as negatively charged alcoholates. This property was abused by a few research units for partition, trailed by the ensuing indirect UV, conductivity (Carvalho et al., 2003), or electrochemical detection. After stoichiometric cleavage of the glycosidic bonds, the monosaccharide components of plant gums could be isolated by CE at pH 12.1 and identified with indirect UV absorbance. The CE method is as yet a zone of improvement to overcome the trouble experienced with the investigation of carbohydrates that lack detection focuses on their structures. A CE technique utilizing 2,6-pyridinedicarboxylic acid

(PDC) as the chromophoric background electrolyte at a high pH was depicted by Soga and Imaizumi (2001) by adding cetyltrimethylammonium bromide (CTAB) to the running electrolyte. This framework was adjusted for the evaluation and characterization of sugar monomers and uronic acid contained in various plant gums, matrices that were not examined by CE. The constituent sugars were broken down, either by direct detection in light of chelation with Cu(II) after derivatization to acquire UV-absorbing species, or by deciding the underivatized sugars through complexation utilizing direct UV identification. Complexation of the saccharides with borate-based electrolytes, a negative charge important for electrophoresis is bestowed. In this system, the plant gums are basically hydrolyzed and investigated by CE, and because of the utilization of backhanded UV detection, no derivatization of the sugars must be performed. The technique focuses essentially on the recognizable proof of the gum family, which is essential enthusiasm for the restorer and conservator and does not consider parts of chemical taxonomy.

Größl et al. (2005) examined different plant gums. They got monosaccharides after hydrolysis with 2M trifluoroacetic acid, and lyophilization of the hydrolysate were arabinose, galactose, mannose, rhamnose, xylose, fructose, and glucose, and the two sugar acids, galacturonic and glucuronic acids, as per the literature. They isolated these sugars in a foundation electrolyte comprising of NaOH to modify the pH, 20 mM PDC as a chromophore for detection, and 0.5 mM CTAB as an added substance to invert the electroosmotic flow. In light of their electropherograms, they recognized the plant gums by their typical composition as follows: a pinnacle of glucuronic acid, together with that of rhamnose, is demonstrative for GA. Pinnacles of galacturonic acid and fructose indicate gum tragacanth. Locust bean gum demonstrates a noteworthy peak for mannose (with the attendant galactose peak in proportion, 4:1), though a glucuronic acid and a mannose top, together with a noticeable arabinose peak, indicate cherry gum. These writers applied the strategy to recognize the plant gums in tests such as water colors and in a few paint layers, such as gum-based paint or those with egg white or drying oils as added substances (Größl et al., 2005).

Mechanical stability

Mechanical stability is determined by stress–strain relations in order to know the possible deformation or stability of the solid structure. The mechanical solidness of the networks is an imperative element, giving a sign about the applicability of further handling or bundling steps, and it fills in as determinant criteria for a protected application by the patient. A high mechanical resistance can be an incredibly preferred standpoint contrasted with other quick-dissolving oral dose shapes arranged by lyophilization, where quick discharge rates are regularly identified with low hardness and a decreased stability (Saigal et al., 2008). Instantgum AA, a shower-dried form of GA, paracetamol, and phenazone, are utilized. A disintegration media of 0.1 N HCl is utilized to stimulate gastric liquid. Phosphate buffer pH 6.8 is utilized to stimulate intestinal liquid.

Determining the organic elements

GA contains the elements C, H, and O as the main constituents of its carbohydrate moiety and C, H, O, S, P in its protein moiety, while it contains cationic species such as Mg, Ca, K, and Na, and trace amounts of other metals such as Cu, Zn, Fe, Ni, and Pb (Bulajic et al., 2002). Basic investigation is performed with a specific end goal to determine the carbon, hydrogen, nitrogen, and sulfur contents. The outcomes demonstrated that the isolated characteristic polysaccharides are comprised of a specific rate of carbon, nitrogen, sulfur, and hydrogen in every one of the gums. The basic composition shows that polysaccharides can be utilized as a part of a mucoadhesive medication delivery system, as they contain sulfur and hydrogen in copious amounts (Bothara and Singh, 2012).

Laser-induced breakdown spectroscopy (LIBS) is a generally exploited atomic emission spectroscopic method appropriately imagined for the examination of the elemental structure of an extensive assortment of materials (solid, fluid, and gas samples) (Singh and Thakur, 2007). The LIBS capacity of giving the basic composition of practically any sort of sample with no specimen pretreatment and its potential for in situ and remote examination, even in harsh conditions, together with its small scale of damaging tendency and comparatively minimal cost, renders this procedure especially reasonable for the investigation of an extensive variety of samples. Yousif (2017) characterized GA by using the LIBS technique. This method gives a clear picture of the components of GA and does not change the nature of the materials. It is also useful to note the presence of higher ionization states of some of the elements present in the gum samples, such as Fe^{+3}, Fe^{+2}, Cr^{+3}, Th^{+2}, Ca^{+2}, Cr^{+5}, Ti^{+2}, and Ti^{+3}. Almuslet et al. (2016) utilized LIBS to determine the constituent of GA (*A. senegal*) collected from five different locations in Sudan. Gum samples were irradiated with 80 mJ pulse energy of Nd:YAG laser (1064 nm), and the Atomic Spectra Database was utilized for the spectral investigation of the plasma emitted from these samples. The outcomes showed that all the five samples contain Fe, Na, Ca, Mg, K, S, C, N, O, Cr, Br, Ti, Ar, and H in different amounts. Some elements, like Br, Ti, and Ar, were recorded for the first time.

The X-ray fluorescence (XRF) has the benefit of being nondestructive, multielemental, and quick strategy; in this manner, it is utilized as an investigative apparatus for subjective and quantitative examination of K, Ca, Fe, Cu, Rb, Sr, and other heavy metals. The last outcomes are investigated by quantitative X-ray analysis system (QXAS) programming. Moreover, XRF gives a genuinely uniform recognition restriction over an extensive portion of the periodic table and is applicable to an extensive variety of concentrations, from 100% to few parts per million (ppm). The significant burdens are that examinations are, by and large, limited to components heavier than fluorine, and cost-effective. At that point, the material to be analyzed is irradiated with a follow of X-rays with a shorter wavelength to discharge M, L, or even K electrons.

Adam et al. (2013) used XRF to analyze GA; their results indicated that GA (*A. senegal* and *Acacia seyal*) contained calcium (Ca), iron (Fe), copper (Cu), lead (Pb), and strontium (Sr). The FTIR spectra of both types of GA showed broad and strong absorption bands in the range of 600–3000 cm^{-1}. These absorptions were assigned to the different stretching vibrations. These authors concluded that conductivity was found to increase with frequency, while the opposite behavior was observed for the dielectric. The absorption was found to increase with the increase of concentration. Mohamed (2010) used X-ray diffraction (XRD) technique to identify heavy elements in samples of GA, using the XRF line energies (keV) as: K (Ka 3.312), Ca (Ka 3.690), Cr (Ka 5.411), Mn (Ka 5.895), Fe (Ka 6.400), Cu (Ka 8.041), Zn (Ka 8.631), Br (Ka 11.907), Rb (Ka 13.375), Sr (Ka 14.142), and Pb (La 10.540).

Disintegration studies

The generation of thin layer matrices is an intriguing choice to give a quick drug delivery to the oral cavity. High mechanical strength consolidated with quick disintegration circumstances can be a great advantage for the development of orodispersible tablets (Kipping and Rein, 2016). Controlled discharge matrices in light of GA are set up by applying a nonstop hot-melt extrusion technique: the preblend comprised of GA and the incorporated active pharmaceutical ingredient (API) is plasticized by a co-rotating twin-screw extruder, a middle-of-the-road strand is formed by a round nozzle. Single-dosed matrices are set up by cutting the semielastic strand with a rotating fly cutter. Paracetamol and phenazone are utilized as model medication substances. High diffusion–based loadings up to 70% can be figured out. Matrices are portrayed concerning their crystalline structure, in vitro disintegration, breaking down time, and different physical parameters, including glass transition temperature (T_g). Release characteristic conduct is, for the most part, affected by erosion of the matrices. At higher medication loadings, additionally, diffusion-based transport picks up significance. The solvency of the API demonstrates an impact on the erosion rate of the matrix and ought to, in this manner, be considered during formulation advancement. T_g is, for the most part, impacted by the solubility of the API in the encompassing matrix. High-solvent phenazone demonstrates a decrease, while paracetamol expansion has about no impact on the T_g of the polymeric system. Activation energy (EA) of the glass transition is resolved through unique mechanical examination. The expansion of APIs prompts a reduction of EA, demonstrating an expanded molecular movement at the T_g district contrasted with placebo extrudates. XRD is utilized to decide the crystalline condition of the extruded matrices and communication among matrix and joined APIs. The production of thin-layer matrices is an intriguing alternative to give a quick medication delivery to the oral cavity. High mechanical strength joined with quick disintegration circumstances can be an awesome preferred standpoint for the improvement of orodispersible tablets. An awesome advantage of the assessed processing technology is the straightforward adaptation of the last dosage by differing either the cutting length or the diameter of the nozzle, bringing about a cost-effective production of single-dosed matrices without modification of the base formula (Kipping and Rein, 2016).

Determining viscosity (η)

Viscosity is the resistance to a fluid's flow through a capillary tube. There are various methods used to measure the viscosity of a fluid: vapillary viscometer, Zahn cup, falling sphere viscometer, vibration viscometer, rotational viscometer, falling cylinder viscometer, rising bubble/droplet viscometer, Redwood viscometer, saybolt viscometer, and Ubbelohde, or Ostwald, viscometer. Most gums shape extremely viscous sols or gels at 1–5% concentration.

GA arrangement viscosities are much lower, and the gum is not utilized for swelling or high thickening uses. Its viscosity increments gradually up to 25% concentration and displays Newtonian behavior. The intrinsic viscosity is determined by using the dilute solution viscometer; the relative viscosity η_1 of test solution can be determined by measuring the distance travelled by an iron ball through it, during a defined time, down a column, in comparison to η_2 of a known standard liquid material under similar experimental conditions.

Gómez-Díaz et al. (2008) studied the inherent viscosity and flow conduct of fluid solutions of various concentrations of Arabic gum. They figured the molecular weight utilizing the estimation of the intrinsic viscosity, utilizing Huggins, Kraemer's, and Mark–Houwink equations. The estimation of the obvious viscosity and the impact of the shear rate on the rheological conduct of various polymer concentrations in the watery solution were resolved. The impact of temperature over the flow conduct was additionally examined. The Ostwald model has been utilized to fit the test data, and model parameters have been acquired. They reasoned that Arabic gum watery solutions demonstrated a non-Newtonian and pseudoplastic conduct. The polymer utilized has been described in connection to intrinsic viscosity and normal atomic weight. The power law model has been utilized to portray this conduct under shear in rotational concentric cylinder system. The rheological parameters have been resolved, and the impact of polymer focus and temperature has been assessed (Gómez-Díaz et al., 2008). Onset temperatures associated with the second mass loss were characteristic for the different gum samples, implying different gum structures and different degrees of stability.

Thermogravimetric analysis

Thermogravimetric analysis (TGA) is a basic and exact technique to measure the progressions in physical and chemical properties of materials, as elements of expanding temperature (with consistent heating rate), or as components of time, with constant temperature and additionally consistent mass loss (Liu and Yu, 2006). TGA gives data about physical or chemical phenomena, for example, second-order stage transitions. TGA is ordinarily utilized to decide chosen characteristics of materials that display either mass loss or gain because of deterioration, oxidation, or loss of volatiles (D'Antone et al., 2001; Tikhonov et al., 2009). TGA is an especially useful technique for the study of the decomposition design and the thermal solidness of polymeric materials, including plastic films, composites, elastomers, thermosets, fibers, coatings, thermoplastics, and paints. TGA is being equipped with electronic self-balancing techniques.

Daoub et al. (2016) studied physicochemical and thermal stability of four acacia species. The primary thermograms show the thermal stability data and thermal behavior for the gum samples. Heating of the gum samples from 30°C to 400°C, at a rate of 10°C per minute rate, brings about two mass loss occasions. The primary mass loss is because of the loss of adsorbed and basic water of gums that takes place between 30° and 150°C, as the polysaccharide decomposition is related to the second mass loss with an onset temperature of about 260°C, which resulted in a 60% mass loss. Mass residues of the different gums were ranging from 30% to 35%. The decomposition onset temperature of 260°C suggests that polysaccharide gums are thermally stable.

Differential scanning calorimeter

Onset temperatures associated with the second mass loss were characteristic for the different gum samples, implying different gum structures and different degrees of stability. The differential scanning calorimeter (DSC) is used to accurately determine the amount of heat emitted or absorbed associated with phase onset temperatures associated with the second mass loss were characteristic for the different gum samples, implying different gum structures and different degrees of stability. Transitions from crystalline (T_c) to glass (T_g) to melting temperatures of polymeric materials compared to those of thermally inert reference material. Bothara and Singh (2012) thermally characterized natural gums using DSC, differential thermal analysis, and TGA under nitrogen atmosphere. They decided the significant thermal transitions and also activation energies of the real deterioration stages. These researchers revealed that DSC traces show a noteworthy serious exothermic transition taken after by weaker exotherm(s). Thermogravimetric investigation demonstrated two phases of weight reduction. The main phase has minor weight reduction in tests and is ascribed to the loss of adsorbed and structural water of biopolymers or because of desorption of moisture as hydrogen-bound water to the saccharide structure. The second weight reduction occasion might be ascribed to the polysaccharide decomposition. They presumed that the natural gums are thermally steady and can be utilized as discharge modifiers in different dose forms. Mothe and Rao (2000) studied the thermal behavior of GA and cashew gum by DSC and thermogravimetry (TG). They found that DSC thermal profiles for GA with low water content (0 ± 40%) showed an endothermic event at about 908°C, and with increasing water content (50% ± 80%), multiple melting endotherms and associated enthalpies were observed. Similar behavior was found in the DSC curves of cashew gum with endothermic event of about 978°C. TG curves show two decomposition stages; the major decomposition occurred at 252°C to pure GA and cashew gums. DSC measurements are carried out with samples enclosed in 50 ml aluminum pans, and temperature scans are performed at a heating rate of 150°C/min, in the temperature range of 25–40°C. The use of quick heating rates permits the perception of low energetic transitions found in biopolymers (Liu et al., 2009). These authors concluded that DSC warm profile information propose gel formation and a dissolvable helped melting of crystallites amid heating of the GA and cashew gum. The TG depict a comparative thermal stability for both.

Infrared spectroscopy

Research by infrared (IR) spectroscopy has limited utilization, since various of the spectra overlap. Near-infrared (NIR) spectroscopy gives multicomponent investigation of almost any matrix to give important information about the nature of chemical bonds, functional groups, and molecular structure of organic and organometallic molecules. The technique is characterized by an ease of rapid and nondestructive sample preparation without any pretreatments, the probability of separating the test estimation position by utilization of fiber optic probes, and the desire of chemical and physical example parameters from one single spectrum of *A. seyal* and *Acacia polyacantha.* Adam et al. (2013) performed FTIR on powdered samples of two types of GA (*A. senegal* and *A. seyal*). The samples illustrated broad and strong absorption band in the range of 600–3000 cm^{-1}, and these absorptions were assigned to the different stretching vibrations. Podder et al. (2014). used FTIR to characterize their samples, and most of their observed stretching vibrations are similar to Adam et al. (2013) results.

There are some current procedures utilizing particular frequencies of infrared radiation that are consumed by the functional groups characteristics for water. The concentration of moisture in the specimen conversely corresponds to the energy consumed; thus, the measure of water substance is controlled by measuring the energy that is reflected or transmitted by the sample.

Dong et al. (2017) developed a rapid and reliable method for GA authentication based on NIR spectroscopy and chemometric methods. The two major *Acacia* gum species, *A. Senegal* and *A. seyal*, were assigned perfectly by the NIR spectroscopic method. In addition, a partial least squares regression model was calibrated to predict the blending percentage of the two pure gum types, producing an accurate, root mean square error of cross-validation of 2.8%. These authors concluded that NIR spectroscopy is a very powerful and reliable method for authenticity testing of GA species.

Nuclear magnetic resonance (H NMR)

Over the years, numerous publications have appeared on the nuclear magnetic resonance (NMR) of food polysaccharides, and NMR has become a routine method of analysis. ^{1}H and ^{13}C NMR are the best for the investigation of GA adulteration with lower-grade materials (Figs. 8.1 and 8.2). In the examples below, we show the contrasting spectra of *A. senegal* compared to *A. seyal*, *A. polyacantha*, and *combretum.* Adulteration can be readily observed and quantified by analyzing relative intensities of methyl resonances in the ^{13}C spectrum (http://www.process-nmr.com/adulteration of Acacia senegal.htm).

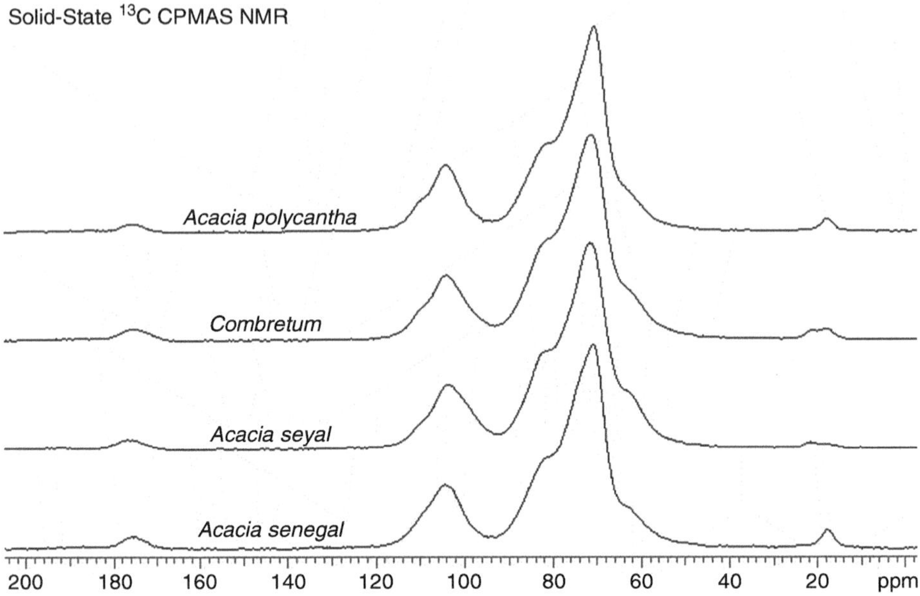

Figure 8.1 Solid-state ^{13}C cross-polarization magic angle spinning (CP-MAS) NMR of different gums.

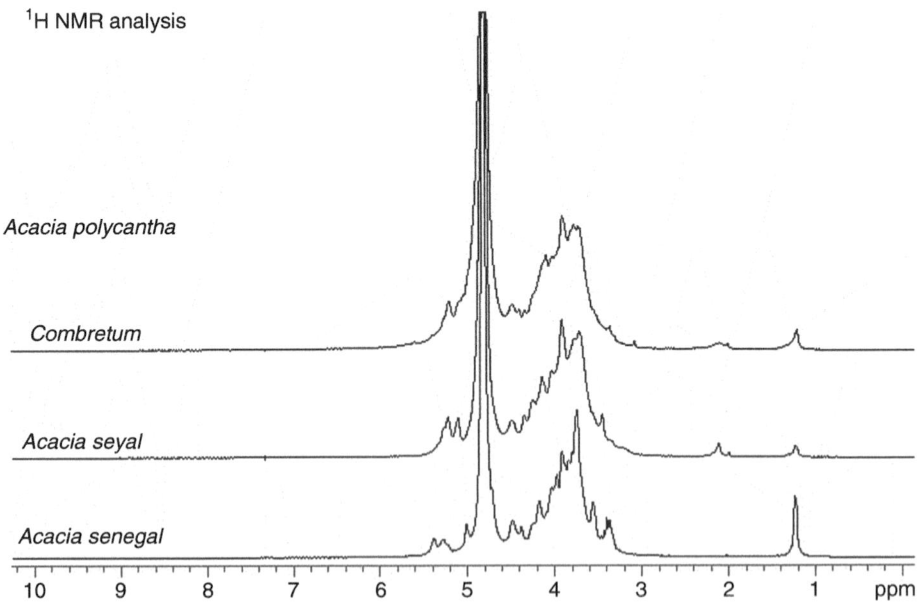

Figure 8.2 ^{1}H NMR of different gums.

A. senegal, *A. seyal*, *Acacia mellifera*, and *Acacia tortilis* were analyzed by Daoub et al. (2016) using NMR. The ^{13}C and ^{1}H NMR spectra of gum samples showed similarity in individual sugar components, but characteristic patterns of each gum were observed. Authors showed crowded signals in the ^{1}H NMR spectrum between 3 and 6 ppm, which is typical of polysaccharides and reflects the presence of similar sugar residues, for each gum an up field peak at around 1.4 ppm assigned to the methyl group of rhamnose sugar which on high resolution shows triplet of triplets.

Optical activity

Certain substances, for example, some organic molecules (carbohydrates), liquid or solutions, are characterized by optical rotating structure, which is considered as the most important criterion of purity and identity of such compounds. Therefore, it is used to identify the type of gum. The optical rotational angle of specific rotation is measured by a polar meter. Its value depends on the nature of substance, solvent, concentration, temperature, wavelength, and path length (Daoub, 2016).

Electric conductivity (Λ)

Conductivity is used to study the molecular or charge mobility in the amorphous phase structure. Mokhtar (2013) studied the electric conductivity of GA samples, where samples of gum solutions of different concentrations were prepared by dissolved gum powder in distilled water. The study emphasized the effects of temperature, distance between capacitor plates, added electrolytes (NaCl), and different concentrations of gum solutions on the conductive properties. The electric conductivity was 0.0 mS at lowest concentration and minimum temperature of 0.0°C; it increased to 0.5 mS when the temperature was increased to 90°C. When the concentration was increased to 0.3% at 0.0°C, the conductivity increased 4.8 mS. The conductivity reached its maximum, 19.5 mS, at 90°C and 0.3% concentration. The electric current reached its maximum value 8.0mA at the concentration of 0.3% at room temperature and 1 cm distance between the capacitor plates. The electric current increased with an increase in temperature to reach the maximum of 63mA at 0.3% concentration and 90°C. The electric current passing through the capacitor plates was reported to be 0.0mA at the lowest concentration at room temperature, regardless of the distance between the two plates. An addition of 2% concentration of NaCl solution to gum solution of different concentrations increased the conductivity, even at the lowest concentration. Increasing NaCl concentration showed proportional increase in the electric conductivity of gum solution. Mokhtar (2013) recommends more research in this field to make use of GA as materials that can be used in techniques of modern electronic manufacture instead of the routine use of gum in food. Also, the researcher recommends the use of direct current on the sun research to reveal the properties and possibility of using GA in transistors, solar cells, and different types of capacitors with different capacities and use of GA in its solid state or crystals.

Determining dielectric relaxation (ε)

Dielectric spectroscopy (DS) is utilized as an instrument for exploring complex materials. The interesting ability of DS is being connected in a wide time scale range to recognize general relaxation phenomena and to find new ones for the concentrated complex systems, independent of the distinction of the material. DS makes it conceivable to cover a wide continuous range of tentatively controlled parameters, an element that no other existing spectroscopic method today can offer. In addition, this method, in contrast with traditional radio spectroscopic methods, for example, electron paramagnetic or atomic attractive reverberation, is considerably more straightforward as far as equipment acknowledgment. It has, as of now, been specified that the properties of a dielectric test, in spite of the morphological contrasts, are elements of many tentatively controlled parameters. The DS technique possesses an exceptional place among the various present-day methods utilized for physical and chemical examination of materials, since it empowers examination of relaxation procedures in a, to a great degree, extensive variety of characteristic times (105–1013 s). Despite the fact that the strategy does not have the selectivity of NMR or electron spin resonance (ESR), it offers imperative and, now and then, one-of-a-kind data about the dynamic and basic properties of substances. DS is particularly touchy to intermolecular collaborations, and along these lines, agreeable procedures might be checked. It gives a link between the properties of the bulk and individual constituents of an unpredictable material.

Ishage et al. (2015) prepared a sample of GA doped with black carbon (soot) and potassium bromide ($KBrO_3$) using the solid-state chemical method. The outcomes of the diffusion coefficient and dielectric indicated direct proportionality with the logarithmic frequency. The porosity of the sample was observed to be specifically corresponding to the dielectric consistent. The impact of temperature was significant, as diffusivity and conductivity were found to increase with temperature. Such results might be attributed to the increase in grains surface to volume ratio obtained in the studied sample and being good evidence for porosity formation. The rate of diffusion is proportional to the concentration gradient of the diffusion. General formulations for diffusion and mass transport are outlined in Fick's first law of diffusion, which expresses the steady state flux of diffusion per unit area.

References

Adam, H., Siddig, M.A., Siddig, A.A., Eltahir, N.A., 2013. Electrical and optical properties of two types of gum Arabic. Sudan Med. Monit. 8 (4), 174–178.

Almuslet, N.A., Ahmed, M.E., Mohammed, M.A., 2016. Characteristics of gum Arabic (*Acacia senegal*) using laser induced breakdown spectroscopy 2 (6), 208–215.

AOCS, 2001. American Oil Chemist's Society, Physical and Chemical Characteristic of Oils, Fats and Waxes. AOCS Press, Champaign, Illinois, USA.

Baldwin, T.C., 2003. Gum Arabic: past, present and future. In: Majumdar, D.K., Govil, J.N., Singh, V.K. (Eds.), Recent Progress in Medicinal Plants, Vol. VIII–Phytochemistry & Pharmacology II. Sci. Tech. Publishers, USA, pp. 403–410.

Bothara, S.B., Singh, S., 2012. Thermal studies on natural polysaccharide. Asian Pac. J. Trop. Biomed., S1031–S1035.

Bulajic, D., Corsi, M., Cristoforetti, G., Legnaioli, S., Palleschi, V., Salvetti, A., Tognoni, E., 2002. A procedure for correcting self-absorption in calibration free-laser induced breakdown spectroscopy. Spectrochim. Acta Part B 57, 339–353.

Carvalho, A.Z., da Silva, J.A.F, do Lago, C.L., 2003. Determination of mono- and disaccharides by capillary electrophoresis with contactless conductivity detection. Electrophoresis 24, 2138–2143.

Colombini, M.P., Ceccarini, A., Carmignani, A., 2002. Ion chromatography characterization of polysaccharides in ancient wall paintings. J. Chromatogr. A 968, 79–88.

D'Antone, S., Bignotti, F., Sartore, L., D'Amore, A., Spagnoli, G., Penco, M., 2001. Thermogravimetric investigation of two classes of block copolymers based on poly(lactic-glycolic acid) and poly(ε-caprolactone) or poly(ethylene glycol). Polym. Degrad. Stab. 74, 119–124.

Daoub, R.M.A., 2016. Rheology and Stability of Emulsions of Some Acacia Gums and Their Blends (Ph.D. thesis). Sudan University of Science and Technology, Khartoum, Sudan.

Daoub, R.M.A., Elmubarak, A.H., Misran, M., Hassan, E.A., Osman, M.E., 2016. Characterization and functional properties of some natural Acacia gums. J. Saudi Soc. Agric. Sci., (online).

Dong, Y., Sørensen, K.M., He, S., Engelsen, S.B., 2017. Gum Arabic authentication and mixture quantification by near infrared spectroscopy. Food Control 78, 144–149.

Elzain, E.M.I., Mobarak, L.M., Dirar, M., 2012. Investigating the electric conductivity, magnetic inductivity, and optical properties of gum Arabic crystals. J. Basic Appl. Chem. 2 (6), 35–49.

FAO, 1990. JECFA specifications for identity and purity of certain food additives. Food Nutr. Pap. 49 (23), Rome.

Gómez-Díaz, D., Navaza, J.M., Quintáns-Riveiro, L.C., 2008. Intrinsic viscosity and flow behaviour of Arabic gum aqueous solutions. Int. J. Food Prop. 11 (4), 773–780.

Größl, G., Harrison, S., Kaml, I., Kenndler, E., 2005. Characterization of hatural polysaccharides (plant gums) used as binding media for artistic and historic works by capillary zone electrophoresis. J. Chromatogr. A 1077, 80–89, http://www.process-nmr.com/adulteration of Acacia senegal.htm.

Ishage, A.H.M., Elkhair, H.M., Siddig, M.A., Elbadawi, A.A., 2015. Diffusivity and electrical properties of gum Arabic, carbon black/$KBrO_3$ composite material. Int. J. Eng. Sci. Res. Technol. 4 (4), 263–269.

Kipping, T., Rein, H., 2016. Continuous production of controlled release dosage forms based on hot-melt extruded gum Arabic: formulation development, in vitro characterization and evaluation of potential application fields. Int. J. Pharm. 497 (1–2), 36–53.

Lelon, J.K., 2008. Uptake of Micronutrients by *Acacia senegal* Varieties and It Possible Effect on Gum Arabic Quality (Ph.D. thesis). University of Nairobi.

Lelon, J.K., Jumba, I.O., Keter, J.K., Chemuku, W., Oduor, F.D.O., 2010. Kenya Forestry Research Institute (KEFRI), Nairobi, Kenya. University of Nairobi, Nairobi, Kenya.

Liu, P., Yu, L., Liu, H., Chen, L., Li, L., 2009. Glass transition temperature of starch studied by a high-speed DSC. Carbohydr. Polym. 77, 250–253.

Liu, X., Yu, W., 2006. Evaluating the thermal stability of high performance fibers by TGA. J. Appl. Polym. Sci. 99, 937–944.

Mejanelle, P., Bleton, J., Tchapla, A., Goursaud, S., 2002. Gas chromatography–mass spectrometric analysis of monosaccharides after methanolysis and trimethylsilylation. Potential for the characterization of substances of vegetal origin: application to the study of museum objects. J. Chromatogr. Lib. 66, 845–902.

Mohamed, M.M.A., 2010. Determination of Trace Elements in Gum Arabic Using X-ray Fluorescence (M.Sc. thesis). Faculty of Science University of Khartoum, Khartoum, Sudan.

Mokhtar, L., 2013. Electric conductivity of gum Arabic from *Acacia senegal*. Int. J. Sci. Res. 5 (2), 583–593.

Mothe, C.G., Rao, M.A., 2000. Thermal behavior of gum Arabic in comparison with cashew gum. Thermochim. Acta 357–358, 9–13.

Okalebo, J.R., Gathua, K.W., Woomer, P.L., 2002. Laboratory Methods of Soil and Plant Analysis: A Working Manual, second ed. TSBFCIAT and SACRED Africa, Nairobi, Kenya.

Pitthard, V., Finch, P., 2001. GC-MS analysis of monosaccharide mixtures as their diethyl dithioacetal derivatives: application to plant gums used in art works. Chromatographia 53, 317–321.

Podder, A.K., J.K., Chakrobarty, Faroque, A.B.M., 2014. Qualitative and quantitative analysis of sildenafil in traditional medicines and dietary supplements. Asian J. Pharm. Clin. Res. 7 (1), 25–30.

Saigal, N., Baboota, S., Ahuja, A., Ali, J., 2008. Fast-dissolving intra-oral drug delivery systems. Expert Opin. Ther. Pat. 18 (7), 769–778.

Schneider, U., Kenndler, E., 2001. Identification of plant and animal glues in museum objects by GC–MS, after catalytic hydrolysis of the proteins by the use of a cation exchanger, with simultaneous separation from the carbohydrates. Fresenius J. Anal. Chem. 371 (1), 81–87.

Singh, J.P., Thakur, S., 2007. Laser-induced Breakdown Spectroscopy. Elsevier, Amsterdam, The Netherland.

Soga, T., Imaizumi, M., 2001. Electrophoresis 22, 3418.

Tikhonov, N.A., Arkhangelsky, I.V., Belyaev, S.S., Matveev, A.T., 2009. Carbonization of polymeric nonwoven materials. Thermochim. Acta 486, 66–70.

Tiwari, A., 2007. Gum Arabic – Graft-Polyanaline: an electrically active redox biomaterial for sensor applications. J. Macromol. Sci. Part A Pure Appl. Chem. 44, 735–745.

Yousif, M.A.M., 2017. Characterization of Gum Arabic Using Laser Induced Breakdown Spectroscopy (LIBS) (Ph.D. thesis). Sudan University of Science and Technology, Khartoum, Sudan.

Gravimetric and Calorimetric Analysis of Some Sudanese Acacia Exudate Gums

9

B.I. Beelly Osman, M.A. Hussein**, Elfatih A. Hassan†, Mohamed E. Osman‡, Mahiran Basri§*

*University of Blue Nile, Ad-Damazīn, Sudan; **International University of Africa, Khartoum, Sudan; †University of Khartoum, Khartoum North, Sudan; ‡The Gum Arabic Company Ltd, Khartoum, Sudan; §University Putra Malaysia, Serdang, Malaysia

Introduction

Thermal analysis can be defined as a group of techniques in which a physical property of a substance and/or its reaction products is measured as a function of temperature when the substance is subjected to a controlled temperature program (Mackenzie, 1979). Heat flow into a substance leads to many physical and chemical changes by which we can identify and characterize a sample (Hill, 1991). Carbohydrates represent more than 90% of the dry weight of all biomass, and more than 90% of the carbohydrate mass is in the form of polysaccharides (Zohuriaan and Shokrolahi, 2004). Since these polymers are numerous, hail from renewable sources, and are naturally available, nontoxic, and chemically and biochemically frail and adjustable, it is justifiable that they find extensive and far-flung use.

Acacia gum is a dried exudation obtained from the stems and branches of natural strains of *Acacia senegal* (L.) Willdenow or closely related species of *Acacia* (family: Leguminosae). It consists mainly of high molecular weight polysaccharides and their calcium, magnesium, and potassium salts, which, on hydrolysis, yield arabinose, galactose, rhamnose, and glucuronic acid (Assaf and Phillips, 2005). Exudates gums are produced by different types of trees and shrubs during the gummosis process; this wound response results in plastic sealant that plugs cracks in the bark, which, when drying in contact with air and sunlight, leads to the formation of hard, glasslike lumps that can easily be collected (Renard et al., 2014). Acacia gums exhibit unique properties in a wide variety of applications and have been used in important items of international trade in the food, pharmaceutical, adhesive, paper, cosmetics, textile, and other industries for centuries due to their emulsifying, microencapsulation, thickening, and stabilizing properties (Orozco-Villafuerte and Cruz-Sosa, 2003; Verbeken and Dierckx, 2003).

The term "gum" stands for a group of industrially useful polysaccharides or their derivatives that can hydrate in hot or cold water to form viscous solutions at low concentrations (Whistler, 1982). When used in foods, gums sometimes behave as hydrocolloids (Liu and Ishida, 1996). Regarding the commercial interest in and utility of

Gum Arabic. http://dx.doi.org/10.1016/B978-0-12-812002-6.00009-9

gums, physicochemical characterization of these gum samples is of great importance. According to the covered literature reviewed, there is no methodical relative study that has been done on the thermal description of the present acacia gums; few studies in this field are done on gum Arabic (*Acacia senegal*; ASG) (Mothé and Rao, 2000). The present study deals with thermal descriptions of the investigated gum samples from three acacia trees using differential scanning calorimeter (DSC) and thermogravimetric analysis (TGA).

Classification and botanical description of the investigated samples

Family: *Fabaceae* **Sub family**: *Mimosoideae* **Genus**: *Acacia* **Subspecies:** *Acacia sieberana.Benth.var sieberana*

A. sieberiana var. *sieberiana* is a tree that grows 3–25 m tall with a trunk diameter of 0.6–1.8 m. It is a perennial native to Africa and introduced into Pakistan (Orwa et al., 2009). *A. sieberiana* var. *sieberiana* grows in the savannah and woodland. It was found with various botanical characteristics in the entire Sahel and many other semiarid regions in Africa. The species is especially suited for riverbanks or low ground and replaces *Acacia nilotica* in the south of the latter's natural range. *A. sieberiana* var. *sieberiana* is drought and frost resistant, and it grows in deep, well-drained, heavy clay; light sandy; and medium loamy acidic soils (Figs. 9.1 and 9.2).

Figure 9.1 ***A. sieberiana*** **tree.**

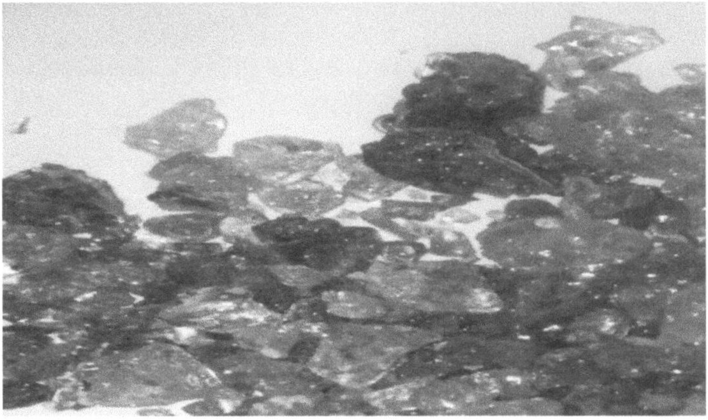

Figure 9.2 ***A. sieberiana*** **gum sample.**

Figure 9.3 ***A. gerrardii*** **tree.**

Acacia gerrardii var. gerrardii

Family: Fabaceae (alternative leguminosae). **Sub family**: Mimosoideae - **Genus:** Acacia **Subspecies**: Acacia gerrardii Benth.Var. gerrardii

It is a tree usually 3–15 m height, with a flat, umbrella-shaped crown and rough, gray bark that is fissured (Figs. 9.3 and 9.4) (EL Amin, 1990).

Acacia tortilis var. raddiana

- *Fabaceae*: *Mimosoideae*: *Acacia*: *Acacia tortilis*.var.*raddiana*.

A. tortilis var. *raddiana* tends to grow in areas where temperature varies from 0 to 50°C and rainfall is anywhere from about 10–100 mm (3.9–39 in) per year (Meyer and Ward, 2005). In Sudan, it is found mainly in northern, central, and northeast regions. Medium umbrella-shaped crown, happed tree, 4–15 m tall, often with several trunks,

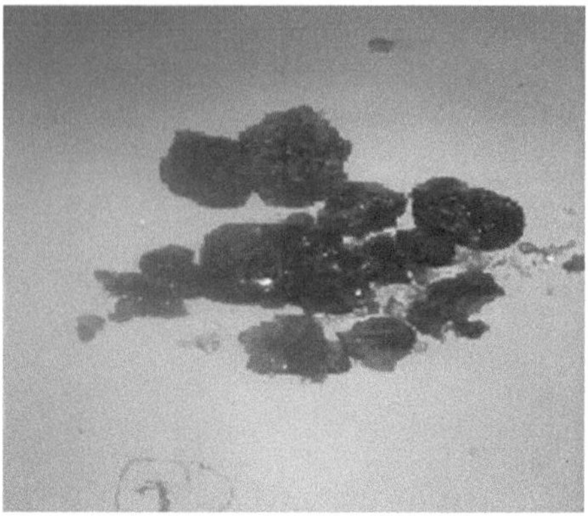

Figure 9.4 ***A. gerrardii*** **gum sample.**

Figure 9.5 ***A. tortilis*** **tree.**

which are reduced to small, wiry shrubs less than 1 m tall under extremely arid conditions (Figs. 9.5 and 9.6) (Table 9.1).

Results and discussion

Thermogravimetric analysis results

TGA is a simple and accurate method used to find out the weight loss, decomposition pattern, and thermal stability of material from heating. Transitions involving mass changes are detected by TGA as a function of temperature and time (Singh and Selvam, 2010). Table 9.2 and Figs. 9.7–9.9 give details of the thermal behaviors according to the primary and derivative thermograms of the studied gums. The early

Figure 9.6 ***A. tortilis gum sample.***

Table 9.1 Sources and general specification of the investigated or studied natural acacia gum samples

Parameter	A. sieberiana	A. gerrardii	A. tortilis
Sources	White Nile state	South Kordofan state	While Nile state
Appearance (color)	White (clear solution)	Light brown	Dark brown
Solubility	Cold and hot water	Cold and hot water	Cold and hot water
Moisture content %	8.28	8.37	7.89
Ash %	2.1	2.3	3.3
pH	4.29	4.32	4.75

Table 9.2 Peak temperatures in the differential scanning calorimeter thermograms, specific heat and enthalpy changes associated with the acacia gums in studies

Gum sample	No. of main enthalpy changes	Initial temp. (°C)	Peak temp. (°C)	Final temp. (°C)	ΔH(j/g)	Mass loss (%)	Specific heat (j/g°C)
A. sieberiana							
	1	40.4	104.97	209.6	−399.1	6.50	−3.19
	2	265.4	312.4	352.2	250.6	66.74	6.08
A. gerrardii							
	1	51.79	104.14	164.44	−383.9	11.88	−4.18
	2	290.54	322.61	343.56	247.3	63.85	6.95
A. tortilis							
	1	48.52	101.33	164.54	−174.5	7.35	−4.54
	2	280.53	315.53	343.34	203.8	63.40	5.39

minor mass loss that appeared in all thermograms of the samples is attributed to desorption of moisture as hydrogen-bound water to the saccharide structure (Kittur and Prashanth, 2002), or it may be attributed to the loss of the adsorbed and the structural water of the gums sample (Nep and Conway, 2010). While *A. sieberiana* and ATR gums exhibited three stages of decomposition patterns, *A.*

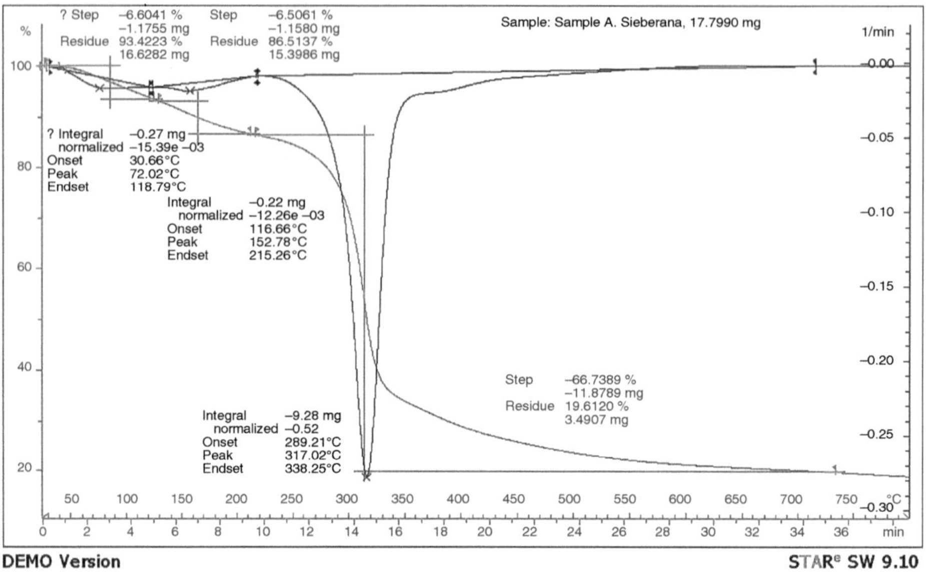

Figure 9.7 DSC profile of *A. sieberana* gum.

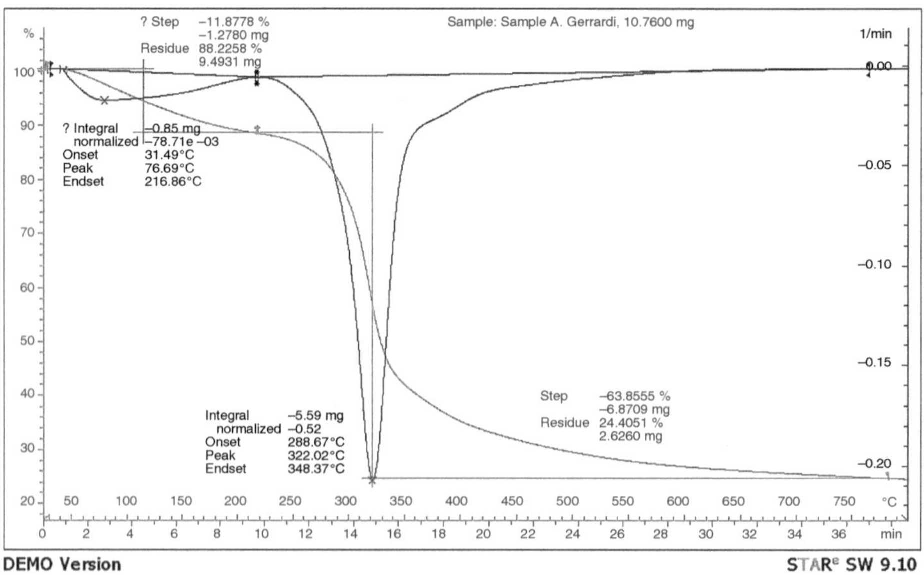

Figure 9.8 DSC profile of *A. gerrardii* gum.

gerrardii exhibited only a double-stage decomposition pattern. Generally, the main decomposition of polysaccharides starts above 200°C and is mostly related to the major second mass loss (Da Silva and de Paula, 2007; Varma and Kokane, 1997). Although *A. sieberiana* gum was resistant to heat and has high initial decomposition

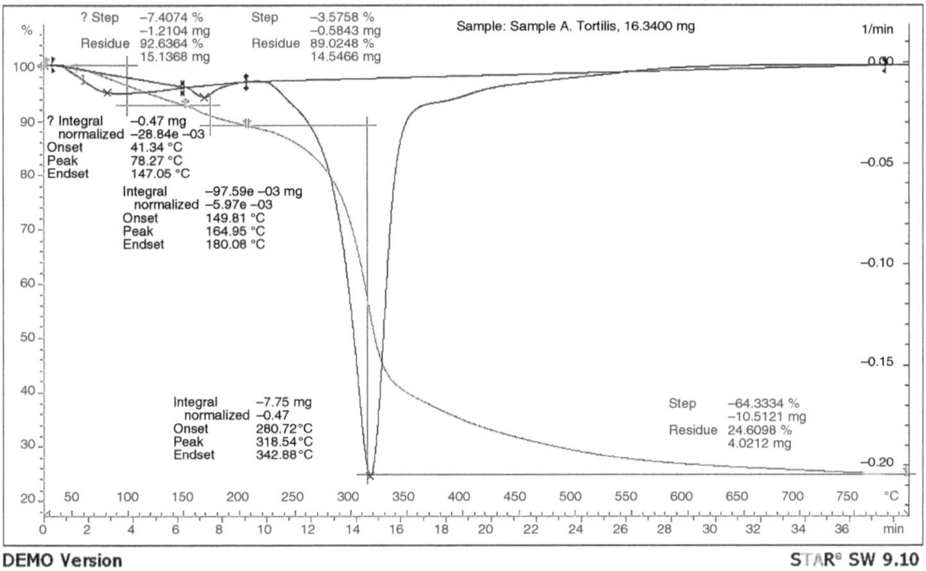

Figure 9.9 DSC profile of *A. tortilis*.

temperature, it decomposed rapidly as soon as it reached its initial decomposition temperature (IDT) to yield low char residue, while ATR gum showed low IDT value and exhibited low decomposition to yield the highest char residue (24.9%) among the studied gums at the final temperature (600°C). The weight percentages of all studied gums fall in the range of 11%–14%) at a temperature range of 180–200 °C. These are low compared to weight loss reported for other polymers, such as starch, and it gives a glimpses that these acacia gums have excellent thermal stability (Singh and Selvam, 2010).

Differential scanning calorimeter analysis

DSC and DTA have stood out as sturdy physical tools to control and observe physical and chemical changes that occur in the polysaccharides during thermal transitions over the course of heating under an inert nitrogen purge. These methods produce unique curves for the studied polysaccharide samples (Bothara and Singh, 2012). Typical DSC thermograms of the given natural *acacia* gums were representatively illustrated in Figs. 9.10–9.12. The peak temperatures for various thermal effects as well as the associated specific heat and enthalpy changes for all gum samples studied were given in Table 9.3.

A. sieberiana and Acacia tortilis gum not (ATR) gum showed early negative glass transition temperatures in the range of −32 to −37°C, while no glass transition was recorded for *A. gerrardii*. All the DSC traces exhibited early intensive endothermic peaking in a close temperature range of 101–105°C, which fall in the same range for ASG,

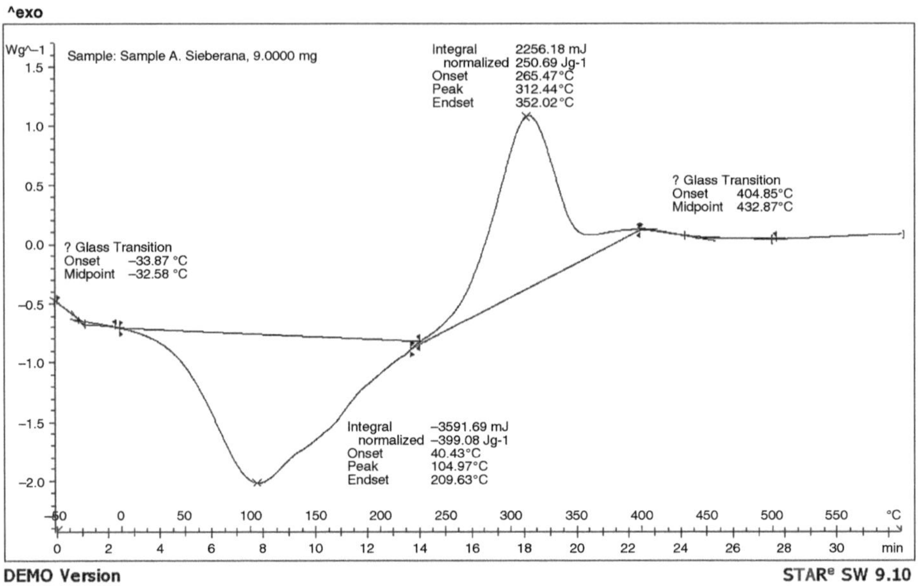

Figure 9.10 TGA and DTG of *A. sieberana*.

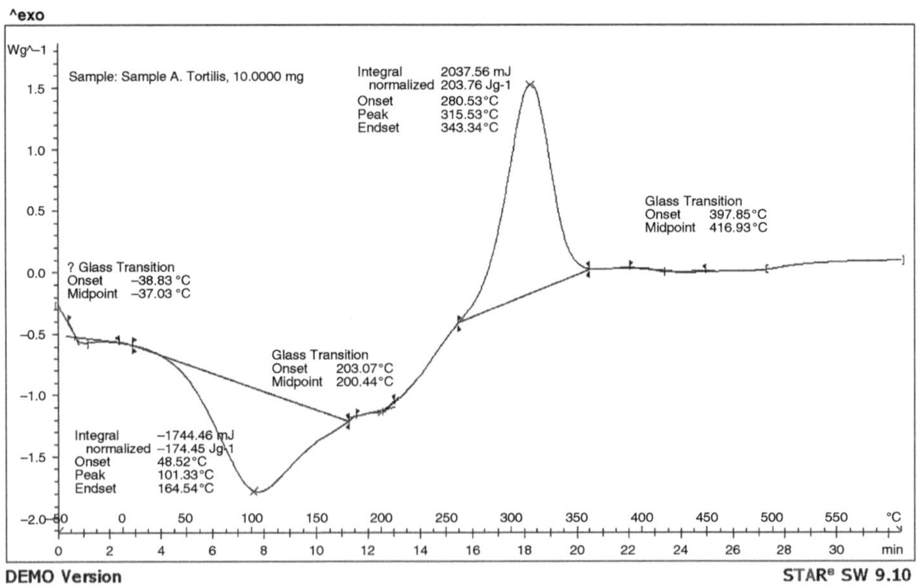

Figure 9.11 TGA and DTG of *A. tortilis*.

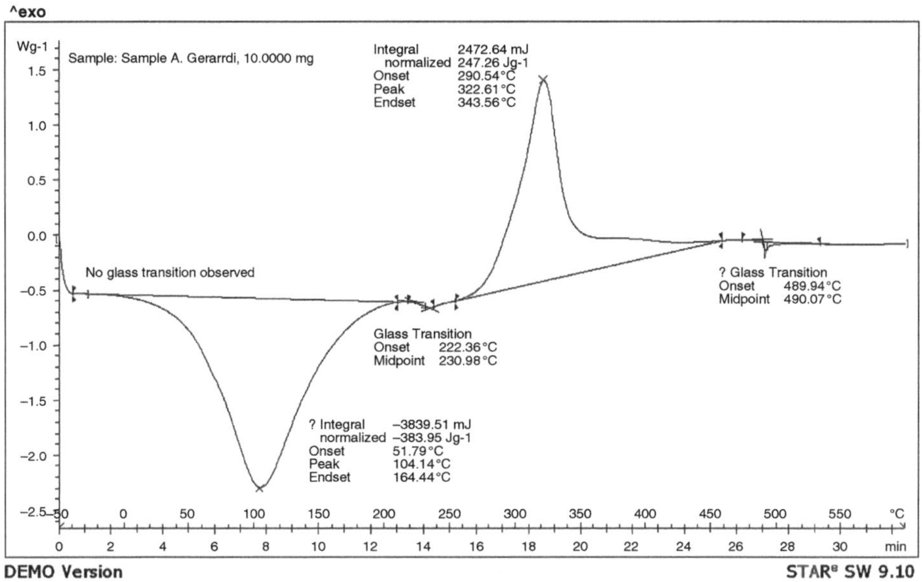

Figure 9.12 TGA and DTG of *A. gerrardii.*

Table 9.3 Thermogravimetric (thermogravimetric analysis and DTA: Differential thermal analysis) data of natural acacia gums

Name	No. of decompositions	Temperature range (°C)	DTG maximum (°C)	Weight loss (%)
A. sieberiana				
	1	30.66–118.79	72.02	6.604
	2	116.19–215.26	152.78	13.486
	3	289.21–338.25	317.02	66.739
A. gerrardii				
	1	31.49–216.86	76.69	11.878
	2	288.67–348.37	322.02	63.856
A. tortilis				
	1	41.34–147.05	78.27	7.407
	2	149.81–180.08	164.95	10.975
	3	280.72–342.88	318.54	64.333

DTA, Differential thermal analysis; DTG, differential thermogravimetric analysis.

Acacia mellifera (AMF), *Acacia seyal* (ASY), and ATR, reported by Daoub et al. (2016). These transitions are related to the loss of water (Zohuriaan and Shokrolahi, 2004).

The second intensive peak recorded in the DSC thermograms is an exothermic transition at 300°C, and this is found to be in the same range for ASG, AMF, ASY, and ATR, reported by Daoub et al. (2016). Generally, dehydration, depolymerization, recrystallization, and pyrolytic decomposition are involved in these high temperature stages, resulting in the formation of H_2O, CO_2, and CH_4. However, the difference in the particles' sizes, structures, and functional groups affect the transitions to some extent, which means that either the degradation routes or the resulting fragments will be different. Since most polysaccharides are composed of carboxylate or carboxylic acid functional groups, thermal cleavage of the carboxylate groups and efflorescence of CO_2 from the corresponding carbohydrate backbone may be a probable mechanism for the thermal transitions (Zohuriaan and Shokrolahi, 2004).

Elemental analysis results

Table 9.5 shows the results of the elemental analysis (CHNSO), which revealed that all acacia gums studied contained carbon, oxygen, nitrogen, hydrogen, and sulfur. Presence of S along with N in these gums possibly suggested the presence of proteins having S amino acids, such as cysteine. The ratios of carbon to hydrogen was just over 7:1, 8:1, and 9:1 for *A. sieberiana, A. gerrardii*, and ATR gums, respectively, pointing out, together with the ratios of carbon to oxygen, a good number of unsaturations due to polysaccharide composition (Singh and Selvam, 2010).

In conclusion, no clear difference regarding the thermal behaviors of the investigated gums is indicated. This suggests that they are more or less similar in their structure and functional groups. Although *A. sieberiana* gum showed the highest initial decomposition temperature, it exhibited a fast disintegration to yield low char residue (19.6%) at the final temperature (600°C), which indicates that it has low activation energy (E_a), while ATR, which showed the lowest initial decomposition temperature, decomposed slowly and produced high char yield (24.9%) at 600°C, indicating that it

Table 9.4 Initial decomposition temperature and char yield of the gum samples (thermal stability indication)

Gum sample	IDT (°C)[a]	Char yield % at (600 °C)[b]	Weight loss (%) at (180–200 °C)
A. sieberiana	289.21	19.6120	13.4863
A. gerrardii	288.67	24.4051	11.8778
A. tortilis	280.53	24.8807	10.9752

[a] Initial decomposition temperature (IDT) is the thermogravimetric analysis (TGA) curve onset.
[b] Char yield is the residual weight left after recording the TGA trace at the highest temperature, i.e., 600°C in this study.

Table 9.5 The CHNSO composition of three acacia species

Element %	C	H	N	S	O
A. sieberiana	51.75	6.903	2.506	0.153	62.08
A. gerrardii	50.71	5.996	2.386	0.144	64.63
A. tortilis	53.20	5.960	3.532	0.151	66.51

has high activation energy. Accordingly to the DSC and TGA results obtained in the present study, the suggested order of the thermal stability for the studied gum samples is as follows: ATR > *A. gerrardii* > *A. sieberiana*. The results of the weight loss percentage at temperature range of 180–200°C (summarized in Table 9.4) also sustained and reinforced the above suggested order for the thermal stabilities of these acacia gums. The overall result of this study is that these acacia gums are thermally stable compared to the thermal characteristics of some natural and modified gums reported in literature by Zohuriaan and Shokrolahi, (2004), and they possess the potential to be used as modifiers in different dosage forms.

Acknowledgments

We are grateful to professor Mahiran Basri for her interest in this study and for her valuable help and support. Our thanks also extend to the Universiti Putra Malaysia for technical assistance during the thermal analysis experiments.

References

Assaf, S.A., Phillips, G.O., 2005. Studies on acacia exudate gums. Part I: the molecular weight of *Acacia senegal* gum exudate. Food Hydrocoll. 19 (4), 647–660.

Bothara, S.B., Singh, S., 2012. Thermal studies on natural polysaccharide. Asian Pac. J. Trop. Biomed. 2 (2), S1031–S1035.

Da Silva, D.A., de Paula, R., 2007. Graft copolymerization of acrylamide onto cashew gum. Eur. Polym. J. 43 (6), 2620–2629.

2007. Tree and Shrubs of the Sudan. Ithaca Press, USA.

Daoub, R.M.A., Elmubarak, A.H., Misran, M., Hassan, E.A., Osman, M.E., 2016. Characterization and functional properties of some natural *Acacia* gums. J. Saudi Soc. Agric. Sci.doi: 10.1016/j.jssas.2016.05.002.

Hill, J., 1991. For better thermal analysis III. Special edition of the International Confederation for Thermal Analysis (ICTA). In: http://www.ictac.org/.

Kittur, F., Prashanth, K.H., 2002. Characterization of chitin, chitosan and their carboxymethyl derivatives by differential scanning calorimetry. Carbohydr. Polym. 49 (2), 185–193.

Liu, J., Ishida, H., 1996. A new class of phenolic resins with ring-opening polymerization. In: Salamone, J.C. (Ed.), The Polymeric Materials Encyclopedia. CRC Press, Florida, USA, pp. 484–494.

Mackenzie, R., 1979. Nomenclature in thermal analysis, Part IV. Thermochim. Acta 28 (1), 1–6.

Meyer, K.M., Ward, D., 2005. Big is not better: small *Acacia mellifera* shrubs are more vital after fire. Afr. J. Ecol. 43 (2), 131–136.

Mothé, C., Rao, M., 2000. Thermal behavior of gum arabic in comparison with cashew gum. Thermochim. Acta 3 (57), 9–13.

Nep, E.L., Conway, B.R., 2010. Characterization of grewia gum, a potential pharmaceutical excipient. J. Excipients Food Chem. I (1), 30–40.

Orozco-Villafuerte, J., Cruz-Sosa, F., 2003. Mesquite gum: fractionation and characterization of the gum exuded from *Prosopis laevigata* obtained from plant tissue culture and from wild trees. Carbohydr. Polym. 54 (3), 327–333.

Orwa, C., Mutua, A., Kindt, R., Jamnadass, R., Simons, A., 2009. Agroforst tree Data base: a tree reference and selection guide version 4.0 (http://www.Worldagroforstry.Org/af/treedb/).

Renard, D., Lepvrier, E., Garnier, C., Roblin, P., Nigen, M., Sanchez, C., 2014. Strucure of glycoprotiens from acacia gum: an assembly of ring – like glycoprotiens modules. Carbohydr. Polym. 99, 736–747.

Singh, A.K., Selvam, R.P., 2010. Isolation, characterization and formulation properties of a new plant gum obtained from *Mangifera indica*. Int. J. Pharm. Biomed. Res. 1 (2), 35–39.

Varma, A., Kokane, S., 1997. Thermal behavior of galactomannan guar gum and its periodate oxidation products. Carbohydr. Polym. 32 (2), 111–114.

Verbeken, D., Dierckx, S., 2003. Exudate gums: occurrence, production, and applications. Appl. Microbiol. Biotechnol. 63 (1), 10–21.

Whistler, R.L., 1982. Industrial gums from plants: Guar and chia. Econ. Bot. 36 (2), 195–202.

Zohuriaan, M., Shokrolahi, F., 2004. Thermal studies on natural and modified gums. Polym. Test 23 (5), 575–579.

Processing of a Highly Purified Grade of Gum Arabic From Gum Dust

10

Nawal Abdel-Gayoum Abdel-Rahman, Inaam Awad Ismail, Muna Mohamed Mustafa
National Food Research Centre, Khartoum North, Sudan

Introduction

Gum Arabic is obtained from *Acacia senegal* trees, and the international specification quality is based on Sudanese gum variety *senegal* (Larson and Bromley, 1991). Those specifications include moisture (13%–15%), ash (2%–4%), internal energy (30%–39%), volatile matter (51%–65%), optical rotation (−26 to −34 degrees), and nitrogen content (0.26%–0.39%). In addition, the cationic composition of total ash, according to FAO (1990), include copper (52–66 ppm), iron (730–2490 ppm), manganese (69–117 ppm) and zinc (45–111 ppm).

The application of the term "gum" is confusing, as it is often employed to describe materials that affect the senses of touch, taste, and sight in a measure summed up as a property of gumminess, which is difficult to define. But visual and manual examination of a material may cause the observer to call it a gum (Ismail, 2000). Gum Arabic is the sap that is exuded from various species of acacia trees when wounded to prevent desiccation of the tissue beneath. Once hardened, the sap is collected by hand and sold at the market, eventually making its way through sorting, cleaning, grading, and processing steps that transform it into a form that is usable by industry (Tadesse et al., 2007).

Sudan is considered one of the largest countries in Africa, with varied ecological zones, from the desert in the north to the tropical rainforest in the south (Abdel-Rahman et al., 2014). Sudan is the largest producer and exporter of gum Arabic throughout the world; the gum Arabic is produced in rainy areas in Korofan and Darfur states. These zones are habitats for abundant plant species; one of these is gum Arabic, with major production annually. The annual exportation decreased within the last years from 25,000 and 30,000 metric tons in 2002 and 2003 to 14,000 metric tons in 2005 (GAC, 2009). The exported amount is derived from *A. senegal* and is exported into different grades, which are handpicked, cleaned, and sifted into siftings and gum dust. Major markets for gum Arabic are the United States, the United Kingdom, Italy, Germany, Japan, France, Belgium, and the Netherlands. These countries take about 9% of the Sudanese exports, or about 76% of the world's supply (GAC, 1991). Internationally, the quality of gum Arabic is based on the Sudanese gum variety *A. senegal* (Macrae and Merlin, 2002).

Gum Arabic. http://dx.doi.org/10.1016/B978-0-12-812002-6.00010-5

Gum dust is the by-product of the operations of gum cleaning and sorting to obtain the above-mentioned grades. In the international market, gum dust fetches only 25% of the price of the other grades. It is purified and spray-dried in other countries for industrial purposes. Local purification will hopefully create added value.

The most important nonwood forest product in the Sudan is gum Arabic, which is exudates of *A. senegal*, known as gum hashab, and *A. seyal*, which is known as talha (NOF, 2007). Sudan produces about 85%–90% of the world demand of gum Arabic (Abdel-Nour, 1999). World demand is increasing for gum Arabic for use in food and other purposes; it is used as food ingredients in several food industries. Also, gum has a wide range of applications in pharmaceutical and cosmetic industries. This is a natural material that is nontoxic, biocompatible, cheap, and easily available (Prajapati et al., 2013). In most of the importing countries, gum Arabic is purified, spray-dried, and used in the powder form. The processing of gum Arabic to a high-quality, pure form for analytical and industrial purposes will create an added value.

Gum dust is the inferior grade of cleaning, grading, and packaging of gum Arabic. Gum dust fetches low prices compared to high-quality grades; <25% (GAC, 1991). It is currently exported to external markets whereby it is purified and used for different purposes.

Material and methods

Preparation of samples

Gum dust was produced from the gum Arabic Company Ltd. in Khartoum. The following experimental steps were applied for purification:

1. Gum dust was dissolved in distilled water at 50°C to obtain a solution of 40% concentration.
2. The extraction was filtered and kept in the refrigerator overnight (15–20°C) for sedimentation.
3. After decantation, the solution was centrifuged at 1000 rpm for 30 min and then pasteurized at 80–85°C for 10 min.
4. The clear solution was then loaded on trays (7 kg/m^2) and then dried.

Triplicate samples of the following variables were prepared:

a. oven dried (sample A)
b. shade dried (sample B)
c. fatherly centrifuged, pasteurized, oven dried (sample C)
d. fatherly centrifuged, pasteurized, shade dried (sample D)
e. pasteurized, shade dried (sample E)
f. treated with charcoal, filtered concentration from 25% to 40%, centrifuged, and then oven dried (sample F).

Analytical methods

1. *Physicochemical analysis*: Prepared samples were subjected to the following tests, according to FAO/WHO (2007), to determine moisture content, specific rotation, relative viscosity

(Phillips et al., 2008), pH value, ash content, total nitrogen content, crude protein and color, OD at 85 nm (AOAC, 2000). The heavy metal (HM) contents were determined by X-ray fluorescence, according to the method of Tertain and Claisse (1992).

2. *Microbiological evaluations*: The prepared samples of purified gum dust were evaluated using the recommended growth media and incubation temperatures. The determination includes total bacterial counts, molds, and yeasts, as well as pathogenic bacteria, Coliform and *Salmonella* (FDA, 2001).

Results and discussion

The results in Table 10.1 show that the purified product is of comparable properties to kibbled gum Arabic, which is a commercial grade of good quality. The results also conform to the Sudanese standards for gum Arabic. The moisture content of the samples ranged from 10.17% (sample C) to 8.20% (sample D); these results were lower the maximum limit recorded by Sudanese standards, as well as close to results obtained by Ismail (2000). In addition, they conform to the values specified by FAO/WHO (2007), which specified that the good-quality gum Arabic sample should not contain more than 15% moisture content.

Table 10.1 Analyses of purified gum and kibbled and crude gum dust

Sample	Moisture (%)	Optical density (degrees)	Ash (%)	Nitrogen (%)	Crude fiber (%)	Heavy metal (mg/kg)
A	9.80^{ab}	-28.00^{de}	2.90^{bc}	0.32^{a}	2.00^{b}	19.65^{c}
B	8.25^{bc}	-28.30^{d}	2.91^{bc}	0.32^{a}	2.00^{b}	19.61^{c}
C	10.17^{a}	-27.80^{c}	2.84^{c}	0.27^{b}	1.69^{d}	18.73^{d}
D	8.20^{bc}	-28.00^{de}	2.88^{c}	0.33^{a}	2.01^{b}	18.66^{d}
E	8.33^{bc}	-28.00^{de}	3.00^{b}	0.30^{ab}	1.88^{c}	20.32^{b}
F	9.71^{ab}	27.80^{b}	2.85^{c}	0.23^{bc}	1.44^{de}	18.76^{d}
K[1]	8.62^{b}	29.30^{a}	3.16^{b}	0.22^{bc}	1.38^{de}	20.34^{b}
G[2]	6.17^{c}	26.17^{c}	8.68^{a}	0.45^{a}	3.01^{a}	23.25^{a}
$Lsd_{0.05}$	0.6541*	0.9217**	0.1456*	0.0183*	0.0226*	0.5422*
SE±	0.0279	0.0084	0.0351	0.0016	0.0034	0.0236
Sudanese standards	Not more than 15%	27 to −39	Not more than 4%	0.27 to 0.40	1.69 to 2.50	–

Any two mean value(s) having same superscript(s) in a column do not differ significantly ($p \leq 0.05$). 1: kibbled gum 2: Crude gum dust.

Moreover, the specific rotation of the purified samples also considerably improved; it was −28.0 to 27.8 degrees. These findings are in agreement with Sudan standards specified for gum Arabic (27 to −39 degrees), while samples A, B, C, D, and E were recorded results closest in agreement to that obtained by Al-Assaf et al. (2005) for clarified dust. It is noticeable that the ash content of crude gum dust sample G (8.68%) is far beyond the specified limit (not more than 4%), but nearly 7.15% reported by Ali (1998), whereas the purification process has succeeded in bringing ash content to the specified limit, which agreed with results reported by Yebeyen et al. (2009) and were higher than the value obtained by Mhinzi (2003).

On the other hand, the total nitrogen and crude protein were recorded at ranges from 0.22% to 0.33% and 1.38% to 2.01%, respectively. These levels are within the range reported by Sudanese standards for gum Arabic, which is 0.27%–0.4% and 1.69%–2.50%, respectively. Investigations on mean amounts of total nitrogen and protein were carried out by Ali (1998) and found to be of 0.28 and 1.75%, 0.32 and 2.00% by Idris et al. (1998), and 0.33 and 2.03% by Jurasek et al. (1993), respectively.

The ranges of nitrogen contents were comparable with ranges of 0.27%–0.40% and 1.88%–2.50%, recommended by FAO/WHO (2007). The process of purification followed could be considered satisfactory in bringing up the quality of gum dust to specified limits in the parameters evaluated so far. The samples of purified gum Arabic dust contain levels of crude fiber that ranged from 2.01% (sample D) to 1.44% (sample F). The low value of the crude fiber fell out of the range reported by Sudanese standards for gum Arabic (1.69%–2.50%), and the high value was within this range. The specific levels of heavy matter give an idea about cationic compositions and quality of gum Arabic. The lowest level of HM was recorded by sample D (18.66 mg/kg), while the highest value was recorded by sample E (20.32 mg/kg). It was clear from the results presented in Table 10.1 that the crude gum dust reported values out of range according to the comparable references mentioned above.

Table 10.2 shows that the pH values of the purified samples (4.82 for sample A and 4.88 for sample F) have higher acidity due to the purity. The crude gum dust (4.873) was comparable, indicating similar acidic nature. The rippled sample K has a lower pH value of 4.87.

The relative viscosity of sample C was improved (2.35), which is similar to A except for centrifugation. The value of sample C is higher than that of other samples. Sample A is similar to sample E in all process steps except that E is pastimes before drying. It seems that lowering the microbial load before drying has a positive effect on drying. This means that more purification has a positive effect on viscosity. Sample F (2.08) is similar to sample C except for the addition of charcoal, which means that treatment with charcoal has an adverse effect on viscosity due to its absorptive characteristics. The application of charcoal for bleaching seems to have an improvement in color due to the observed color intensity (0.287); it is still far beyond that of good commercial grades of light color (0.119 OD at 85 nm).

The results of microbiological analysis are shown in Table 10.3. All samples are free from yeasts and molds. Total bacterial counts are within the permissible level

Table 10.2 Quality criteria of purified gum and kibbled and crude gum dust

Sample	pH value	Relative viscosity (cps)	Color (optical density at 85 nm)
A	4.82^{b}	1.79^{cd}	0.30^{bc}
B	4.87^{a}	1.98^{c}	0.34^{b}
C	4.86^{b}	2.35^{a}	0.36^{a}
D	4.86^{ab}	2.15^{ab}	0.28^{c}
E	4.86^{ab}	2.10^{b}	0.35^{ab}
F	4.88^{a}	2.08^{b}	0.29^{bc}
K[1]	4.87^{a}	2.15^{ab}	0.35^{ab}
G[2]	4.37^{c}	2.17^{ab}	0.12^{d}
$Lsd_{0.05}$	0.0251*	0.1774*	0.0106*
SE±	0.0063	0.00032	0.0351

Any two mean value(s) having same superscript(s) in a column do not differ significantly ($p \leq 0.05$). 1: kibbled gum, 2: Crude gum dust.

Table 10.3 Microbiological evaluation of gum Arabic dust

Sample	Total bacterial count (cfu/g)	Total yeasts and molds	Coliform bacteria (mpn/ml)	*Salmonella* spp.
A	1.5×10^3	NIL	9	− ve
B	5.2×10^3	NIL	7.3×10^2	+ ve
C	2.3×10^3	NIL	NIL	− ve
D	1.4×10^3	NIL	9	+ ve
E	2.7×10^3	NIL	9	+ ve
F	2.0×10^2	NIL	NIL	− ve
K[1]	2.9×10^3	NIL	1.1×10^2	− ve
G[2]	8.5×10^2	NIL	79	+ ve

1: kibbled gum, 2: Crude gum dust.

of raw food material $< 10^6$ cfu/g (SSMO, 2001) specifications, which include nil *Salmonella* and nil *Escherichia coli*/g for gum Arabic.

Sample E, which was subjected to filtration, sedimentation, centrifugation, and pasteurization and finally dried in the oven at 75°C for 9 h, is free from the pathogenic bacteria. On the other hand, sample D was similarly treated, but dried and carried out under the open air, which indicated the presence of *Salmonella* spp. This could be attributed to postprocessing contamination, i.e., environmental or personnel. This indicates that the treatment of sample C was appropriate to reach a hygienic product. Cowan and Steels (1979) reported that most loads of microorganisms can easily be destroyed during thermal treatment of the final products in which gum is added. Those results were in good agreement with results obtained by many researchers (Ali, 1998; Anderson and McDougall, 1987; FAO, 1999; Idris et al., 1998; Ismail, 2000; Khalid, 1988; NEWI, 1987).

As can be shown in Table 10.3, samples C and E are similarly prepared except that E is not centrifuged. The results of microbial load show that centrifugation is effective in decreasing the microbial criteria.

Conclusion

The economic value of the gum dust can be raised after being purified using some laboratory techniques. The oven drying and centrifugation techniques are appropriate if we consider the overall performance of raw gum dust and purified samples. Both techniques are promising to be used in purifying gum dust.

References

Abdel-Nour, H.O., 1999. Gum Arabic in Sudan: Production and Socio-Economic Aspects. Medicinal, Culinary and Aromatic Plants in the Near East. FAO Forestry Department, FAO, Cairo, Egypt.

Abdel-Rahman, N.A., Ismail, I.A., Elshafe'a, E.B.B., 2014. Characterization of some Sudanese edible forest fruits. J. Agri-Food Appl. Sci. 2 (2), 39–44.

Al-Assaf, S., Phillips, O.G., Williams, A.P., 2005. Studies on Acacia Exudate Gums. Part I: the molecular weight of *Acacia senegal* gum exudates. J. Food Hydrocoll. 19, 647–660.

Ali, E.M.O., 1998. Microbiological and Physico-chemical Studies on Gum Arabic Quality and Safety (M.Sc. thesis). University of Khartoum, Sudan.

Anderson, D.M.W., McDougall, F.J., 1987. Degradative studies of gum Arabic *Acacia sengal* L. (Willd.) with special reference to the fate of the amino acids percent. J. Food Addit. Contam. 4 (3), 247–255.

AOAC. 2000. Association of Official Agriculture Chemist. Official Methods of Analysis. Assoc. of Analytical Chemist. 17th. Gaithersburg, DW. Washington DC, USA.

Cowan, S.T., Steels, K.G., 1979. Cowan and Steels Manual for the Identification of Medical Bacteria. Cambridge University Press, London, England.

FAO. 1999. Gum Arabic. Compendium of food additive. Addendum 7, Joint FAO/WHO Expert Committee on Food Additives 53rd session. FAO, Rome, Italy.

FAO. 1990. Specifications for Identity and Purity of Certain Food Additives. Food and Nutrition. Paper No. 49, 23. FAO, Rome, Italy.

FAO/WHO. 2007. Food and Nutrition. Guide to JECFA Specification for Food Additives and Contaminants. FAO joint WHO, Paper No. 49, 23-25. FAO, Rome, Italy.

FDA. 2001. Evaluation of the Health Aspects of Gum Arabic as a Food Ingredient. Report PB: 234-904. Department of Health and Human Services. Food and Drug Administration.

GAC. 2009. Annual Reports for the Years 1970–2004. GAC Research Unit, Gum Arabic Company Ltd. Khartoum, Sudan.

GAC. 1991. Gum Arabic and an Ancient Ingredient for the 21st Century. Gum Arabic Company Ltd. Khartoum, Sudan.

Idris, O.H., Williams, P.A., Phillips, G.O., 1998. Characterization of gum from *Acacia senegal* trees of different age and location using multidetection gel permeation chromatography. J. Food Hydrocoll. 12, 379–388.

Ismail, I.A., 2000. Quality Criteria of Crude and Processed Gum Arabic (Ph.D thesis). University of Khartoum, Sudan.

Jurasek, p., Kosik, M., Phillips, G.O., 1993. Chemometric study of the Acacis (gum Arabic) and related natural gums. J. Food Hydrocoll. 7 (1), 73–85.

Khalid, A.S., 1988. Microbiological quality for raw processed gum Arabic. Gum Arabic Symposium ZDS. Solingen, West Germany.

Larson, B.A., Bromley, D.W., 1991. Natural resource price, export, policies and deforestation: the case of Sudan. World Dev. 19 (10), 1289–1297.

Macrae, J., Merlin, G., 2002. The Prospects and Constraints of Development of Gum Arabic in Sub-Saharan Africa. World Bank, Washington DC., USA.

Mhinzi, G.S., 2003. Intera-species variation of the properties of gum exudates from *Acacia senegal* var. senegal and Acacia var fistula from Tanzania. BCSE 171, 67–74.

NEWI. 1987. Noeth East Wales Institute for Higher Education. Report on Microbiological Examination of Kibbled Gum Arabic. Department of Microbiology.

NOF. 2007. National Organization for Forestry. Ministry of Agriculture and Forestry. Khartoum, Sudan. Annual Report 2007. pp. 42.

Phillips, O., Ogasawara, T., Ushida, K., 2008. The regulatory and scientific approach to defining gum Arabic (*Acacia senegal and Acacia seyal*) as a dietary fiber. J. Food Hydrocoll. 22 (1), 24–35.

Prajapati, V.D., Jani, G.K., Moradiya, N.G., Randeria, N.P., 2013. Pharmaceutical applications of various natural gums, mucilages and their modified forms. Carbohydr. Polym. 92 (2), 1685–1699.

SSMO (2001). Sudanese Microbiological Standards for Foods:Bottled, Natural and Formulated Juices. No. 20. SDS 525. Sudanese Standards and Metrology Organization (SSMO), Khartoum. Sudan.

Tadesse, W., Desalegn, Ali, R., 2007. Natural gum and resin bearing species of Ethiopia and their potential application. Investigaci6n Agraria: Sisremas y Recursos Forestales 16 (3), 200–221.

Tertain, Claisse 1992. Principles of Quantitative X-RAY Fluorescence Analysis. London, England.

Yebeyen, D., Lemenih, M., Feleke, S., 2009. Characterisation and quality of gum Arabic from naturally grown Acacia senegal (Linne) Willd. Trees in the Central Rift Valley of Ethiopia. J. Food Hydrocoll. 23, 175–780.

Processing and Modification of Gum Arabic in Specific Applications

11

Olatunji Ololade
University of Lagos, Akoka, Nigeria

Introduction

Owing to its high solubility and low viscosity in addition to its low toxicity and desirable emulsifying properties, gum Arabic is seen as one of the most widely used polysaccharides in the industry, especially in the food and pharmaceutics industry, as stabilizers, thickeners, emulsifiers, and film-forming and suspending agents (Rayner et al., 2016).

Like all gums, gum Arabic is a hydrocolloid polysaccharide natural exudate from the trunk and branches of the acacia tree. The gum is generally harvested from the tree by creation of incisions/injuries by tearing off the branches. This then causes the tree to release latex-like water sap, which coagulates to form large, waxy crystalline balls, which can then be picked from the tree. Branches over 3 cm wide are suitable for harvesting (Okatahi and Onyibe, 1999).

Although the properties of any given particular batch of gum are dependent on the particular tree source, as it varies with species, location, and season, gum Arabic is generally characterized by relatively high solubility in cold and hot water and low viscosity, which distinguishes it from other polysaccharides. Gum Arabic achieves up to 50% w/w solubility in cold water. The quality of gum Arabic is generally assessed using five key properties: moisture content, specific optical rotation, presence of tannins, ash content, and protein content (Daoub et al., 2016; Hassan, 2000; Karamalla et al., 1998).

For specific applications, it is important to select the ideal variety of gum Arabic suitable for the purpose. Two main species of gum Arabic exist in the markets; these are gum Arabic from the *Acacia senegal* var. *senegal* trees and that from the *Acacia seyal* var. *seyal* trees. Generally, the former is preferred due to its superior properties; it forms better emulsions and suspensions and forms a more colorless solution than the latter (Daoub et al., 2016). Although it has inferior properties, gum Arabic of the *A. seyal* is also used in some applications.

The general process of harvesting, washing, drying, and grinding applies to all forms of gum Arabic. These steps are then followed by specific processing and modifications, which are specific to the end use. The processing referred to herein is the processing applied to either gum Arabic prior to inclusion in the formulation or the

Gum Arabic. http://dx.doi.org/10.1016/B978-0-12-812002-6.00011-7

processing applied to the entire system within which gum Arabic plays an active role in determining the properties of the end material.

In the following sections, this chapter explores each processing technique and discusses the various reported cases where these techniques have been applied to gum Arabic and the specific purpose for which these were done. A description of each process is given to provide readers a guideline for reproducing such.

Compositing and blending

In the simplest case, gum Arabic is mixed with other polymers or materials to form a blend or composite with minimal chemical modification to the gum Arabic structure. Compositing or blending gum Arabic with other polymers results in a new material that combines the properties of the materials in the blend or composite, as it obtains all blends and composites (Olatunji and Olsson, 2016). The chemical, mechanical, and morphological properties of the resulting formulation are therefore dependent on the mixing conditions and proportion of each polymer in the formulation. This applies in applications such as food where gum Arabic is mixed in with other components to serve as a thickener or stabilizer. This could be mixing in already modified gum Arabic or gum Arabic in its unmodified form. Modification of gum Arabic prior to inclusion in a blend or composite could be aimed at improving its compatibility, solubility, behavior in lipophilicity, hydrophilicity, activity at given pH, etc.

Gum Arabic is also applied for delivery of dosage forms orally or transdermally. Gum Arabic dissolved in water or a suitable solvent is mixed in with the active pharmaceutical ingredient (API) followed by dehydrating to achieve a solid or semisolid form as a patch, oral tablet, syrup, or other. In addition to its low toxicity and biocompatibility, gum Arabic is particularly attractive for such applications due to its ability to achieve a relatively lower viscosity of solution at higher concentration, unlike most other water-soluble polymers, which show significant increase in the viscosity of the solution at a relatively lower concentration. This allows optimal drug loading into the dosage form (Kipping and Rein, 2016).

For the advantage of combining properties of more than one polymer, gum Arabic is blended with other polymers such as pectin, gelatin, and chitosan. Various combinations of biopolymers exist for applications such as food, drug delivery, or tissue engineering. In applications such as food, where a combination of biopolymers is blended together to give the active food or drug ingredient that they carry some protection or enhancement, gum Arabic in combination with other polymers or materials can be either applied as coatings or serve as a matrix within which the food or pharmaceutical ingredient is embedded. An example is coating of mango extract with gum Arabic–calcium chloride blend. The results indicated that the gum Arabic–based coating delayed ripening time, inhibited decay, and retained acid content, color, firmness, and overall quality of the fruit (Khaliq et al., 2015). The addition of calcium chloride improves the microstructure of the gum Arabic.

Gum Arabic blends containing xylitol and calcium chloride are generally used as a coating for food, cosmetics, and pharmaceutical products, such as chewing gums,

Table 11.1 Example formulation for preparing gum Arabic–based sugar-free hard coating

Ingredient	Weight (g)
Gum Arabic	55
Xylitol	312
Calcium chloride	6
Water	198
Total	571 g

Source: Huzinec and Graff (1987).

candies, and coated tablets. This technique for coating dates as far back as 1987 (Huzinec and Graff, 1987.

The center used in the particular example is a chewing gum center made using the well-established process from producing chewing gum. The syrup used in the coating is prepared using the formulation listed in Table 11.1.

Gum Arabic content typically ranges between 5% and 10% weight of gum Arabic per weight of coating. In place of calcium chloride, other calcium-based water-soluble salts such as calcium nitrate, calcium iodide, or calcium bromide can be used. Typical concentrations of calcium chloride range between 3% and 8% the weight of gum Arabic used. Similarly, the xylitol could also be replaced by other sweeteners such as fructose, sorbitol, or mannitol. For sugar-containing coatings, sucrose, fructose, glucose, mannose, or galactose could also be used. Sweetener concentrations range between 40% and 60% w/w. Other components could be used as additives in minute quantities, such as moisture absorbers, antiadherents, film formers, or dispersing agents. Drying is achieved using air-drying or a drying agent such as calcium carbonate.

The prepared centers to be coated are placed on a revolving coating pan. The pan-coating technique could be replaced with other suitable coating techniques. The coating syrup is heated to a temperature range of 50°C–60°C and applied onto the centers on the revolving pan for a given distribution time. The distribution time is the time within which the centers are allowed to contact the syrup on the revolving pan; this ranges between 1 min and 3 min. A typical coating process involves a repeated cycle of drying followed by further deposition of the syrup until the target coated weight is achieved. An example of such cycle is given in Table 11.2. The coated pieces are then dried at 16°C, followed by 48 h conditioning at room temperature.

A composite of gum Arabic and tapioca starch results in a composite with improved viscosity, elasticity, and pasting property. However, this is only at the lower concentration of gum Arabic, in the range 0.1%–1.0%. At higher concentrations, gum Arabic tends to play an inhibitory effect as regarding the rheological properties of tapioca starch (Singh et al., 2016) such that for every formulation, there is need to obtain an optimum concentration at which the gum Arabic acts to improve the material property.

In addition to the role it plays in improving the rheological and storage properties of foods and pharmaceutical formulations, gum Arabic is also considered as a dietary fiber possessing therapeutic properties, such as antioxidant properties and treatment of

Table 11.2 **Example coating sequence for chewing gum with hard coating**

Charge	Material applied	Material weight (g)	Distribution time (min)
1	Coating syrup	30	1.5
2	$CaCO_3$	30	1.5
3	$CaCO_3$	30	1
4	Coating syrup	30	1
5	$CaCO_3$	30	2
6	$CaCO_3$	30	1
7	78% Xylitol solution	30	1.5
8	$CaCO_3$	30	2
9	Coating syrup	30	1
10	$CaCO_3$	30	2

Source: Huzinec and Graff (1987).

renal and kidney conditions as well as diabetes (Ahmed et al., 2016). It can therefore serve as a nutritional supplement on its own or added to other foods or supplements.

Compositing or blending of gum Arabic with other polymers or material therefore generally involves good mixing to achieve a homogeneous system whereby either particles are evenly distributed in the continuous phase or the two continuous phases are evenly blended in each other. The sections that follow look at more complex methods of combining gum Arabic with other polymeric and nonpolymeric materials, where the chemical interaction between gum Arabic and other molecules play a significant role.

Protein polysaccharide emulsions and polyelectrolytes formation

Protein polysaccharide emulsions are particularly effective emulsions due to the opposing surface charges of the polymers; proteins carry positive surface charges, while polysaccharides carry negative surface charges (Fig. 11.1). This could result in either hydrogen bonding or electrostatic interactions, which result in formation of complexes. The extent and nature of interaction depend on the conditions of mixing, particularly pH and temperature. For this reason, gum Arabic is often mixed with proteins such as soy protein and gelatine (Costa de Conto et al., 2013). The purpose for which this is done includes emulsion preparation, microencapsulation, or nanoparticle preparation.

Gum Arabic has the ability to form a polyelectrolyte (a polymer that dissolves in water with charged surfaces). Polyelectrolyte complexes can be obtained as a result of combining cationic and anionic polyelectrolytes. An example of such polyelectrolyte complex is a blend of chitosan, pectin, and gum Arabic (Tsai et al., 2014). Chitosan has the ability to form cationic polyelectrolytes at low pH, while pectin and gum Arabic form anionic polyelectrolytes. Such polyelectrolyte complexes can be applied as controlled drug delivery and bioelectrochemical systems where the electrolyte nature of the

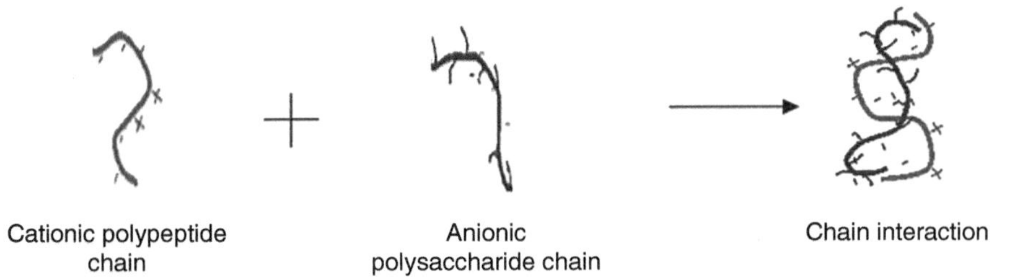

Figure 11.1 Positively charged polypeptide chain and a negatively charged polysaccharide chain.

complex formed play significant roles in the release and transportation within biological systems. The polyelectrolyte complex of chitosan/pectin/gum Arabic has been proven to achieve the complete release of insulin over a 6-hour period. The polyelectrolyte complexes act as carriers for the active drug and release it into the body. A protein-based drug such as insulin interacting with a polysaccharide-based complex aids in preserving its bioactivity during transportation into the target cell. It also aids the transportation through hydrophilic or lipophilic membranes and varying pH.

Gum Arabic plays the role of stabilizer in polyelectrolyte complexes due to its unique surface property and ability to attain lower viscosity at a higher concentration in solution compared to other polysaccharides. The charged nature of the polyelectrolyte also serves as a mechanism to maintain stability in an emulsion. In the case of lipid-based emulsions, coating the emulsion with a protein provides a positively charged surface. Including a polysaccharide in the formulation will result in the anionic polyelectrolyte of the polysaccharide attaching onto the protein-coated surface due to opposite charges attracting. This provides a negative surface of the lipid droplet, which prevents them from aggregating. Including two other anionic polyelectrolytes would result in a competitive adsorption where the two polysaccharides compete to attach onto the surface of the droplets. An example of this is illustrated in Fig. 11.2,

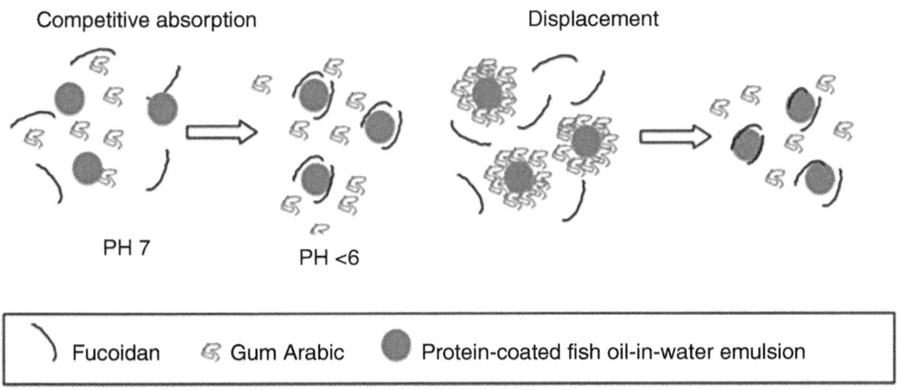

Figure 11.2 Schematics of polyelectrolytes of fucoidan and gum Arabic being electrostatically being deposited onto protein coated lipid droplets at varying pH.

where gum Arabic competes with fucoidan for the positively charged site (Chang et al., 2016). In such a case, the fucoidan has a stronger negative charge than the gum Arabic; it not only competes for the chance to attach to the site, but it also is strong enough to displace already attached gum Arabic from the surface. This should be carefully considered when including other polysaccharides or polyelctrolytes in a complex with gum Arabic.

Here we see that the ability of gum Arabic to form anionic polyelectrolytes in water plays a key role in the formation of complexes with polypeptides and other polysaccharides. The presence of another polysaccharide within a formulation could play an inhibitory role in the activity of gum Arabic within the complex; this therefore serves as a means to adjust the properties of the target formulation using parameters such as concentration of each polymer as well as system conditions such as temperature and mixing rate.

Cross-linking of gum Arabic with other polymers

Gum Arabic, as a branched-chain polysaccharide, can form a cross-linked network through a reaction between the end group of the branches with other end groups. This cross-linking can either occur through direct reaction, or it could involve intermediate steps. A number of studies have reported cross-linking of gum Arabic with other biopolymers such as gelatine, chitosan, and pectin. Cross-linking of polymers results in a new polymeric material with improved properties and functionality. Cross-linked polymers have been shown to have improved mechanical properties, for example, cross-linking of poly(methyl vinyl ether-co-maleic acid) with polyethylene glycol, resulting in a cross-linked hydrogel with desirable swelling properties. The hydrogels formed could absorb water up to 300% of their dry mass (Thakur et al., 2009). Such polymers are attractive in biomedical applications, as they combine good mechanical properties with equally good water absorption capacity. The level of cross-linking can also be controlled by varying the ratio of components and operating conditions.

The cross-linking generally involves creating a network structure from branched polymers through a chemical reaction between the branches, resulting in the formation of a network, which alters the physical and chemical properties of the polymers; the most pronounced of which is their mechanical and swelling properties. In most cases, the cross-linking is achieved by the use of a cross-linking agent, which initiates the reaction between the branches of the polymers.

Here, we illustrate with the case of cross-linking of gum Arabic with gelatine. Simply mixing gelatine and gum Arabic would result in a blend with little or no strong chemical bonds between the protein and carbohydrate structure, but would not result in a cross-linked polymer. To achieve cross-linking, the gum Arabic is first converted to gum Arabic aldehyde by oxidizing with periodate. Unlike the unmodified gum Arabic, the gum Arabic aldehyde can undergo a Schiff base reaction with the gelatine due to the reaction between the aldehyde group and the amino groups of the gelatine (Sarika et al., 2014). This reaction is illustrated in Fig. 11.3, and an outline of the cross-linking of gelatine with gum Arabic is shown in Fig. 11.3.

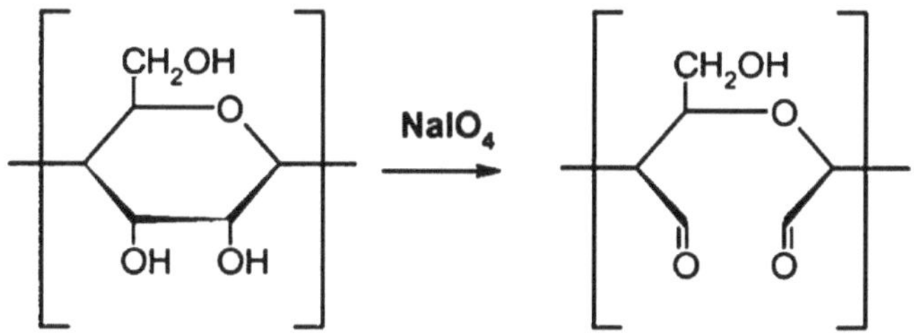

Figure 11.3 Oxidation of a carbohydrate unit with sodium periodate to form an aldehyde.

From Fig. 11.4, we can see that the alcohol group of the carbohydrate unit has been converted into an aldehyde unit. This makes this region more reactive, allowing it to react with the amino unit of the protein (Halsall and Hirst, 1947; Stefano et al., 2003).

Coacervation-based cross-linking of gum Arabic–gelatin complex can also be achieved using other cross-linkers and without prior modification of the gum Arabic. Fish gelatine–gum Arabic hydrogels have been achieved using both oxidized and nonoxidized tannic acid as cross-linking agent (Anvari and Chung, 2016). Fig. 11.5 further provides a flow chart of the processes for obtaining gelatin–gum Arabic hydrogel system, which combines coacervation with cross-linking prossess. The fish gelatin and gum Arabic are separately dissolved in water and then mixed together at 40°C and pH 3.55. This is then followed by inclusion of a cross-linker, which could be either an oxidized or nonoxidized tannic acid. This is then further mixed for 2 h at specified temperature and mixing rate. The system is then cooled to 10°C, which results in phase separation after 24 h where the cross-linked polymers separate from the water phase. However, the water molecules remain trapped within the polymer network. Gelation then occurs when the temperature is further dropped to 3°C (Sarika et al., 2015).

In this cross-linked form, application of gum Arabic can be extended to applications such as scaffold in tissue engineering and drug delivery systems for controlled release. These gum Arabic aldehyde–gelatin cross-linked polymers can also be loaded with active agents for therapeutic applications; for example, nanogels of gelatine–gum

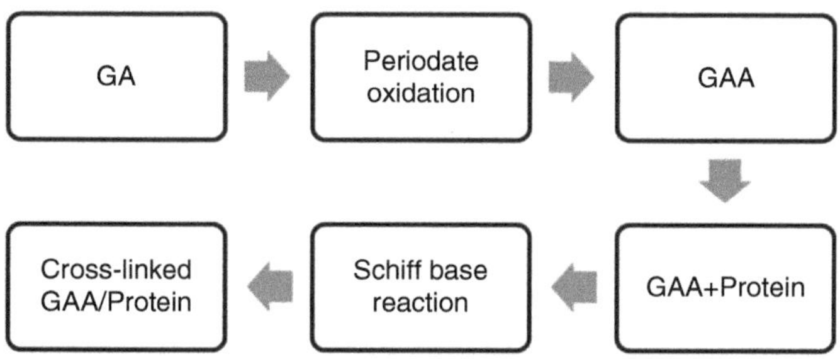

Figure 11.4 Modification of gum Arabic (GA) to form gum Arabic aldehyde (GAA) and cross-linking with gelatin.

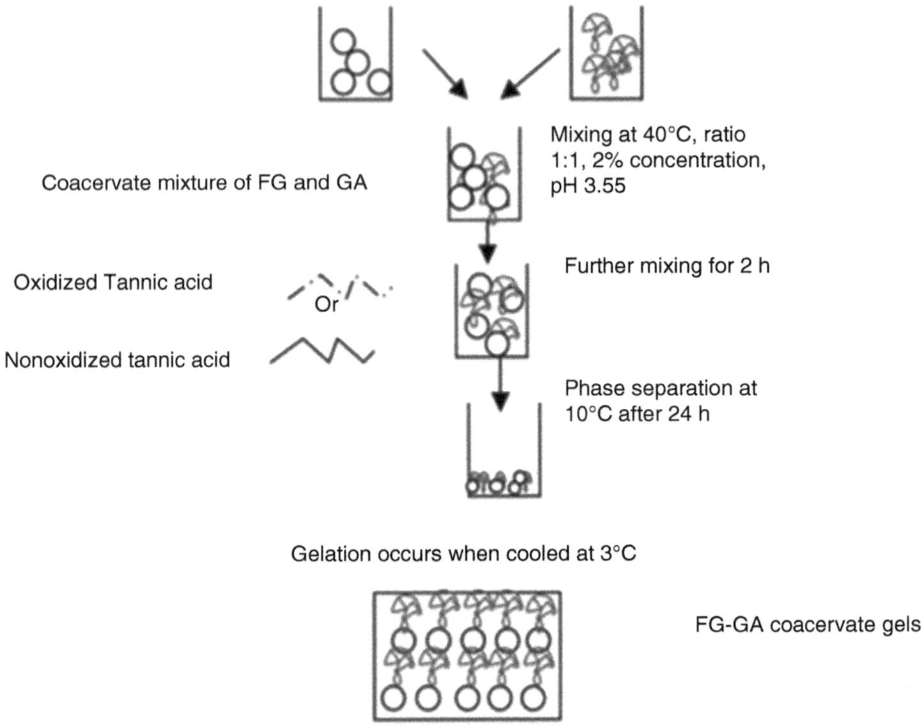

Figure 11.5 Coacervation-based cross-linking of fish gelatine (FG) and gum Arabic (GA).

Arabic aldehyde complex loaded with curcumin have been investigated for breast cancer treatment (Sarika and Nirmala, 2016).

Chemical modification with octenyl succinic anhydride for improved emulsification

Although gum Arabic is said to possess good emulsifying properties, the emulsifying properties are further improved through modification with octenyl succinic anhydride. For this purpose, gum Arabic that has been extracted and purified is reacted with octenyl succinic anhydride in an esterification reaction as outlined in the flow chart in Fig. 11.6. The esterification results in the attachment of lipophilic groups to the polysaccharide chains of gum Arabic (EFSA Panel on food additives and Nutrient sources added to food, 2010).

The emulsifying properties of gum Arabic are also improved through modification with dodecenyl succinic anhydride. This modification takes a similar route to modification with octenyl succinic anhydride. In this case, derivatization of the polymer occurs and is followed by incorporation of a hydrophobic group onto the polymer chain.

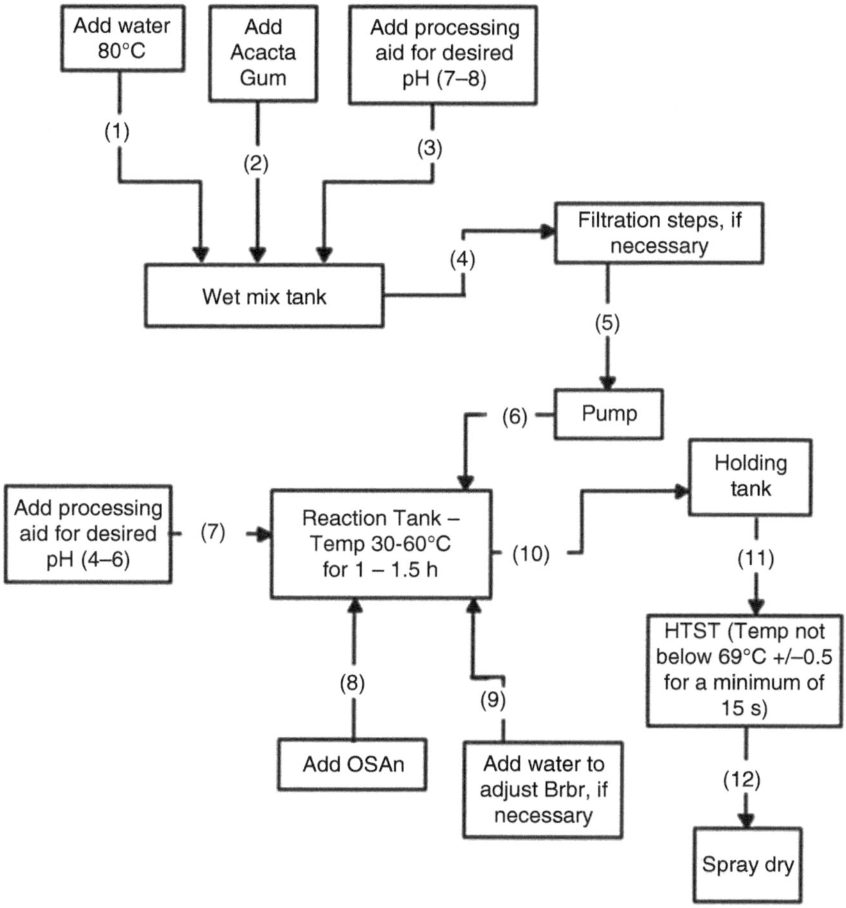

Figure 11.6 Flow chart for modification of gum Arabic with octenyl succinate.
Source: EFSA Panel on food additives and Nutrient sources added to food, 2010. Scientific Opinion on the use of Gum Arabic modified with Octenyl Siccinis Anhydride (OSA) as food additive. EFSA J. 8(3), 1539–1562.

Microencapsulation using gum Arabic

The emulsifying property of gum Arabic makes it very useful in preparation of emulsions and microencapsulation of other substances. Microencapsulation is of particular importance in food and pharmaceutics where the bioactivity of active agents needs to be protected to withstand processing conditions or to enable them dissolve in conditions (such as pH, temperature) that they normally would not dissolve in.

The general methods for the preparation of emulsion involves incorporating fat droplets within an aqueous system and inclusion of an emulsifier to maintain the distribution of the droplets within the system, hence the stability of the emulsion.

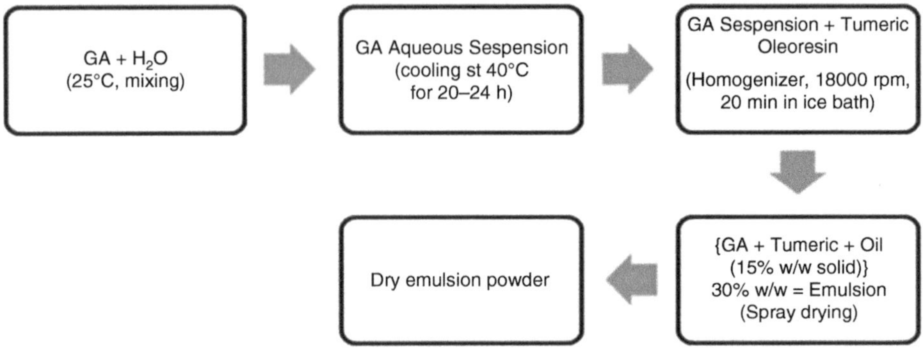

Figure 11.7 Flow chart of microencapsulation of turmeric oleoresin.

Gum Arabic, due to its solubility and surface properties, has proven to possess remarkable emulsifying properties for various applications, used alone or in combination with other emulsifiers. In the production of microencapsulations, gum Arabic serves to provide a wall within which the active ingredients are embedded. The wall is achieved as a porous system created as a result of drying of the oil–water emulsion, leaving gaps within the matrix within which the dry active ingredient is embedded such that the dry emulsion creates a wall around the active ingredient to be protected, resulting in a microencapsulation.

An example of this is the preparation of microencapsulated turmeric oleoresin. The microencapsulation was achieved by spray drying of the emulsions produced. The process is described in Fig. 11.7. Gum Arabic was used in combination with two other polymer emulsifiers, maltodextrin and modified starch, in order to combine the attributes of all three, as it has been shown that using a combination of three polymers result in improved microencapsulation properties, giving improved protection to cardamom, turmeric, cinnamon, and cumin, relative to when gum Arabic is used solely (Cano-Higuita et al., 2014; Kanakdande, 2007; Krishnan et al., 2005; Vaidya, 2006).

The effectiveness of the encapsulation was measured as a ratio of the total amount of turmeric with active ingredient curcumin present inside the microcapsules to the amount of curcumin present on the surface of the microcapsules (i.e., that was not encapsulated). This was done using UV spectrometry. Since the curcumin is soluble in methanol while the emulsifiers are not, dissolving the microcapsule powder in methanol allowed the surface curcumin to dissolve off in methanol and be quantified. To obtain total curcumin content, the whole sample is dissolved in water to release the encapsulated curcumin from within the water-soluble biopolymers. The higher content of gum Arabic and modified starch yielded better encapsulation and retention (Cano-Higuita et al., 2014).

An alternative method for processing of gum Arabic for encapsulation application is using coacervation. This involves formation of colloidal systems using two polymers of opposing charges, resulting in encapsulation of the active agent within the droplets formed. A particular example is the microencapsulation of sulforaphane within gelatine–gum Arabic complex (Garcia-Saldana and Campas-Baypoli, 2016).

Sulforaphane is easily degraded during storage and processing as a result of temperature and pH changes. Encapsulation protects this phytochemical from degradation. Fourier-transform infrared spectroscopy and UV spectrometry results from the aforementioned study showed successful encapsulation of sulforaphane in gelatine–gum Arabic coacervation complex. Although the encapsulation efficiency was higher for gelatine–pectin complex, gelatine–gum Arabic complex also achieved up to 80% encapsulation efficiency, thus indicating gelatine–gum Arabic complex as a suitable polymer for microencapsulation of sulforaphane.

The general processing of gum Arabic in microencapsulation therefore involves dispersion in a suitable solvent to form a suspension; this is then followed up by introduction of the active compound to be encapsulated and homogenizing the mixture. An oil–water emulsion is then formed by addition of oil. The microcapsule is finally achieved by drying.

Electrospinning of gum Arabic

The electrospinning process is generally used to obtain fibers from polymers from their dissolved states using electricity to align the polymer molecules into fibers. The electrospinning processing technique has been applied to gum Arabic for a variety of applications ranging from packaging to scaffolds for tissue engineering. Generally, gum Arabic, in its unmodified form, does not possess sufficient physical property to be electrospun; however, when dissolved in a suitable solvent and mixed with another polymer, it can be processed using electrospinning.

In a study of three polysaccharide gums: gum Arabic, gum Karaya, and Kondagogu gum, it was found that mixing these gums with synthetic polymer polyvinyl alcohol (PVA) resulted in formulations suitable for electrospinning. To begin with, the respective gums were dissolved in their solvents; the additive PVA was also dissolved in a suitable solvent. The electrospinning apparatus includes a syringe with a needle nozzle connected to a high-voltage source. By controlling the electrospinning parameters such as the voltage, pump pressure, and distance between plate and nozzle, the property of the resulting fiber can then be controlled. In the said study, an optimum formulation of gum Arabic to PVA ratio of 90:10 weight ratio achieved beadless nanofibers of 220 (±40) nm average diameter with good uniformity (Padil et al., 2016). The specific area, water contact angle, surface porosity, roughness, and hydrophobicity can be further improved using methane plasma treatment.

Nanosizing using gum Arabic

In preparing nanoscale formulations either in gel, powder, or liquid forms, a template or carrier system is usually needed to prevent aggregation and to control the nanoparticle distribution. In other cases, the nanofibers are achieved through processes such as electrospinning, which require dissolution of the polymer into a solvent, in which case the ease and effectiveness of the processing may be limited by the solubility of the polymer or the properties such as viscosity of the polymer in the dissolved state. The surface attribute of gum Arabic makes it useful in such cases.

A particular application in tissue engineering, nanofibers from chitosan–gelatin using less toxic solvent as acetic acid are improved by the addition of gum Arabic. This allowed a higher concentration of the chitosan–gelatin blend in the solvent prior to electrospinning into nanofibers, as a higher concentration of the polymer in the solvent could be reached while maintaining moderate viscosity for processing (Tsai et al., 2015). This is of particular importance in biomedical applications where scaffolds and implants can be fabricated from biocompatible and eco-friendly materials, eliminating the need for toxic solvents.

Gum Arabic is also applied in nanoformulation production. The ability of gum Arabic to form networks with other polymers makes it suitable in production of semiinterpenetrating networks (Pater, 1992). Poly(sodium acrylate) cross-linked with gum Arabic has been used as template for silver nanoparticles prepared from extract of *Syzygium aromaticum* (clove) (Bajpai and Kumari, 2015). In this process, the extracted silver nanoparticles are mixed with the low-viscosity monomer, sodium acrylate. This is then polymerized using solvent-based polymerization with the monomer dispersed in gum Arabic in solution using potassium persulfate as initiator. This allows the sodium acrylate to polymerize around the silver nanoparticles, forming a uniform composite. The network is then formed by cross-linking of the sodium acrylate and the gum Arabic using N,N′-methylenebisacrylamide as a cross-linker. This is illustrated in Fig. 11.8.

Just as the opposing charges of protein-based polymer gelatine and gum Arabic come in handy in preparing nanoparticles through coacervation technique, the anionic polyelectrolyte nature of gum Arabic combined with the cationic polyelectrolyte of chitosan comes in handy in producing polysaccharide-based nanoparticles. When gum Arabic is mixed with chitosan in the ratio 1:1 at a pH of 4, monodisperse, hydrophilic colloidal nanoparticles are formed. These nanoparticles carry high positive charges and can be loaded with active ingredients for drug delivery applications (Tan et al., 2016).

Drying and molding

Following the blending, compositing, and/or chemical processing stage, it is necessary to further process the derived gum Arabic–incorporated material into the final desired shape or form such as sheets, films, or powders. Various processes exist for processing of polymer materials, such as hot-melt extrusion, drum drying, film casting and electrospinning. Previous sections in this chapter have discussed the electrospinning process. Here we look at the processes that are applied mainly to convert the polymeric material to powders and flat films: drum drying, spray drying, and extrusion.

Gum Arabic–based formulations can prepared in dry powder form using the conventional drying processes used for biopolymer-based materials, such as drum drying, as illustrated in Fig. 11.9. The drum drying process typically involves passing the

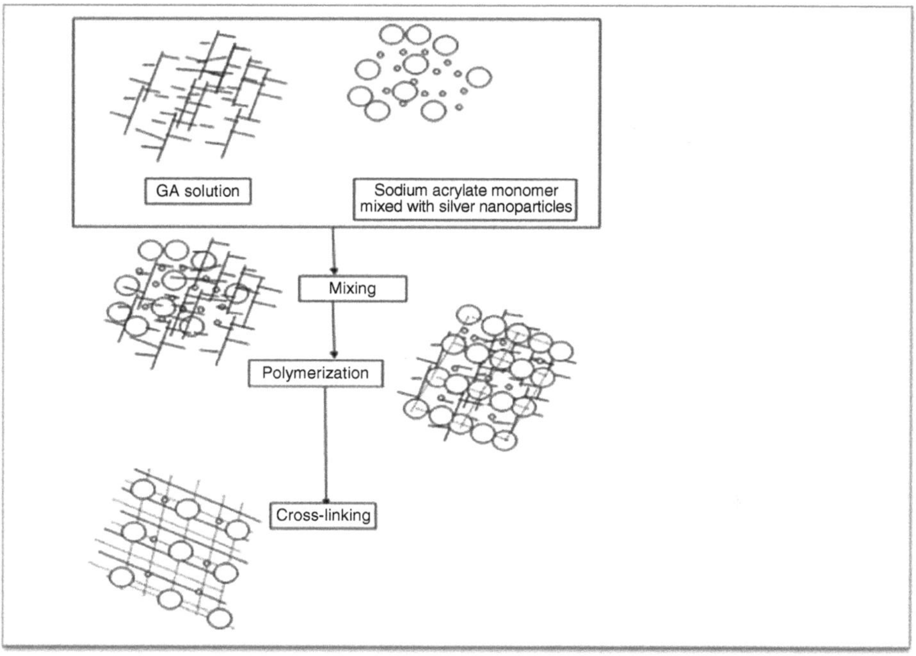

Figure 11.8 Semipenetrating network of gum Arabic and sodium acrylate as template for silver nanoparticles.

puree or slurry to be dried through two heated drums rotating at set speed and spacing. The drums are usually operated at moderate temperatures around 120°C. The setup includes parts for collecting or releasing the evaporated moisture and a collector plate for dried product on either side of the drums. Fig. 11.9 illustrates an example process for drying jackfruit puree, which contains soy lecithin and gum Arabic as emulsifiers and a stabilizing agent to protect the nutrients in the jackfruit (Pua et al., 2007).

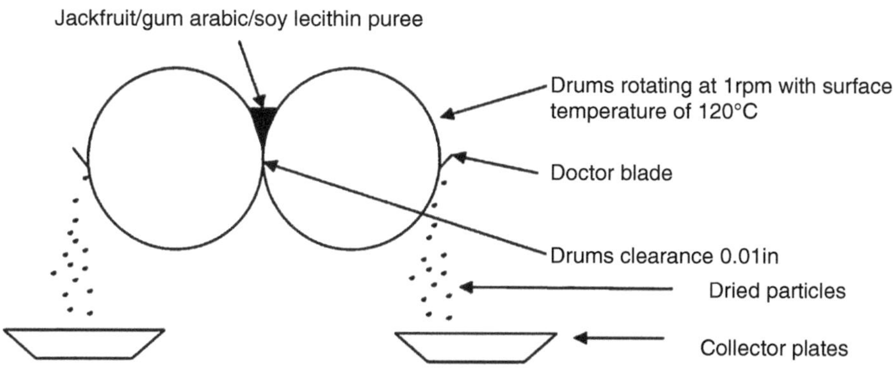

Figure 11.9 Schematics of drum drying process for Jackfruit puree containing gum Arabic and soy lecithin.

Such a drying process is applied in food processing, and gum Arabic is included here to act as a protector for the nutrients in the food throughout the drying process.

To achieve flat films from formulations of gum Arabic, processes applied to thermoplastic can also be applied to such. In such processes, care must be taken not to raise the operating temperature to reach heat that would degrade gum Arabic. On the other hand, the material must have just enough thermoplastic property to withstand the hot-melt process. A continuous hot-melt extrusion process has been investigated for production of gum Arabic–based dosage forms containing API. The dosage formulation developed achieved up to 70% drug loading (Kipping and Rein, 2016). Ability to produce dosage forms using a continuous extrusion process with a single polymer as the carrier is attractive, since extrusion is the most versatile and economic polymer processing technique that achieves uniformity of the product.

Conclusion

Gum Arabic can be processed in a variety of ways in various applications, such as food or biomedical applications. The processing could take different approaches, such as modifying its chemical structure by reacting with nonpolymeric groups, such as is seen in formation of gum Arabic aldehyde. Processing of gum Arabic could also be in the form of the polysaccharide itself being included in the formulation in its unmodified form in order to impose its physical or chemical properties, where it acts to, for example, reduce viscosity, stabilize droplets, or form nanoparticles.

In combination with other polymers, gum Arabic can simply be mixed, cross-linked, or formed into semiinterpenetrating networks with other polymers. These systems have been widely applied in developing microcapsules, nanoparticles, and emulsions for drug-loading applications.

The unique properties of gum Arabic, coupled with an ever-increasing number of processing techniques of polymers, makes it a versatile biopolymer with seemingly endless possibilities of applications.

References

Ahmed, A., et al., 2016. Gum Arabic supplementation improved antioxidant status and alters expression of oxidative stress gene in ovary of mice fed high fat diet. Middle East Fertil. Soc. J. 21, 101–108.

Anvari, M., Chung, D., 2016. Dynamic rheological and structural characterization of fish gelatin-gom arabic coacervate gels cross-linked by tannic acid. Food Hydrocoll. 60, 516–524.

Bajpai, S., Kumari, M., 2015. A green approach to prepare silver nanoparticles loaded gum acacia/poly(acrylate) hydrogels. Int. J. Biol. Macromol. 80, 177–188.

Cano-Higuita, D., Velez, H., Telis, V., 2014. Microencapsulation of tumeric oleoresin in binary and ternary blends of gum Arabic, maltodextrin and modified starch. Cienc. Agrotecnol Lavras 39 (2), 173–182.

Chang, Y., Hu, Y., McClements, J., 2016. Competitive adsorption and displacement of anionic polysaccharides (flucoidan and gum Arabic) on the surface of protein-coated lipid droplets. Food Hydrocoll. 52, 820–826.

Costa de Conto, L., Grosso, F., Goncalves, G., 2013. Chemometry as applied to the production of omega-3 microcapsules by complex coacervation with soy protein isolate and gum Arabic. LWT-Food Sci. Technol. 53, 218–224.

Daoub, R., Elmubarak, A.H., Misran, M., Hassan, A.E., Osman, M.E., 2016. Characterization and functional properties of some natural Acacia gums. J. Saudi. Soc. Agric. Sci.

EFSA Panel on food additives and Nutrient sources added to food, 2010. Scientific Opinion on the use of gum Arabic modified with Octenyl Siccinis Anhydride (OSA) as food additive. EFSA J. 8 (3), 1539–1562.

Garcia-Saldana, J., Campas-Baypoli, O., 2016. Microencapsulation of sulforaphane from broccoli seed extracts by gelatin/gum Arabic and gelatin/pectin complexes. Food Chem. 201, 94–100.

Halsall, T., Hirst, E., 1947. Oxidation of carbohydrates by periodate ion. J. Chem. Soc. (Resumed) 269 (0), 1427–1432.

Hassan, E., 2000. Characterization and Fractionation of *Acacia Seyal* Gum, s.l.: University of Khartoum, Sudan.

Huzinec, R.J., Graff, A.H., 1987. Coatings for Chewing Gums Containing Gum Arabic and a Soluble Calcium Salt. US Patent No. 4,681,766.

Kanakdande, D., 2007. Stability of cumin oleoresin microencapsulated in different combination of gum Arabic, maltodextrin and modified starch. Carbohydr. Polym. 67, 536–541.

Karamalla, K., Saddig, N., Osman, M., 1998. Analytical data for *Acacia senegal* var. *senegal* gum samples collected between 1993 and 1995 from Sudan. Food Hydrocoll. 12 (4), 373–378.

Khaliq, G., et al., 2015. Effect of gum Arabic coating combined with calcium chloride on physico-chemical and qualitative properties of mango (*Mangifera indica* L.) fruit during low temperature storage. Sci. Hortic. 190, 187–194.

Kipping, T., Rein, H., 2016. Continuous production of controlled release dosage forms based on hot-melt extruded gun arabic: formulation development, in vitro characterization and evaluation of potential application fields. Int. J. Pharm. 497 (1–2), 36–53.

Krishnan, S., Bhosale, R., Singhai, R., 2005. Microencapsulation of cardamon oleoresin: evaluation of blends of gum Arabic, 2005 and a modified starch as wall materials. Carbohydr. Polym. 61, 95–102.

Okatahi, S., Onyibe, J., 1999. Production of Gum Arabic, Zaria: National Agricultural Extension and Research Laison Services. Ahmadu Bello University & Rubber research Instittue of Nigeria.

Olatunji, O., Olsson, R., 2016. Processing and characterization of natural polymers. In: Olatunji, O. (Ed.), Natural Polymers, Industry Technique and Application. Springer International Publishing, Stockholm, pp. 19–57.

Padil, V., Senan, C., Waclawek, S., Cernik, M., 2016. Electrospun fibers based on Arabic, Karaya and Kondagogu gums. Int. J. Biol. Macromol. 91, 299–309.

Pater, R.H., 1992. Semi-interpenetrating polymer network for tougher and more microcracking resistant high temperature polymers. United States, Patent No. 5149746.

Pua, C., Hamid, S.A., Rusul, G., AbdRahman, R., 2007. Production of drum-dried jackfruit (*Artocarpus heterophyllus*) powder with different concentration of soy lecithin and gun arabic. J. Food Eng. 78 (2), 630–636.

Rayner, M., Ostbring, K., Purhagen, J., 2016. Application of Natural Polymers in Food. In: Olatunji, O. (Ed.), Natural Polymers: Industry techniques and Applications. Springer International Publishing, Stockholm, pp. 115–157.

Sarika, P., Nirmala, R., 2016. Curcumin loaded gum Arabic aldehyde-gelatin nanogels for breast cancer therapy. Mater. Sci. Eng. C Mater. Biol. Appl. 65 (1), 331–337.
Sarika, P., et al., 2014. Modified gum Arabic cross-linked gelatin scaffold for biomedical applications. Mater. Sci. Eng. C Mater. Biol. Appl. 43, 272–279.
Sarika, P., Pavithran, A., James, N., 2015. Cationized gelatin/gum Arabic polyelectrolyte complex: study of electrostatic interactions. Food Hydrocoll. 49, 176–182.
Singh, A., Gevekea, D., Yadav, M., 2017. Improvement of rheological, thermal and functional properties of tapioca starch by using gum Arabic. LWT-Food Sci. Technol. 80, 155–165.
Stefano, T., Fabiana, S., Attilio, C., 2003. The kinetics of periodate oxidation of carbohydrates 2. Polymeric substrates. Carbohydr. Res. 338 (10), 1083–1095.
Thakur, R., McCarron, P., Woolfson, A., Donnelly, R., 2009. Investigation of swelling and network parameteres of poly(ethylene glycol)-crosslinked poly (methyl vinyl ether-co-maleic acid) hydrogels. Eur. Polym. J. 45, 1239–1249.
Tsai, R., et al., 2014. Chitosan/pectin/gum Arabic polyelectrolyte complex: process-dependent appearance, microstructure analysis and it's application. Carbohydr. Polym. 101, 752–759.
Tsai, R., et al., 2015. Use of gum Arabic to improve the fabrication of chitosan-gelatin based nanofibers for tissue engineering. Carbohydr. Polym. 115, 525–532.
Tan, C., et al., 2016. Polysaccharide-based nanoparticles by chitosan and gum Arabic polyelectrolyte complexation as carriers for curcumin. Food Hydrocoll. 57, 236–245.
Vaidya, S.E.A., 2006. Microencapsulation of cinnamon oleoresin by spray drying using different wall materials. Drying Technol.: Int. J. 24, 983–992.

Enhancement of Color Stability in Foods by Gum Arabic

12

Abdalbasit A. Mariod[*,**]
[*]University of Jeddah, Alkamil, Kingdom of Saudi Arabia; [**]Indigenous Knowledge & Heritage Center, Ghibaish College of Science & Technology, Ghibaish, Sudan

Introduction

Gum Arabic (GA) is predominantly collected from two related species, namely *Acacia senegal* and *Acacia seyal*. *A. senegal* (L.) is a tree (or shrub) of 2–6 m, occasionally up to 8 m in height, with an umbrella-shaped crown. It is very branched, with many upright twigs. Producers harvest the gum commercially from wild trees, mostly in Sudan (80%) and throughout the Sahel, from Senegal to Somalia (von Maydell, 1990). *A. senegal* var. *senegal* (hashab) is a variety that grows in the Sudan and is the main source of commercial GA. On the other hand, both varieties of *A. seyal*, i.e., var. *seyal* and var. *fistula*, are found in Sudan. The former is characterized by normal spines and green, white, or red bark, while var. *fistula* is characterized by inflated spines and a whitish bark. Both varieties produce a commercial gum referred to as talha (Chikami et al., 1997).

Chemistry

GA is a complex mixture of glycoproteins (GPs) and polysaccharides. It is the original source of the sugars arabinose and ribose, both of which were first discovered and isolated from it, and are named after it (Muller and Okoro, 2004). In recent years, several investigations have been conducted in order to reveal the molecular structure of GA and relate it to its exceptional emulsifying and rheological properties. The chemical composition of GA is complex and consists of a group of macromolecules (97%), which are predominantly characterized by a high proportion of carbohydrates, composed of D-galactose and L-arabinose units and a low proportion of protein (<3%).

The chemical composition of GA may vary slightly depending on its origin, climate, harvest season, tree age, and processing conditions, such as spray drying. The two varieties of Acacia gum, *A. senegal* and *A. seyal*, are hyperbranched polysaccharides rich in arabinose and galactose, mainly formed by chains of 3,6-linked β-D-Gal*p* substituted in position 6 by side chains of 3-linked α-L-Ara*f* (Fig. 12.1). Beyond this common structure, *A. senegal* presented the highest degree of branching (78.2% vs. 59.2%), with more branched galactopyranoses, shorter arabinosyl side branches, and more rhamnopyranoses in terminal position. Circular dichroism experiments evidenced that both acacia gums were partially structured into polyproline type II helices,

Gum Arabic. http://dx.doi.org/10.1016/B978-0-12-812002-6.00012-9

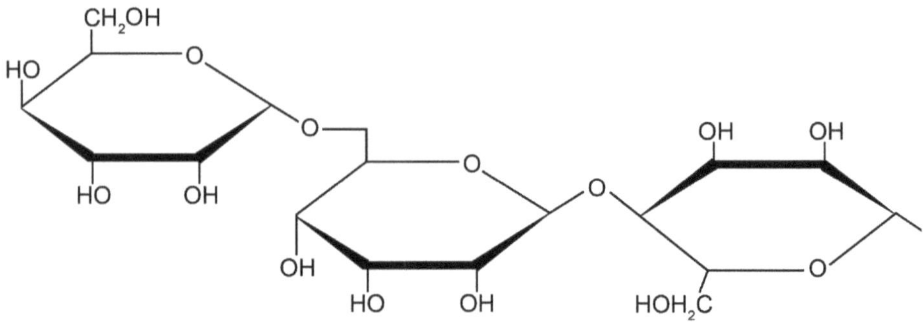

Figure 12.1 Chemical structure of gum Arabic.

but in a lesser extent for A. *senegal* gum, suggesting its less structured organization. The average molecular weight of *A. seyal* macromolecules was larger than that of *A. senegal* (8.2×10^5 g/mol vs. 6.8×10^5 g/mol). The most compact structure of *A. seyal* gum can be partly explained by the lowest concentration of charge sugars that induces less electrostatic repulsion, and the highest content of long arabinose side chains that may self-organize and interact between them (e.g., hydrogen bonding, steric effect, etc.). For both acacia gums, the anisotropy of macromolecules increased with the molecular weight; however, *A. senegal* macromolecules were the most anisotropic ones (Al-Assaf et al., 2005).

Gashua et al. (2016) studied the molecular composition of *A. senegal* and *A. seyal* gum exudate using transmission electron microscopy. The molecules observed in both samples were found to have diameters of either ~20 μm, ~60 μm, or ~10 μm. These most likely represent the arabinogalactan, arabinogalactan–protein (AGP), and GP molecules present in acacia gum exudates. Micrographs obtained for gum solutions that had been left to stand for up to 5 days indicated that molecular aggregation had occurred; this was particularly evident for the *A. senegal* sample. These authors determined the thickness of the gums adsorbed onto polystyrene latex particles using dynamic light scattering. They found that the adsorbed layer thickness increased over time and, after 14 days, had a value of 61 nm for the *A. senegal* gum sample. For the *A. seyal* gum sample, the adsorbed layer thickness was only ~3 nm and did not increase over time. Transmission electron microscopy revealed the presence of a distinct, thick adsorbed layer for the *A. senegal* gum and the presence of a much thinner, more diffuse layer for the *A. seyal* gum sample (Gashua et al., 2016).

Menzies et al. (1996) compared the physiochemical properties of GA from *A. senegal* and *A. seyal* var. *seyal*. They reported an optical rotation (ca. −30 degrees) and the contents of rhamnose (12%–14%), arabinose (24%–29%), galactose (36%–42%), glucuronic acid (16%–17%), nitrogen (0.327%–0.365%), and protein (2.16%–2.41%). These contents of the GA samples were consistent with the FAO (1990) specification for acacia gum. In contrast, the gum tahl (*A. seyal*) samples had positive (alpha) *D* values (+45 degrees to +54 degrees), lower rhamnose (3%–4%), higher arabinose (41%–45%) contents, and lower nitrogen (0.147%–0.175%) and, hence, protein (0.97%–1.15%), contents. Therefore, there are some physicochemical and structural differences between *A. senegal* gum and *A. seyal* gum (Menzies et al., 1996).

Food uses

GA is used primarily in the food industry as a stabilizer. It is edible and has E number E414. GA is a key ingredient in traditional lithography and is used in printing, paint production, glue, cosmetics, and various industrial applications, including viscosity control in inks and in textile industries, though less expensive materials compete with it for many of these roles (Smolinske, 1992). GA is a mixture of polysaccharides, and GP gives it the properties of a glue and binder that is edible by humans. It remains an important ingredient in soft drink syrups; gummy candies such as gumdrops, marshmallows, chocolate candies; and edible glitter. Also, GA is used as an emulsifier and a thickening agent in icing, fillings, chewing gum, and other confectionery treats (Rinsky and Rinsky, 2009). The inherent characteristics of GA exemplified by its emulsification, acid stability, low viscosity at high concentration, adhesive and binding properties, and good mouth feel made it widely used in pharmaceuticals, insecticides, ceramics, lithographic plates, soft drinks, candies, and countless other applications (Karama, 2002).

GA is mainly used in the confectionery industry, where it is incorporated in a wide range of products. It has a long tradition of use in wine gums, where it produces a clarity that is higher than can be obtained with other hydrocolloids. Furthermore, it prevents sucrose crystallization, provides a controlled flavor release, and slows down melting in the mouth, making the wine gum long-lasting. It also provides the appropriate texture to these candies, which are easily deformed in the mouth, but do not adhere to the teeth (Arja et al., 2011).

Gum Arabic as food color preserver

Color is the first characteristic that a person notices when judging a food. Although other characteristics of food (e.g., flavor, texture, nutrition value) are important in the overall development, perhaps none can influence how a person perceives a food as much as color (Wrolstad and Smith, 2010). The color is an indicator of pigment concentration that is measured immediately. Color can be changed during heat processing based on the different reactions of pigments, such as pigment degradation, browning reactions like Maillard reactions, enzymatic browning, and oxidation of ascorbic acid (Maskan, 2006). Color, as it is perceived by humans, is the reflection or transmission of various wavelengths of light off of or through an object that are detected by the eye (Barry, 2013). GA is used in many applications in the food industry, mainly as stabilizer, thickener, emulsifier, and for anticaking and other uses as a color preservative (Duke, 1981). At present, GA is one of the most widely used biopolymers in foods and beverages (Garti and Reichman, 1993). Its AGP component is responsible for its unequaled emulsifying properties, including the ability to form a stable emulsion over a wide pH range and in the presence of electrolytes (Buffo et al., 2001). The combination of maltodextrin and GA produced colored particles with the longest half-life and spherical particles that were less than 50 μm in size. An ethanolic anthocyanin extract from blueberries (*Vaccinium ashei*) was successfully encapsulated using mesquite gum as the wall material (Jiménez-Aguilar et al., 2011).

The potential of GA to improve the stability of anthocyanins in commercial beverages was investigated by Chung et al. (2016). These authors proceeded the degradation of anthocyanin in beverage systems containing L-ascorbic acid with a first-order reaction rate during storage (40°C for 5 days in light). They found that the addition of GA significantly enhanced the color stability of anthocyanin, and when they increased the concentration, they noticed an efficacy reduction due to a change in the conformation of the GA molecules that hindered their exposure to the anthocyanins. Fluorescence quenching measurements showed that the anthocyanin could have interacted with the GP fractions of the GA through hydrogen bonding, resulting in enhanced stability (Chung et al., 2016).

Idham et al. (2012) signifies that the encapsulation process could stabilize and extend the shelf-life of anthocyanins content. These authors microencapsulated anthocyanins with a combination of maltodextrin and GA; they found that it had the highest encapsulation efficiencies. Their storage stability results supported that combination both of maltodextrin and GA as wall material gave the longest shelf-life and the smallest change in the pigment color. Storage period significantly affects the color changes of the spray-dried powder.

Achouri et al. (2012) studied the ability of GA as an emulsifier to stabilize oil-in-water emulsions at high oil concentrations. Parameters investigated included droplet size distribution and color (ΔE). They reported that at 10% (w/v) emulsifier-to-oil ratio, emulsions stabilized with GA had average oil droplet sizes lower than 1 μm. GA-stabilized emulsions had lower ΔE. Droplet size strongly affected color. GA emulsions showed good stabilities over the 15-day storage period.

Khaliq et al. (2015) investigated the effect of 10% GA and 3% calcium chloride (CA) on the physiological and biochemical properties of mango fruits that stored for 5 days. They observed significant ($p \leq 0.05$) differences in fruits treated with 10% GA and 3% CA as compared to control. The combined treatment of 3% CA + 10% GA significantly alleviated the increase in decay incidence. Furthermore, 10% GA alone or combined with 3% CA effectively reduced weight loss, color changes, soluble solid concentration, respiration rate, and ethylene production, and maintained high firmness, titratable acidity, and ascorbic acid. These authors suggested that application of 10% GA coating combined with 3% CA might be a simple and effective technique for preserving mango fruit quality during low-temperature storage.

The efficacy of GA as a color preservative and inhibitor of nonenzymatic browning of dehydrated tomato during storage was studied by Idris et al. (2013); they compared GA with sodium metabisulfite. They treated tomato homogenates with 1, 2.5, 5, 7.5, and 10.0% GA and 1.0% sodium metabisulfite, respectively. The authors monitored dehydrated tomato, and color changes during 4 months of storage. They found that water activity (Fig. 12.2) was decreased significantly ($p \leq 0.05$), and Brix values were increased significantly ($p \leq 0.05$) in GA-treated samples.

The pH of metabisulfite-treated samples was found significantly ($p \leq 0.05$) higher than that of control and GA-treated samples (Fig. 12.3).

Lightness (L*) value was increased significantly ($p \leq 0.05$) with the addition of 5%–10% GA compared to the dehydrated control sample. Browning index of dehydrated tomato was decreased significantly ($p \leq 0.05$) with the treatments, 1.0 and 2.5% GA; 1% metabisulfite; 5, 10, and 7.5% GA, respectively (Fig. 12.4). GA at 5%–10% (w/w) was more suppressed to browning than sodium metabisulfite. GA preserved color of dehydrated tomatoes for up to 4 months of storage (Idris et al., 2013).

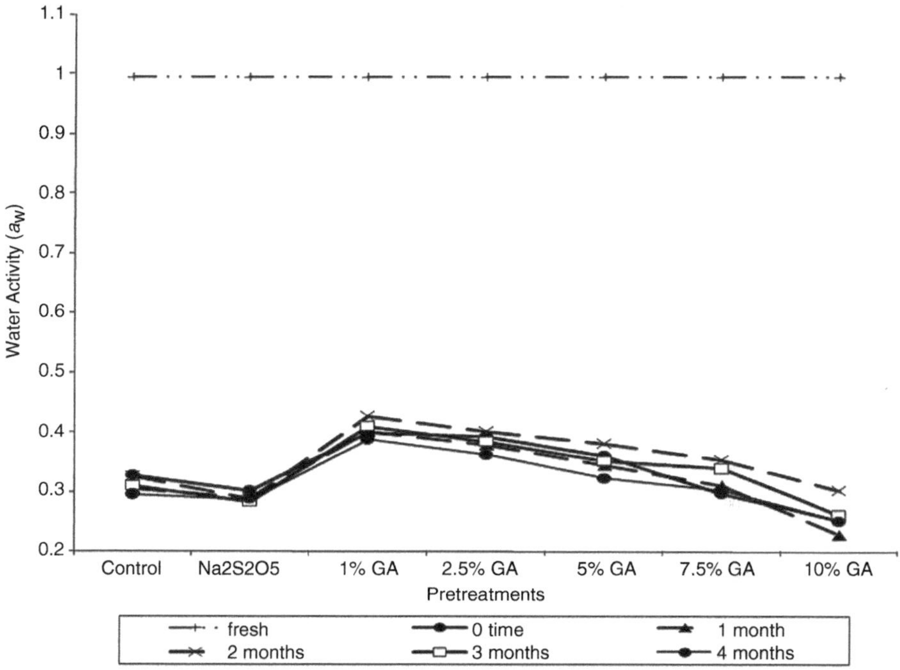

Figure 12.2 Water activity (a_W) of dehydrated tomato pretreated with gum Arabic at different storage periods. Values expressed are mean ± standard deviation ($N = 2$), at room temperature (25°C).
Source: Idris, Y.M.A., Ibrahim, Y.A., 2011. Effects of gum Arabic on dehydrated tomato color. unpublished data.

Khazaeia et al. (2014) studied anthocyanin stability and color of encapsulated freeze-dried saffron petal extract with various matrices consisting of GA and maltodextrin (M7 and M20). They measured the total anthocyanins of powders and color parameters (a*, b*, L*, C, H, and total color differences) immediately after production and during storage of up to 10 weeks by pH differential method and computer vision, respectively. All powders had a high value of color parameter a*, which was attributed to high anthocyanins content. The powder with 100% AG showed the biggest total color differences, and the sample with a wall formula of the M20, M7, AG: 66.7, 16.7, and 16.7 represented the lowest total color differences. They found no significant differences between different compounds of wall materials in terms of stabilizing anthocyanins ($p < 0.01$), and no significant decrease in anthocyanin content of the powders was observed after storage. The efficiency order of wall materials considering total color differences was AG > M20 > M7. These authors recommended microencapsulation by freeze-drying as a suitable method for stabilizing anthocyanins of saffron petal extract. Guan and Zhong (2015) studied GA to improve the thermal stability of anthocyanins. Solutions with 0.51 mg/mL anthocyanins at pH 5.0 were heated at 80°C and 126°C up to 80 min with and without 10 mg/mL GA. The half-life of thermal degradation of anthocyanins at 80°C and 126°C increased by 2.0 and 1.8 times, respectively,

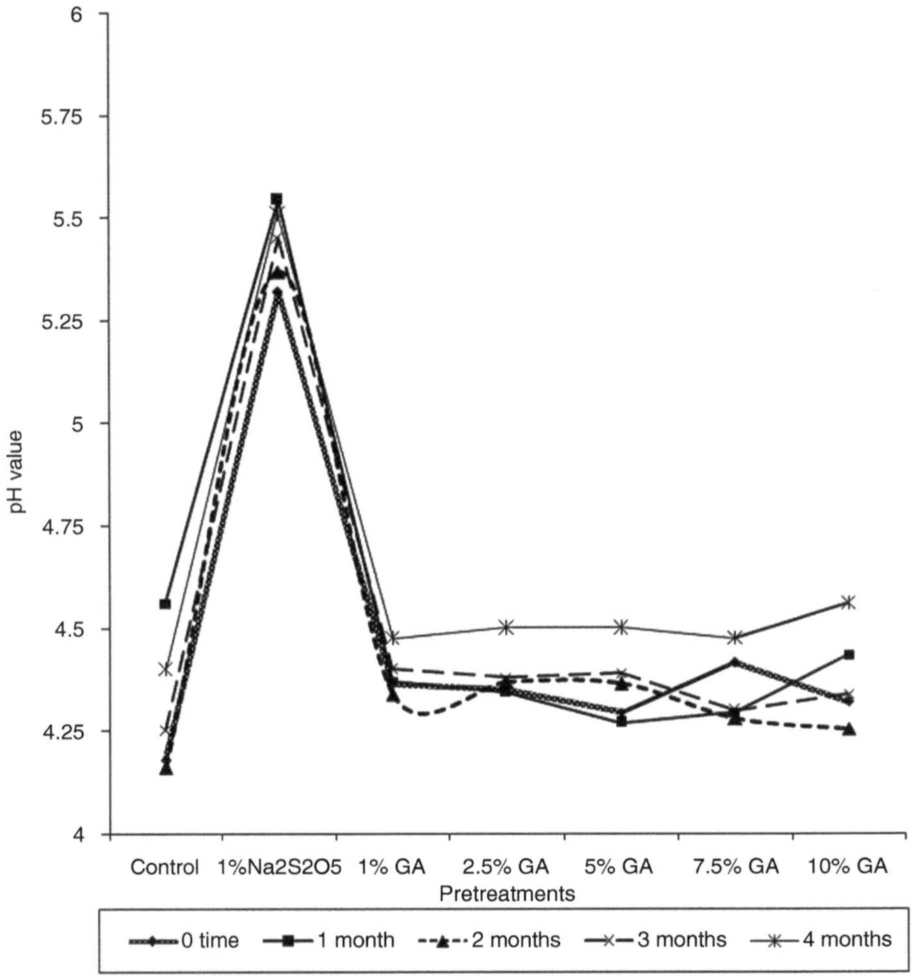

Figure 12.3 pH of dehydrated tomato pretreated with gum Arabic at different storage periods. Values expressed are mean ± standard deviation ($N = 2$), at room temperature (25°C). *Source:* Idris, Y.M.A., Ibrahim, Y.A., 2011. Effects of gum Arabic on dehydrated tomato color. unpublished data.

after adding GA. The residual concentration of anthocyanins with GA was 1.02 and 1.35 times higher than that without GA after 30-min heating at 80°C and 126°C, respectively.

Conclusion

Dehydration and subsequent storage promoted discoloration and nonenzymatic browning of tomatoes. Sodium metabisulfite increased the initial and during storage pH, and its effects as a color preservative decreased with time. GA is a potential substitute for metabisulfite as a color preservative for dehydrated tomatoes.

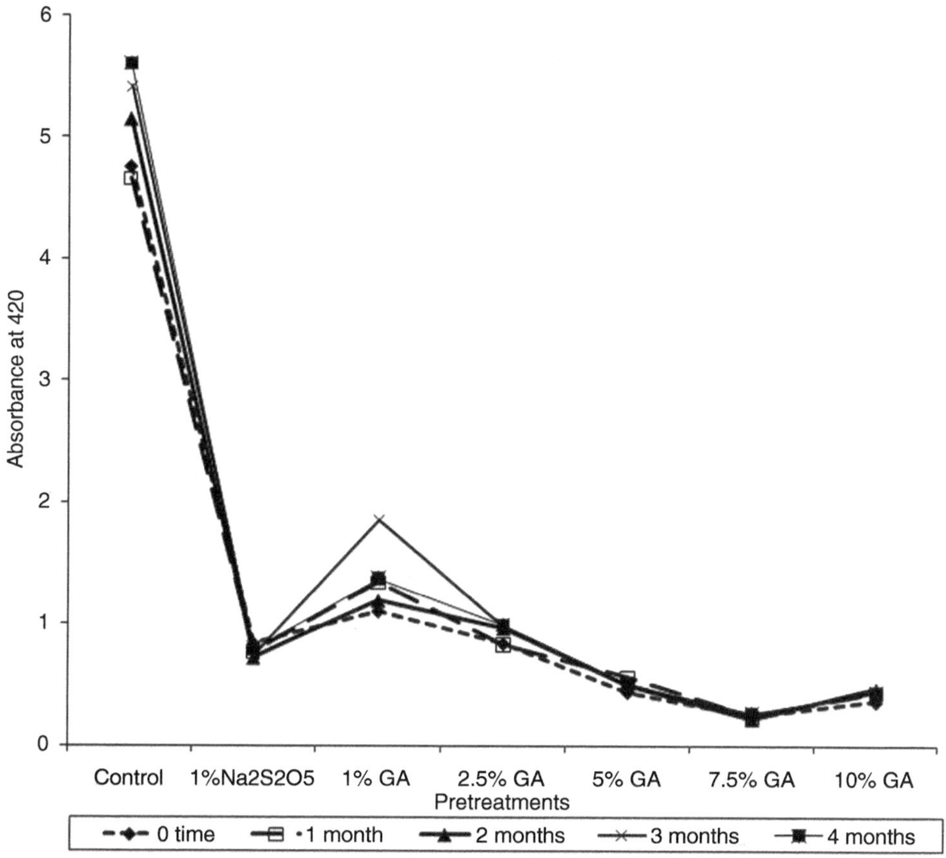

Figure 12.4 Browning index (A_{420}) of dehydrated tomato pretreated with gum Arabic at different storage periods. Values expressed are mean ± standard deviation ($N = 2$), at room temperature (25°C).
Source: Idris, Y.M.A., Ibrahim, Y.A., 2011. Effects of Gum Arabic on dehydrated tomato color. unpublished data.

References

Achouri, A., Zamani, Y., Boye, J.I., 2012. Stability and physical properties of emulsions prepared with and without soy proteins. J. Food Res. 1 (1), 254–267.

Al-Assaf, S., Phillips, G.O., Williams, P.A., 2005. Studies on acacia exudate gums. Part I: the molecular weight of Acacia senegal gum exudate. Food Hydrocoll. 19 (4), 647–660.

Arja, V., Maija, S., Kaija, L., 2011. Gum Arabic as a cause of occupational allergy. J. Allergy, 1–5, ID 841508.

Barry, A.M., 2013. Encapsulation, Color Stability, and Distribution of Anthocyanins from Purple Corn (*Zea mays* L.), Blueberry (*Vaccinium* sp.), and Red Radish (*Raphanus sativus*) in a Cold-Setting Pectin-Alginate Gel (Msc. thesis). The Ohio State University, USA.

Buffo, R.A., Reineccius, G.A., Oehlert, G.W., 2001. Factors affecting the emulsifying and rheological properties of gum acacia in beverage emulsions. Food Hydrocoll. 15, 53–66.

Chikami, B.N., Casadei, E., Coppen, J.J.W., Abdel Nour, H.O., Cesareo, D., 1997. A Reivew of Production, Markets and Quality Control of Gum Arabic in Africa. FAO. Technical Cooperation Programme. TCP/RAF/4557, Rome.

Chung, C., Rojanasasithara, T., Mutilangi, W., McClements, D.J., 2016. Enhancement of color stability of anthocyanins in model beverages by gum Arabic addition. Food Chem. 201, 14–22.

Duke, J.A., 1981. Handbook of Legumes of World Economic Importance. Plenum Press, New York, USA.

Garti, N., Reichman, D., 1993. Hydrocolloids as food emulsifiers and stabilizers. Food Struct. 12, 411–426.

Gashua, I.B., Williams, P.A., Baldwin, T.C., 2016. Molecular characteristics, association and interfacial properties of gum Arabic harvested from both *Acacia senegal* and *Acacia seyal*. Food Hydrocoll. 61, 514–522.

Guan, Y., Zhong, Q., 2015. The improved thermal stability of anthocyanins at pH 5.0 by gum Arabic. LWT – Food Sci. Technol. 64 (2), 706–712.

Idham, Z., Muhamad, I.I., Sarmidi, M.R., 2012. Degradation kinetics and color stability of spray-dried encapsulated anthocyanins from *Hibiscus sabdariffa* l. J. Food Process. Eng. 35, 522–542.

Idris, Y.M.A., Ibrahim, Y.A., 2011. Effects of Gum Arabic on dehydrated tomato color. unpublished data.

Idris, Y.M.A., Ibrahim, Y.A., Mariod, A.A., 2013. Color of dehydrated tomato: effects of gum Arabic. Int. J. Food Prop. 16, 838–851.

Jiménez-Aguilar, D.M., Ortega-Regules, A.E., Lozada-Ramírez, J.D., Pérez-Pérez, M.C.I., Vernon-Carter, E.J., Welti-Chanes, J., 2011. Color and chemical stability of spray-dried blueberry extract using mesquite gum as wall material. J. Food Compos. Anal. 24, 889–894.

Karama M. Musa, 2002. Gum Arabic Marketing Issues: Marketing Model, Buffer- Stocking, Processing and Financing International workshop on Promotion of Gum Arabic Production Processing & Marketing.

Khaliq, G., Mohamed, M.T.M., Alib, A., Ding, P., Ghazali, H.M., 2015. Effect of gum Arabic coating combined with calcium chloride onphysico-chemical and qualitative properties of mango (*Mangiferaindica* L) fruit during low temperature storage. Sci. Horticult. 190, 187–194.

Khazaeia, K.M., Jafaria, S.M., Ghorbania, M., Kakhki, H.A., 2014. Application of maltodextrin and gum Arabic in microencapsulation of saffron petal's anthocyanins and evaluating their storage stability and color. Carbohydr. Polym. 105, 57–62.

Maskan, M., 2006. Production of pomegranate (*Punica granatum* L.) juice concen-trate by various heating methods: colour degradation and kinetics. J. Food Eng. 72 (3), 218–224.

Menzies, A.R., Osman, M.E., Malik, A.A., Baldwin, T.C., 1996. A comparison of the physicochemical and immunological properties of the plant gum exudates of *Acacia senegal* (gum Arabic) and *Acacia seyal* (gum tahla). Food Addit. Contam. 13 (8), 991–999.

Muller, D., Okoro, C., 2004. Production and Marketing of Gum Arabic. Network for Natural Gums and Resins in Africa (NGARA), Nairobi, Kenya.

Rinsky, L.H., Rinsky, G., 2009. The Pastry Chef's Companion: A Comprehensive Resource Guide for the Baking and Pastry Professional. John Wiley & Sons, Chichester, pp. 1–134.

Smolinske, S.C., 1992. Handbook of Food, Drug, and Cosmetic Excipients. CRC Press, ISBN 0-8493-3585-X, p. 7.

von Maydell, J.H., 1990. Trees and Shrubs of The Sahel. pp. 525.

Wrolstad, R.E., Smith, D.E., 2010. Color analysis. In: Nielsen, S.S. (Ed.), Food Analysis. Boston, MA, Springer US, pp. 573–586.

Gum Arabic as Thickener and Stabilizing Agents in Dairy Products

13

Abdel Moneim El-Hadi Sulieman
University of Hail, Kingdom of Saudi Arabia

Introduction

Food hydrocolloids are high–subatomic weight hydrophilic polysaccharides that can control microstructure, texture, flavor, and time span of foods and drinks. In this manner, hydrocolloids as functional ingredients can be found in many foods. Gum Arabic (acacia; GA), tragacanth, xanthan, and certain changed gums have been used as food added substances, as they have exceptional capacities, for example, impediment of precipitation of scattered strong particles and a mixture of oil beads. Part of these gums where emulsions stayed darkened to a few degrees (Garti, 1999, Garti and Reichman, 1993). Food hydrocolloids have three major types:

1. Stabilizer, which keeps up the homogeneous dispersion of at least two immiscible substances in a food;
2. Thickener, which increases the consistency of the constant stage. The structure and measure of these polysaccharides molecules make thickness at genuinely low concentrations; and
3. Gelling agent which is useful in the formulation of stable dispersed systems gel by chain–chain, chain-dissolvable cross-link, utilized as a bit of a cerebrum boggling, basically masterminded in the food area.

Be that as it may, there are similarly broad nonfood applications. GA is comprehensively used for current purposes as a stabilizer, a thickener, an emulsifier, and, to a lesser degree, in textiles, earthenware production, lithography, and in the cosmetic and pharmaceutical industries. In the food industry, GA is, on a very basic level, used in candies, baked goods, dairy, beverages, and as a microencapsulating agent (Montenegro et al., 2012).

GA, which is an exudate of *Acacia senegal* and *Acacia seyal*, has many benefits, including:

1. multifunctional: great emulsifier, used in film processing and as a texturizer and low-consistency water cover and bulking agent;
2. high source of fiber: contains less than 85% soluble dietary fiber (dry basis);
3. high percentage purity: no added substances; free from drugs and polluting influences; has, to a great degree, low bacterial levels;
4. quick hydration and simplicity of scattering: accessible in prehydrated or agglomerated form; and
5. "normal" labeling: gum Arabic is not synthetically adjusted and fits the bill for "characteristic" labeling, or "no manufactured added substances" claims (Fig. 13.1).

Gum Arabic. http://dx.doi.org/10.1016/B978-0-12-812002-6.00013-0

Figure 13.1 Gum Arabic on *A. senegal* tree shoot.

Emulsion stability is a standout among the most essential components deciding the time span of usability of numerous food products (Bai et al., 2016; Yang et al., 2013). In addition, it gives a profitable pointer of the adequacy of various emulsifiers at settling emulsions. Emulsion thickness is expanded with expanding polysaccharide concentration. The capacity tests demonstrated that the GA–balanced out emulsions had great dependability to bead accumulation and gravitational partition fundamentally because of less oil drop estimate as well as high emulsion thickness.

Mechanism of food stabilizers

A stabilizer is a substance added to food to help protect its structure. Typical uses include preventing oil, water emulsions from separating in products, such as salad dressing; avoiding formation ice crystals in frozen food such as ice cream; and preventing fruit from settling in products, such as jam, yoghurt and jellies (Imeson, 2011).

There is a conspicuous distinction among emulsifiers and stabilizers in food systems. As per Dickinson and Stainsby (1988): "An emulsifier is a single chemical or mixture of components having the capacity for promoting emulsion formation and short-term stabilization by interfacial action." While a stabilizer is a single chemical component or a blend of parts that can offer long-term dependability in an emulsion, potentially by a system including adsorption; however, this is not really so. The main applications of food stabilizers include: alteration of the rheological properties of watery frameworks, consistency developers, jellification specialists, and stabilizers (BeMiller, 1988; Dickinson, 1988).

Those functionalities are hard to clarify regarding consistency and water gum associations. It was necessary to consider the emulsions of oil in the water adsorption or interaction with the oil phase.

However, some polysaccharides, such as GA, xanthan, and tragacanth, have been noted to show the particular surface tension and stability of the dispersed particles of oil beads in water frameworks (Achouri et al., 2012).

When considering the stability with respect to the mixture and flocculation, it is imperative to recognize adsorbing polymers and nonadsorbing polymers. Covering particle surfaces with a thick, cautious layer of adsorbing polymer hinders blending. The covered particles emphatically repulse each other at short proximity through a mix of polymeric flexible and osmotic impacts, and then the colloid is said to be well balanced out. It must be set combining a solid association with the surface of the low-fondness constant stage, and the other way around for the chains dangling. Also, there must be sufficient buildup of polymers to cover the surface of the particles flawlessly, and there ought to be adequate thickness of the layer to keep particles from detaching of a couple of particles (van der Waals). Therefore, it appears to be evident that the known proteins have water for no incredible adaptability; it can be effortlessly consumed by the particles or beads of oil, and in this manner, it will be considered, and emulsions with labor stability in the short term on separate molecules (Dickinson and Stainsby, 1988).

On the other hand, hydrocolloids will be considered as stabilizers with limited absorption, stability, and long-term stability (which is called "depletion stabilization mechanism"). Recent reviews have concentrated on the role of GA as an emulsifier; these reviews essentially have a specific end goal to comprehend the solid adsorption of the oil beads in beverage emulsions (Randall et al., 1988; Williams et al., 1990).

Gum Arabic

It is demonstrated that while GA adsorbs strongly and effectively onto the oil droplets via its proteinaceous moieties, guar gum and locust bean gum adsorb weakly and, for the most part, only "precipitate" on the oil surface, and form birefringent layers of the polymer oriented with its hydrophobic mannose backbone facing the oil.

Among all the gums advertised around the world, GA is, without a doubt, the most comprehended and the most looked for, both at the levels of the conveying districts and universally. This routinely causes disintegration, since a few providers blend different components into the product, keeping in mind the end goal to satisfy their commitments. It likewise brought about the overabuse of assets as creators' assets as makers attempt to grow the accessibility of GA (Lelon et al., 2010). GA is unquestionably antiquated and surely an understood gum type. It AU: Please clarify wording of this sentence. is an emulsion of the trunks and branches of acacia trees, and it is a composite material of polysaccharides, appears as a white substance tends to orange cocoa and separates the glass pieces, containing food and other medicines, and if kept in a proper way remains for quite a long time without change, because of the utilization of around five thousand years. It is a consumable, thick fluid. Dissolvable strands occur under anxiety-causing conditions, for example, dry spells, poor soil, and damage (Williams and Phillips, 2009).

Origin and production areas of gum Arabic

As indicated by the Sudanese National Forestry Corporation, the gum belt's gross area is estimated to cover 520,000 km^2, roughly one-fifth of Sudan's total area. More than 5 million individuals work in field of GA production. The Sudanese Gum Arabic Belt, with its acacia trees, the most well-known such belt on the planet, has lost a large number of its assets accordingly because of carelessness and different variables. Specialists affirmed that Sudan had been, from the early 1950s up until the late 1990s, producing roughly 80% of GA globally, and the production has dwindled to less than 50% today. Many Sudanese farmers in central, eastern, and southwestern Sudan rely on GA production. Sudan is the world biggest single maker of GA, and it was utilized to contribute to 15% of the nation's income before the investigation of Sudan's petroleum (siteresources.worldbank.org).

Chemical composition, structure, and properties

Gum Arabic is complex, comprised of a gathering of macromolecules described by a high amount of starches (97%), which are made out of predominantly D-galactose and L-arabinose units and a low proportion of protein (<3%) (Verbeek, 2012). Slight variation could be found in the chemical composition of GA; it depends on its source, atmosphere, collection season, tree age, and processing conditions (Flindt et al., 2005; Hassan et al., 2005; Siddig et al., 2005). The chemical composition of the GA obtained from *A. senegal* differs slightly from that obtained from *A. seyal*. The different characteristics of GA are shown in Table 13.1.

Table 13.1 Characteristics of gum Arabic

Moisture (%)	13–15
Ash (%)	2–4
Internal energy (%)	30–39
Volatile matter (%)	51–65
Optical rotation (degrees)	–26 to –34
Nitrogen content (%)	0.26–0.39
% Galactose	44.00–38.00
% Arabinose	27.00–46.00
% Rhamnose	13.00–4.00
Glucuronic acid	14.50–6.50
4-O-methyl glucuronic acid	1.50–5.50
% Nitrogen	0.36–0.15
Specific rotation/degrees	−30.00 to 51.00
Average molecular mass (MW)	380,000–850,000

Source: Dauqan, E., Abdullah, A., 2013. Am. J. Appl. Sci. 10 (10), 1270–1279.

The major portion of GA is a profoundly expanded polysaccharide comprising of a galactose spine with connected branches of arabinose and rhamnose, which terminate in the glucuronic acid found in nature as magnesium, potassium, and calcium salt.

The excellent solubility of GA in aqueous solutions and its low viscosity, bland taste, neutral odors, use as a fiber source, and low calorie content has lent itself to many nutritional applications, as indicated in Table 13.2.

Emulsifying properties of gum Arabic

Oil-in-water emulsions are fundamental parts of numerous items in the food and drink industries, including milk, creams, dressings, treats, sauces, plunges, and refreshments (Dickinson, 2015; McClements, 2015). They are thermodynamically unsteady colloidal frameworks comprised of little oil beads scattered inside a fluid. Emulsifiers along these lines are required to give active solidness by covering the oil beads and creating unpleasant connections (steric or electrostatic) that repress total drops. Moreover, emulsifiers encourage emulsion development by adsorbing to bead surfaces and lessening the interfacial pressure, in this way encouraging bead separation (Kralova and Sjoblom, 2009).

Various food-grade emulsifiers are legitimately affirmed for usage in the food business (Kralova and Sjoblom, 2009). In any case, they change in their costs,

Table 13.2 Some applications of gum Arabic

No.	Application	Function	Benefit
1	Essential oil emulsions: carbonated and noncarbonated, soft drinks	Preparation and stabilization	Improved emulsion
2	Encapsulation: flavors in powder and liquid form	Encapsulation of essential oils	Increased stability of the flavor
3	Instant beverages Powder form	Suspension of pulp Provides fiber	Good long-term stability
4	Flavored beverage with added pulp	Suspension of pulp Provides fiber	Good long-term stability
5	Wine fining and stabilization (oenology)	Clarification Tannin fixation	Improved clarity and improved color
6	Beer	Clarification Foam stabilization	Improved clarity Improved foam stability
7	Snacks: Dry roasted nuts Spicy nuts	Adhesive Glazing	Improved flavor adhesion and texture
8	Bakery glazing	Film forming Adhesive Emulsification	Improved glaze

Source: http://www.agrigum.com

usability, and usefulness, which makes it trying for food producers to choose the most proper emulsifier for a specific application. Hence, it is vital to describe the execution of various emulsifiers under institutionalized conditions to give the data producers need to settle on educated decisions while choosing emulsifiers.

Purchasers are progressively requesting label-friendly products manufactured from characteristic and feasible ingredients, which has fortified enthusiasm for supplanting engineered surfactants with normal emulsifiers in numerous food and refreshment applications (De Malik et al., 2015; McClements and Gumus, 2016; Ozturk and McClements, 2016).

Biopolymer-based emulsifiers, for example, amphiphilic proteins and polysaccharides, are especially encouraging natural candidates to frame and settle emulsions in food and different applications. Protein-based emulsifiers are, as of now, ordinarily utilized as a part of modern applications because of their extensive variety of sources and great emulsifying capacity (Dickinson, 2015). By the way, there are vital impediments to utilizing protein-based emulsifiers in numerous food items; for instance, protein-covered oil beads are typically flimsy at pH values around their isoelectric point, at high ionic qualities, and at raising temperatures (McClements and Gumus, 2016). Certain sorts of polysaccharide-based emulsifiers are likewise normally utilized as a part of the nourishment business as characteristic emulsifiers, for example, GA, corn fiber gum, and beet pectin (Ngouemazong et al., 2015).

GA is a standout among the most broadly utilized amphiphilic polysaccharides in the food industry (Reiner et al., 2010). Not at all like protein-based emulsifiers, GA is not exceptionally delicate to collect at intermediate pH values, high ionic qualities, and high temperatures because of the development of thick steric balancing-out layers around the oil beads. However, one significant issue is that it is hard to utilize low levels of polysaccharides to create emulsions with little bead sizes (Charoen et al., 2011). Also, at high emulsifier focuses, consumption flocculation brought on by nonadsorbed polysaccharide particles may contrarily influence the time frame of realistic usability of completed items.

The usage of GA secluded from the *A. senegal* tree has likewise been constrained because of variances in the accessibility, quality, and cost of the crude material (Williams et al., 2009). In this way, distinguishing and growing new sorts of polysaccharide-based emulsifiers that can be utilized to create consumable emulsions is essential.

Gum Arabic in dairy production

Colloidal dispersions are included in numerous food products. In most of these products, proteins and polysaccharides are available together, and both macromolecules add to the structure, surface, and stability (Dickinson, 1992). The proteins are the principle stabilizers, and the polysaccharides add to the stability by means of their thickening and steric balancing-out practices, thus of their surface movement or their collaboration with proteins. A few polysaccharides, including gums, have been utilized to upgrade emulsion dependability (Garti, 1999; Yilmazer et al., 1991).

The structure of GA likewise contains proteinaceous material (ca. 2 wt%) covalently appended to the polysaccharide moiety. This protein part makes GA's surface dynamic, and it is, for the most part, connected with a high molecular weight portion representing less than 30% of the aggregate gum. The major molecular fraction (F1) representing ~88% of the total acacia gum mass is an arabinogalactan peptide with a weight-average molecular weight of 2.86 × 105 g/mol. The molecular structure of F1 is really obscure. From small-angle neutron scattering experiments in charge screening conditions, F1 appeared to be a dispersion of two-dimensional structures with a radius of gyration of ~6.5 nm and an inner dense, branched structure (Sanchez et al., 2008). Therefore, GA is used in production of many dairy products such as ice cream, ice pops and water ices, yogurt, and sherbet.

The GA uses are considered minor and of little benefit dwarfed by the uses of the maximum benefits that many people did not expect that the GA do it. So far, initial research has shown that gum has a big role in helping telemedicine and helping many ills, and GA may be a catalyst for the treatment of many, especially kidney, urinary tract, and stomach diseases. Its uses include: stabilizer, sticky flavor and aroma, binder, emulsifier, stiffener, thickener, coater, glazier, carrier, creamer, foaming agent, suspensor, sensitizer, and base, and it works on keeping food free of iron; works on the conservation of germs and microbes; prevents change to the color of milk because it is not subject to oxidation; prevents crystallization of sugar in fluid milk when cooling; reduces the amount of cholesterol in the blood; is used in the sweets, jams, and beverages industries; and is used in the food of the diet industry to help people maintain weight.

Robe is the major fermented milk item in the Sudan. It is mostly made from surplus milk, and its utilization is a piece of Sudanese culture. Hamad et al. (2013) utilized GA as a stabilizer in processing of Robe. They processed gum Robe utilizing three levels of gum powder: 5%, 7.5%, and 10%. The examination demonstrated that there was no observable change in the chemical composition of Robe supplemented with 5% and 7.5% gum powder compared to the control Robe. In any case, the protein substance of the control sample was higher than that of Robe supplemented with 10% GA. The total soluble solids were profoundly expanded in gum Robe. Also, the increase of gum powder brought about diminishing of the moisture and relative increase of the ash content. In addition, the whey partition diminished with the expansion of gum. The sensory evaluation showed acceptance of Robe produced by adding gum powder because of the change of Robe texture. Bashir (2013) studied the impact of GA on enzymatic coagulation of camel milk with addition of calcium chloride and calcium phosphate independently. GA was added at six concentrations, 0.1%, 0.2%, 0.4%, 0.6%, 0.8%, and 1.0%, to full-fat camel milk. There was a significant difference ($p \leq 0.05$) in milk coagulation time among control and GA-treated samples, and among GA–treated samples. Whey volume isolated from the milk increases with expanding GA concentration in the case of calcium chloride and calcium phosphate additions.

GA is utilized as a stabilizer in frozen milk products such as ice cream due to its properties of water absorption. The role of GA in these products is that it causes the delicate tissue development by repressing the arrangement of ice crystals, which is achieved through mixing a lot of water for hydration, the highest melting point of the main attraction of ice cream (Montenegro et al., 2012).

GA is added to milk and other dairy products to produce the desired dish or beverage. This includes applications where GA is used to suspend the ingredient and impart richness and body to the drink. Gum Arabic is, to a great extent, utilized in the soft drinks industry for the stabilization of scattered flavor oils. It has been demonstrated that the protein-containing division adsorbs unequivocally at the oil–water interface (Randall et al., 1988). In this manner, GA not only improves the consistency of the fluid medium like most other polysaccharide gums, but it additionally frames a thick macromolecular layer around the emulsion beads, bringing about great steric stabilization (Montenegro et al., 2012).

GA can be utilized as a part of numerous food items in a mixture with food proteins. In bakery toppings, it counteracts overabundance dampness ingestion by icings or glazing. In frozen dairy items, for example, ice cream, it helps in the development and maintenance of small ice crystals. In refreshments, it works as an emulsion and froth stabilizer. It has a flavor fixative trademark in powdered sustenance details. Since the surface activity of GA is somewhat low, it can be utilized as a soundness enhancer in protein-balance out emulsions which are required to display low consistency (Montenegro et al., 2012).

Niamah et al. (2016) added five different concentrations of GA to yogurt product by probiotic culture starter, which included *Lactobacillus acidophilus*, *Bifidobacterium bifidum*, and *Streptococcus thermophilus*. The best concentration of added GA was 1%. After 1 day of fermentation, pH values decreased to 4.52, while the total acidity was 0.91%. The water-holding capacity (WHC), susceptibility to syneresis (STS), and whey drainage (WD) were 54%, 68%, and 0 mL, respectively. The cells viability of *L. acidophilus* was 7.1 Log CFU/g, 7.15 Log CFU/g of *B. bifidum*, and log 7.72 Log CFU/g of *S. thermophilus* after fermentation. At 21 days (the end of the storage period), the results of yogurt with 1% GA, pH values decreased to 4.05, while total acidity increased to 1.15%. The WHC, STS, and WD were 50.5%, 71%, and 1.2 ml. The cell viability of starter bacteria decreased with *B. bifidum* being affected. When GA was added to the yogurt, the viability of probiotic bacteria was increased after production.

Use of gum Arabic in other food industries

Hydrocolloids, for example, GA, are widely used in many food formulations to improve quality attributes and shelf-life. These include thickening and gelling, emulsifying, stabilizing, and controlling the growth of crystals of ice and sugar, although the basic characteristics of those hydrocolloids allow them to have widespread use in thickening and gelling. As thickening agents, they are used in soups, gravies, salad dressings, sauces, and toppings, while as gelling agents, they are extensively used in products like jam, jelly, marmalade, restructured foods, and low-sugar/calorie gels. Hydrocolloids scattered in the water to give the impact of thickness, or consistently delivered. This water-thickening property is normal for all hydrocolloids and is the primary purpose behind use when all is said and done. Over the thickness fluctuates with the sort and nature of hydrocolloids, with a couple give a low consistency at high concentration to a specific degree, but the majority of them give a high thickness in the center, less than 1% (Saha and Bhattacharya, 2010).

GA has wide mechanical uses: as a stabilizer, a thickener, an emulsifier, and in added substances in the food industry as well as in textiles, earthenware production, lithography, and restorative and pharmaceutical industries (Verbeken et al., 2003). GA has numerous properties that prompted its utilization in production of many food products. These include:

a. Multifunctional: extraordinary emulsifier, film-previous, texturizer, and low-consistency water folio and building agent;
b. High source of fiber: contains over 85% soluble dietary fiber (dry basis);
c. High percentage purity: no added substances; free from the dregs and polluting influences; has, to a great degree, low bacterial count;
d. Fast wetting and simplicity of dispersion: accessible in prehydrated or agglomerated frame; and
e. Natural labeling: GA is not chemically changed and meets all requirements for "common" naming or "no counterfeit added substances" claims (http://www.preparedfoods.com).

Flavor encapsulation

GA is utilized widely in the flavor and drink industry, where its normal properties are used completely. Emerging from its minimal structure, GA is likewise phenomenal for the embodiment of flavors, as it emulsifies, has low thickness and flat flavor, and secures against flavor oxidation. GA is the perfect transporter in flavor packaging on account of normal emulsifying properties and elements at first glance, and great maintenance of unpredictable flavor parts, high dissolvability in water (up to half), corrosive strength, and acid stability. Extra points of interest include: a wonderful flavor, low hygroscopicity, and the capacity to shield the flavors from oxidation. GA is used to enhance flavor of many spray-dried products, such as TIC gums, TIC Pretested products, and FT GA powder. It is perfect for making epitomized splash-dried flavors with the claim of non-GMO. These are low in sediment and have very low bacteria counts due to TIC's manufacturing process, which follows a Hazard Critical Control Points program (https://www.google.com/patents/US20060088574).

Gum Arabic in confectionery

GA offers a wide range of applications in confectionery as a texturing agent and a film-forming agent, as indicated in Fig. 13.2A and 13.2B; the major property utilized is the natural ability of GA to produce and stabilize emulsions. This usefulness emerges from the high-molecular-weight arabinogalactan–protein complex, which is an inherent part of the GA. The most popular confectionery store depends on acacia gum pastilles. In addition to having a nonsticking effect on teeth, enhancing flavor discharge, and amplifying the life of flavors, acacia gum effortlessly works with sugarless detailing, subsequently flawlessly noting the expanding interest for more

Figure 13.2 Gum Arabic powder.

benefits, sugarless and safe-for-teeth confectionery. GA is regularly utilized as a part of delicately formed confections, such as gum drops, and chewy assortments of candy as well. Without sugar, hard confections and low-fat frozen yogurt can likewise be made with GA (http://www.nexira.com/Confectionery_22.html).

Gum Arabic in snack foods

GA is a lubricant and a binder in extruded cereal snacks, at levels from 2%–5%. At 15%–40% levels, GA provides adhesion of dry flavors in peanuts and similar products. In low-fat cake and muffin mixes, GA functions as a partial oil replacer as well as a moisture binder (Table 13.3).

Bakery products

GA is utilized in bakery and sweet roll glazing, forming improved sugar adhesives. Also, it is used in bread (80% concentration) as a baking improver, glazing stabilizer, and in soft cakes as softener and water-retention agent. It is used in dough and baking mixes, flat icing, cake batters and mix, cake filling and toppings, doughnut glaze, and meringue. GA functions in confections as an emulsifier, crystallized sugar, a bulking agent, and film former. In caramel dessert products of GA, the characteristics of emulsification aid in the reduction of fat, and thus the droplet size distribution provide stability for this product (Dauqan and Abdullah, 2013).

Table 13.3 Some gum Arabic confectionary products

No.	Product	Characteristics	Benefit
1	Diabetic sugarless confectionery	Main ingredient Noncaloric fiber source	Good texture
2	High-sugar confectionery	Prevention of sugar crystallization Main binding ingredient Provides fiber	Good texture
3	Reduced-sugar and no-sugar confectionery	Main binding ingredient Provides fiber	Good texture Low calorie
4	Caramels - Toffees	Improves chewiness Fat distribution	Improved texture Increased resistance to oxidation
5	Compressed sugar confectionery	Binder	Improved shape retention
6	Coated products: drageés, chewing gum, chocolate centers, nut centers, coated almonds, peanuts	Coating, glazing, adhesive engrossing, film forming, integration of oil–water interface	Improved glaze Harder coat
7	Chewing gum	Softening of the chews Retention of water Retention of flavor	Improved texture and body
8	Snacks: Dry roasted nuts Spicy nuts	Adhesive Glazing	Improved flavor adhesion and texture

Source: http://www.agrigum.com

Fruit juice and beverages

GA is broadly utilized to balance out emulsions, dietary fiber, and suspended solids and to enhance the grip of the froth glass while drinking. GA is utilized in preparation of soft drinks with fruit pulp, soft drinks (it binds the sugar to the drink; essentially, it's an edible glue), fining wines, juices, vinegar beer, artificially sweetened beverages, and dry beverage mixes (Jensen et al., 1993).

Nutritional products

GA is a low-calorie substance and cholesterol reducer. Numerous items require weight reduction additionally be utilized to expand the strength in huge amounts in the stomach by swelling of the food. GA has a lot of involvement in the field of weight control items. It has been utilized for over 3000 years, and is, for the most part, viewed as protected esteem. Engineered diabetic syrup and sugarless confections are produced (Mohamed et al., 2015).

Cooking products

GA can be consolidated in cooking items, for example, soups, meat items, and bean stew sauce. GA is an exceptionally successful water-binding agent. Fundamental qualities emulsification for all intents and purposes ineffectual in marketable items poured standard and low oil. Dynamic properties of the surface of the gums and emulsion stability of oil droplets provide a creamy mouth feel (*https://en.wikipedia.org/wiki/Gum_arabic*).

The legal requirements for gum Arabic

Various studies connected the quality prerequisites to the item's adequacy by buyers (Bernués et al., 2003). These studies concentrated on the identification of signs that convey information about the quality attributes of the item.

The buyer's prerequisites for GA to effectively market the product include:

- food safety: traceability, hygiene, and control
- contamination
- adulteration
- classification, labeling and packaging
- substances allowed in the European Union
- extraction solvents
- liability
- food safety certification
- Kosher and Halal certification
- documentation
- samples
- payment and delivery terms
- sustainability
- certification of organic production
- quality
- packaging
- labeling

Labeling requirements

It is of prime significance to mark the GA on the off-chance that it will be traded. The label on the item assists the purchasers and customs authorities to trace its starting points. Another reason is to guarantee security during transport and storage. The labels must include:

- product name (for instance, GA from *A. senegal*)
- batch number
- whether the product is destined for use in food products (food grade or not)
- place of origin
- the name and address
- date of manufacture

- best before date
- net weight
- recommended storage conditions

Packaging requirements

GA pulls in dampness and requires bundling in waterproof material. Papers or plastic (polypropylene) bags with a plastic (polyethylene, etc.) lining are used. (https://www.cbi.eu/market-information/natural-food-additives/gum-arabic).

In conclusion, as imperative food additives, hydrocolloids are used in an increasing number of applications in numerous food products, such as stability and thickening agents. Many natural products, such as GA, provide the effects of thickness. Because of its exceptional utilitarian properties; GA can be utilized as an additive in many food products as a good emulsifier, film former, texturizer, and low-viscosity water binder and bulking agent. In addition to its use as texturing additive, the combined functionality of GA is an advantage over other edible biopolymers that do not show antioxidant activities.

Fine emulsions could be created, utilizing the greater part of the polysaccharide-based emulsifiers. However, GA is effective at creating smaller bead sizes at low-emulsifier concentrations. This impact is credited to three noteworthy contrasts between the emulsifiers: (1) the higher surface activity and lower surface heap of GA; (2) the lower interfacial pressure for GA, and (3) the smaller atomic weight of GA.

References

Achouri, A., Zamani, Y., Boye, J.I., 2012. Stability and physical properties of emulsions prepared with and without soy proteins. J. Food Res. 1 (1), 254–267.

Bai, L., Huan, S., Gu, J., McClements, D.J., 2016. Fabrication of oil-in-water nanoemulsions by dual-channel microfluidization using natural emulsifiers: saponins, phospholipids, proteins, and polysaccharides. Food Hydrocoll. 61, 703–711.

Bashir, A.Y.M., 2013. Effect of Gum Arabic on Enzymatic Coagulation of Camel Milk. MSc. Thesis. Sudan University of Science & Technology, Khartoum, Sudan.

BeMiller, J.N., 1988. Some challenges for gums and gum research. Phillips, G.O., Wedlock, D.J., Williams, P.A. (Eds.), Gums and Stabilizers for the Food Industry, 4, IRL Press, Oxford, UK, pp. 3–14.

Bernués, A., Olaizola, A., Corcoran, K., 2003. Extrinsic attributes of red meat as indicators of quality in Europe: an application for market segmentation. Food Qual. Prefer. 14 (4), 265–276.

Charoen, R., Jangchud, A., Jangchud, K., Harnsilawat, T., Naivikul, O., McClements, D.J., 2011. Influence of biopolymer emulsifier type on formation and stability of rice bran oil-in-water emulsions: whey protein, gum Arabic, and modified starch. J. Food Sci. 76 (1), 165–172.

Dauqan, E., Abdullah, A., 2013. Utilization of gum Arabic for industries and human health. Am. J. Appl. Sci. 10 (10), 2013.

De Malik, S., Ghosh, A., Saha, R., Saha, B., 2015. A review on natural surfactants. RSC Adv. 5 (81), 65757–65767.

Dickinson, E., 1988. The role of hydrocolloids in stabilizing particulate dispersions and emulsions. Phillips, G.O., Wedlock, D.J., Williams, P.A. (Eds.), Gums and Stabilizers for the Food Industry, 4, IRL Press, Oxford. UK, pp. 249–263.

Dickinson, E., 1992. An Introduction to Food Colloids. Oxford University Press, Oxford, England, U.K, (140–170).

Dickinson, E., 2015. Colloids in food: ingredients, structure, and stability. Annu. Rev. Food Sci. Technol. 6, 211–233.

Dickinson, E., Stainsby, G., 1988. Advances in Food Emulsions and Foams. Elsevier Applied Science Publishers. LTD, London, UK, pp. 397.

Flindt, C., Al-Assaf, S., Phillips, G.O., Williams, P.A., 2005. Studies on acacia exudate gums. Part V. Structural features of *Acacia seyal*. Food Hydrocoll. 19 (4), 687–701.

Garti, N., 1999. Hydrocolloids as emulsifying agents for oil-in-water emulsions. J. Dispers. Sci. Technol. 20, 327–355.

Garti, N., Reichman, D., 1993. Hydrocolloids as food emulsifiers and stabilizers. Food Struc. 12 (4), 411–426.

Hamad, H.E., Sulieman, A.E., Salih, Z.A., 2013. Quality aspects of the Sudanese fermented milk (robe) supplemented with gum Arabic powder. Disc. J. Agric. Food Sci. 1 (1), 8-7.

Hassan, E.A., Al Assaf, S., Phillips, G.O., Williams, P.A., 2005. Studies on Acacia gums: Part III molecular weight characteristics of *Acacia seyal* var. seyal and *Acacia seyal* var *fistula*. Food Hydrocoll. 19 (4), 669–677.

Imeson, A., 2011. Food stabilizers. Thickeners and Gelling Agents. John Wiley & Sons, Minneapolis, MN55426, USA.

Jensen, C.D., Spiller, G.A., Gates, J.E., Miller, A.F., Whittam, J.H., 1993. The effect of acacia gum and a water-soluble dietary fiber mixture on blood lipids in humans. J. Am. Coll. Nutr. 12 (2), 147–154.

Kralova, I., Sjoblom, J., 2009. Surfactay: a review. J. Dispers. Sci. Technol. 30 (9), 1363–1383.

Lelon, J.K., Jumba, I.O., Keter, J.K., Wekesa, C., Oduor, F.D.O., 2010. Assessment of physical properties of gum Arabic from *Acacia Senegal* varieties in Baringo District. Kenya. Afr. J. Plant Sci. 4, 95–98.

McClements, D.J., 2015. Food Emulsions: Principles, Practices, and Techniques. CRC Press, Boca Raton, Florida, USA.

McClements, D.J., Gumus, C.E., 2016. Natural emulsifiers and biosurfactants, phospholipids, biopolymers, and colloidal particles: molecular and physicochemical basis of functional performance. Adv. Coll. Inter. Sci. 234, 3–26.

Mohamed, R.E., Mohammed, O., Gadour, M.O., Adam, I., 2015. The lowering effect of Gum Arabic on hyperlipidemia in Sudanese patients. Front Physiol. 6, 160.

Montenegro, M.A., Boiero, M.L., Valle, L., Borsarelli, C.D., 2012. Gum Arabic: More than an Edible Emulsifier. In: Verbeek, Dr., J. (Ed.), Products and Applications of Biopolymers. InTech, InTech, Janeza Trdine 9, 51000 Rijeka, Croatia.

Ngouemazong, E.D., Christiaens, S., Shpigelman, A., Van Loey, A., Hendrickx, M., 2015. The emulsifying and emulsion-stabilizing properties of pectin: a review. Compre. Rev. Food Sci. Food Saf. 14 (6), 705–718.

Niamah, A.K., Al-Sahlany, S.T.G., Al-Manhel, A.J., 2016. Gum Arabic uses as prebiotic in yogurt production and study effects on physical, chemical properties and survivability of probiotic bacteria during cold storage. World Appl. Sci. J. 34 (9), 1190–1196.

Ozturk, B., McClements, D.J., 2016. Progress in natural emulsifiers for utilization in food emulsions. Curr. Opin. Food Sci. 7, 1–6.

Randall, R.C., Phillips, G.O., Williams, P.A., 1988. The role of the proteinaceous component on the emulsifying properties of gum Arabic. Food Hydrocoll. 2, 131–140.

Reiner, S., Reineccius, G., Peppard, T., 2010. A comparison of the stability of beverage cloud emulsions formulated with different gum Acacia-and starch based emulsifiers. J. Food Sci. 75 (5), 236–246.

Saha, D., Bhattacharya, S., 2010. Hydrocolloids as thickening and gelling agents in food: a critical review. J. Food Sci. Technol. 47 (6), 587–597.

Sanchez, C., Schmitt, C., Kolodziejczyk, E., Lapp, A., Gaillard, C., Renard, D., 2008. The acacia gum arabinogalactan fraction is a thin oblate ellipsoid: a new model based on small-angle neutron scattering and ab initio calculation. Biophys. J. 94 (2), 629–639, 15.

Siddig, N.E., Osman, M.E., Al-Assaf, S., Phillips, G.O., Williams, P.A., 2005. Studies on acacia exudates gum, part IV. Distribution of molecular components in *Acacia seyal* in relation to *Acacia senegal*. Food Hydrocoll. 19 (4), 679–686, siteresources. worldbank. org/ INTAFRMDTF/Resources/Gum_Arabic_Policy_Note. pdf.

Verbeek, C.J.R., 2012. Products and Applications of Biopolymers. InTech, Janeza Trdine 9, 51000 Rijeka, Croatia, Dr. Johan Verbeek (Ed.), ISBN: 978-95351-0226-7.

Verbeken, D., Dierckx, S., Dewettinck, 2003. Exudate gums: occurrence, production, and applications. Appl. Microbiol. Biotechnol. 63 (1), 10–21.

Williams, P.A., Phillips, G.O., 2009. Gum Arabic. Handbook of Hydrocolloids, second ed. Woodhead Publishing, Cambridge, UK, 173.

Williams, P.A., Phillips, G.O., Randall, R.C., 1990. Structure-function relationships of gum Arabic. Phillips, G.O., Wedlock, D.J., Williams, P.A. (Eds.), Gums and Stabilizers for the Food Industry, 5, IRL Press, Oxford, pp. 25–36.

Yang, Y., Leser, M.E., Sher, A.A., McClements, D.J., 2013. Formation and stability of emulsions using a natural small molecule surfactant: *Quillaja saponin* (QNaturale®). Food Hydrocoll. 30 (2), 589–596.

Yilmazer, G., Carrillo, A.R., Kokini, J.L., 1991. Effect of propylene glycol alginate and xanthan gum on stability of o/w emulsions. J. Food Sci. 56, 513–517.

Further Reading

Howlett, B., MCcarthy, M., 2002. An examination of consumers' perceptions of organic yoghurt. Agribusiness Discussion Paper 38. Department of Businesses and Development, University College, Cork, S.O. http://www.agrigum.com http://www.nexira.com/Confectionery_22. html http://www.preparedfoods.com/articles/106138-focusing-on-gum-arabic. https:// en.wikipedia.org/wiki/Gum_arabic. https://www.cbi.eu/market-information/natural-food-additives/gum-arabic https://www.google.com/patents/US20060088574.

Effect of Gum Arabic (Prebiotic) on Physicochemical and Organoleptic Properties of Yogurt (Probiotic)

14

Mohamed A. Talib, Osama A. Rayis, Emadeldin H.E. Konozy, Mohamed A. Salih
Africa City of Technology, Khartoum Sudan

Introduction

Probiotics are microorganisms that give health advantages when used. The term probiotic is presently utilized to name ingested microorganisms related with useful impacts to people and different animals (Rijkers et al., 2011). Prebiotics stimulate the growth and activity of beneficial bacteria of the intestinal flora. Gum Arabic (GA) is one of the best-studied of the nonabsorbable carbohydrates. GA is therefore a prebiotic. Probiotics have received renewed attention recently from product manufacturers, research studies, and consumers. The history of probiotics can be traced back to the first use of cheese and fermented products that were well known to the Greeks and Romans, who recommended their consumption (Gismondo et al., 1999). The fermentation of dairy foods represents one of the oldest techniques for food preservation (Azizpour et al., 2009).

Acacia senegal and *Acacia seyal* trees are the primary sources of GA. These species develop actually in the semiarid sub-Saharan districts of Africa. There are more than 1000 types of acacia, and a synopsis of their botanical classification was accounted for by Phillips and Williams (1993). GA is, by and large, generally utilized for industrial purposes, for example, as a stabilizer, a thickener, an emulsifier, and an embodying in the food business, and to a lesser degree in textiles, ceramics, lithography, cosmetic, and pharmaceutical industries. In the food business, GA is basically utilized in candy, pastry, dairy, refreshment, and as a microencapsulating agent (Verbeken and Dierckx, 2003). GA has high water solubility and a moderately low consistency contrasted to other gums. Most gums can't break up in water in concentrations over 5% because of their high viscosity. Rather, GA can get broken up in water in a concentration of half w/v, forming a liquid solution with acidic properties (pH 4.5). The very branched structure of the GA particles prompts minimized moderately little hydrodynamic volume, and therefore, GA will just turn into a viscous solution at high concentrations. Solutions containing less than 10% of GA have a low viscosity and react to Newtonian behavior (Williams et al., 1990).

GA tests created by spray drying and drying rollers have preference over crude gum, as they are, for all intents and purposes, free of microbial contamination and break up

Gum Arabic. http://dx.doi.org/10.1016/B978-0-12-812002-6.00014-2

substantially quicker (Mariana et al., 2012). There are a few contrasts between the chemical composition of the GA taken from *A. senegal* and *A. seyal*. Truth be told, both gums have a similar sugar deposits; however, *A. seyal* gum has a lower substance of rhamnose and glucuronic acid and a higher substance of arabinose and 4-O-methyl glucuronic acid than *A. senegal* gum. Rather, *A. seyal* gum contains a lower extent of nitrogen, and specific rotations are also completely different. The evaluation of the last parameters may plainly detect the distinction between the two species (Osman et al., 1993).

Chemically, GA is an intricate blend of macromolecules of various sizes and composition (mostly sugars and proteins). Today, the properties and components of GA have been broadly investigated and created, and GA is being utilized as a part of an extensive variety of industrial sectors, for example, textiles, ceramics, lithography, cosmetics, pharmaceuticals, encapsulation, food, etc. As to the food industry, it is utilized as a stabilizer and a thickener as well as an emulsifier (e.g., soft drink syrup, gummy candies, and creams) (Verbeken and Dierckx, 2003).

Sales of probiotic products have a rising trend from 2010 to 2014, increasing globally by 35% from 23.1 billion USD to 31.3 billion USD. Some regions have increased their use by even more than the average, including Eastern Europe (67%), Asia Pacific (67%), and Latin America (47%), comprising nearly half of probiotics sold globally in 2014. By geographic region, the leading consumers of probiotics in 2014 were Western Europe (8.3 billion USD), Asia Pacific (7 billion USD), Japan (5.4 billion USD), Latin America (4.8 billion USD), North America (3.5 billion USD), and Eastern Europe (2.3 billion USD). Live probiotic cultures are available in fermented dairy products and probiotic fortified foods (Mariana et al., 2012). The objective of the research is to study the effect of GA as prebiotic on the physicochemical and organoleptic properties of yogurt (probiotic).

Materials and methods

Experimental design

The experiment was conducted using 1 × 3 × 1 factorial arrangement in a completely randomized design with three replications. The factors are 1 (one type of cow's milk), 3 (concentration of GA: 0.5, 1.0, and 1.5%), and 1 sampling times (1 day after processing).

Study area

The study was conducted at the Dairy Technology Department, Faculty of Animal Production, Sudan University of Science and Technology, Khartoum, Sudan.

Source of milk

Milk was obtained from a herd of cross (Kenana Frisian) cattle at Dairy Technology Department. The milk will also be collected using 20-L-capacity metallic containers and transported to the laboratory.

Source of gum Arabic

GA samples were obtained from shopping centers of Khartoum, collected in sterile glass bottles with screw caps, and then sent to laboratory.

Yogurt manufacturing

According to the method suggested by OConnor (1995).

Clotting time and curd formation of yogurt

The clotting time of the yogurt was determined by visual observation at 45°C using a constant temperature incubator. Curd formation was observed by rotating the container.

Organoleptic properties of yogurt

Organoleptic properties of yogurt samples were determined using a 10-member trained test panel as described by Williams (1982) (modified) considering the total score of 100 points, which includes flavor 20 points; body and texture, 20 points; color, 20 points; appearance, 20 points; and taste, 20 points. A score below 40 should not be considered acceptable.

Results and discussion

Organoleptic properties of yogurt

Table 14.1 summarizes the average score of the organoleptic properties of yogurt produced with three different concentrations of GA.

The overall percentage scores of the organoleptic parameters recorded increased gradually (71.8, 80.4, 82.3, and 85.3%) for yogurt produced with three different concentrations of GA (control, 0.5, 1.0, and 1.5%), respectively. This means that as the concentration of GA increased organoleptic parameters increased, too. The result also show that yogurt produced using (1.5%) GA had acceptable characteristics (color,

Table 14.1 Average score of organoleptic properties of yoghurt produced with three different concentration of gum Arabic

Property (points scale)	Gum Arabic concentrations (%)			
	Control	**0.5**	**1.0**	**1.5**
Appearance (20)	13.9	16.2	16.2	16.9
Color (20)	15.2	16.2	16.7	17.3
Body and texture (20)	13.0	15.7	16.5	16.9
Flavour (20)	15.1	16.3	16.7	17.4
Taste (20)	14.6	16.0	16.2	16.8
Percentage	**71.8**	**80.4**	**82.3**	**85.3**

Table 14.2 Clotting time (min) and curd firmness of milk during incubation at 45°C

Property	Gum Arabic concentrations (%)			
	Control	**0.5**	**1.0**	**1.5**
Clotting time	120	110	100	90
Curd firmness	Jelly	Semifirm	Firm	Firm

17.3; body and texture, 16.9; flavor, 17.4, appearance, 16.9; and taste, 16.8). The curd obtained with high concentrations of GA was found to be firmer than curd obtained with low concentrations (0.5% and control), which had a weak and jellylike texture. The results of this investigation indicated that 1.5% GA satisfied all the previous characteristics in the manufacture of yogurt, being scored by 85.3% of the respondents.

Effect of gum Arabic concentration on milk clotting time

Table 14.2 shows the effect of different concentrations (0.5, 1.0, and 1.5%) of GA on the physicochemical properties of yogurt. Significant differences were recorded on clotting time due to the GA treatments. Increasing the amount of GA concentrations in the milk significantly decreased the clotting time in all treatments. Yogurt produced with low concentrations of gum had longer coagulation time (100–110 min) than that produced with high concentration (90 min). The curd obtained with high concentrations of gum (1 and 1.5%) was found to be firmer than curd obtained with a lower concentration (0.5%), which had a weak jelly texture. The results revealed that increasing the concentration of gum decreased clotting time. The observed activity was directly proportional to the gum concentration.

Based on the obtained results, it can be concluded that the overall percentage scores of the organoleptic parameters recorded increased gradually for yogurt produced with three different concentrations of GA. The results of this investigation indicated that 1.5% GA satisfied all the previous characteristics in the manufacture of yogurt, being scored by 85.3% of the respondents. Since live probiotic cultures (bacteria) are available in fermented dairy products and probiotic-fortified foods, combinations of probiotic and prebiotic in capsules, powders, and sachets in freeze-dried form are highly recommended.

References

Azizpour, K., Bahrambeygi, S., Mahmoodpour, S., 2009. History and basic of probiotics. Res. J. Biol. Sci. 4 (4), 409–426.

Gismondo, M., Drago, L., Lombardi, A., 1999. Review of probiotics available to modify gastrointestinal flora. Int. J. Antimicrob. Agents 12 (4), 287–292.

Mariana, A., María, L., Boiero, L., Claudio, D.B., 2012. Gum Arabic: More than an edible emulsifier, Products and Applications of Biopolymers, Dr. Johan Verbeek (Ed.), ISBN: 978-953-51-0226-7.

OConnor, C.B., 1995. Rural Dairy Technology manual. International Livestock Research Institute, Addis Ababa, Ethiopia.

Osman, M., Williams, P., Menzies, A., Phillips, G., 1993. Characterization of commercial samples of gum Arabic. J. Agric. Food Chem. 41 (1), 71–77.

Phillips, G., Williams, P., 1993. In: Nishinari, K., Doi, E. (Eds.), Food Hydrocolloids; Structures, Properties and Functions. Plenum Press, New York, pp. 45–61.

Rijkers, G., de Vos, W., Brummer, R., Morelli, L., Corthier, G., Marteau, P., 2011. Health benefits and health claims of probiotics: bridging science and marketing. Br. J. Nutr. 106 (9), 1291.

Verbeken, D., Dierckx, S., 2003. Exudate gums: occurrence, production, and applications. Appl. Microbiol. Biotechnol. 63 (1), 10–21.

Williams, A., 1982. Scoring Methods used in the Sensory Analysis of Foods and Beverages at Long Ashton Research Station. Journal of Food Technology 17, 163–173.

Williams, P., Phillips, G., Stephen, A., 1990. Spectroscopic and molecular comparisons of three fractions from *Acacia senegal* gum. Food Hydrocolloids 4 (4), 305–311.

Further Reading

Adesiyun, A., Webb, I., Rahaman, S., 1995. Microbial quality of raw cow's milk at collection centre in Trinidad. J. Food Prot. 58 (2), 139–146.

Allgeyer, L., Miller, M., Lee, S., 2010. Sensory and microbiological quality of yogurt drinks with prebiotics and probiotics. J. Dairy Sci. 93, 4471–4479.

Birkett, A.M., Francis, C.C., 2010. Short-chain fructo-oligosaccharide: a low molecular weight fructan. In: Cho, S.S., Finocchiaro, E.T. (Eds.), Handbook of Probiotics and Probiotic Ingredients: Health Benefits and Food Applications. CRC Press, USA, pp. 13–42.

Horne, D.S., 2003. Casein micelles as hard spheres: limitations of the model in acidified gel formation. Colloids Surf. A Physicochem. Eng. Aspects 213, 255–263.

Ibrahim, A.H., Khalifa, S.A., 2015. The effects of various stabilizers on physiochemical properties of camel's milk yoghurt. J. Am. Sci. 11 (1), 15–24.

Meilgaard, M., Civille, G.V., Carr, B.T., 1999. Sensory Evaluation Techniques, third ed. CRC Press, Bacon Raton, Florida, pp. 387.

Nauta, A., Bakker-Zierikzee, A.M., Schoterman, M.H., 2010. Galacto-oligosaccharides. In: Cho, S.S., Finocchiaro, E.T. (Eds.), Handbook of Prebiotics and Probiotics Ingredients: Health Benefits and Food Applications. CRC Press, USA, pp. 75–94.

Niamah, A.K., Al-Sahlany, S.H., Al-Manhel, A.J., 2016. Gum Arabic uses as prebiotic in yogurt production and study effects on physical, chemical properties and survivability of probiotic bacteria during cold storage. World Appl. Sci. J. 34 (9), 1190–1196.

Paineau, D., Respondek, F., Bouhnik, Y., 2010. Inulin and oligosaccharides: a special focus on human studies. In: Cho, S.S., Finocchiaro, E.T. (Eds.), Handbook of Prebiotics and Probiotics Ingredients: Health Benefits and Food Applications. CRC Press, USA, pp. 43–74.

Patel, S., Goy, A., 2015. Applications of natural polymer gum Arabic: a review. Int. J. Food Prop. 18, 986–998.

Phillips, G.O., 1998. Acacia gum (Gum Arabic): a nutritional fiber; metabolism and calorific value. Food Addit. Contam. 15, 251–264.

Piggot, J.R., 1988. Sensory Analysis of Foods. Elsevier Applied Science, New York, pp. 426.

Saint-Eve, A., Levy, C., Martin, N., Souchon, I., 2006. Influence of proteins on the perception of flavored stirred yogurts. J. Dairy Sci. 89, 922–933.

Tamime, A.Y., Robinson, R.K., 1999. Yoghurt Science and Technology. CRC Press, New York, pp. 129–131.

Extending Cloud Stability of *Tamarindus indica* L. Juice Using Sodium Alginate and Gum Arabic During Storage in the Refrigerator

15

Hanan S. Fagury, Mohamed A. Talib, Osama A. Rayis, Khalid H. El-Hag
Africa City of Technology, Khartoum, Sudan

Introduction

Fruits and vegetables are extremely thick sources of bioactive compounds and are essential for a healthy eating plan. Currently, fruit and vegetable juices are increasing much interest as nonalcoholic drinks and are utilized as substitutes for caffeine-containing beverages, for example, coffee, black and green tea, and carbonated soft drinks (Azam et al., 2013). Cloud in natural product juice is coveted, and also, almost no cloud is viewed as a quality defect. The cloud contributes a significant part of the color, flavor, turbidity, and the aroma of citrus juice. It is made out of protein, pectin, lipid, hemicelluloses, cellulose, and other minor constituents. Turbidity, or a cloud in juice, is a physical characteristic of good quality in citrus juice and its items. The deterioration of citrus that occurred in the purification procedure was referred to the activity of pectolytic proteins, bringing about the development of a precipitate of calcium (Baker and Cameron, 1999). The utilization of cloudy juice has high nutritional significance in light of the fact that there is a high intent for the greater part of the nutrients to stay in the colloidal suspensions, that is, the capacity of cloudy carrot juice. Juices containing distinctive measures of cloud particles offer the consumer the best understanding of an alternate taste compared with filtered juices (Telesphore and He, 2009). The decrease in juice stability is an indicator of low quality. Ordinarily, the decline of juice stability is because of the precipitation of juice. Also, the precipitate may extremely influence the nutritional value of juices by taking flavor components and clouding substances (Sinchaipanit et al., 2013).

Gum Arabic is the dried gummy exudate from the stems and branches of *Acacia senegal* or of other similar species of *Acacia* (Fam: Leguminosae) (Dondain and Phillips, 1999). Gum Arabic is certified to be utilized as a food additive by the US Food and Drug Administration and is on the list of substances that are generally recognized as safe, with specific limitations (FDA, 1974). Gum Arabic is the most established and best known of all the tree gum exudates from specific types of *Acacia*, family Leguminosae. The chemical and physicochemical properties of this gum can

Gum Arabic. http://dx.doi.org/10.1016/B978-0-12-812002-6.00015-4

change and are altogether contingent upon the source, tree age, time of exudation, sort of storage, and climatic conditions (Islam et al., 1997). Gum Arabic is broadly utilized as a part of food processing, basically to give alluring qualities through its impact on thickness, body, and surface. In fluid frameworks, it is utilized as an emulsifier, stabilizer, thickener, glue, and to counteract crystallization of sugar and the arrangement of ice crystals. Different hydrocolloids are likewise utilized for these reasons (Mothe and Rao, 1999).

Gum Arabic promptly breaks down in cold and hot water in concentration up to half. In view of the reduced, expanded structure and, in this way, little hydrodynamic volume, gum Arabic solutions are portrayed by a low viscosity, permitting the utilization of high gum concentrations in different applications. Solutions display Newtonian behavior at concentrations up to 40% and progress toward becoming pseudoplastic at higher concentrations (Verbeken et al., 2003). Past reviews indicated addition of hydrocolloids at various concentrations was minimum and relied on upon the adaptability of juice formulations and the method for juice extractions. More prominent levels of hydrophilic colloids give hurtful impacts to the structure of the product (Sinchaipanit et al., 2013).

Hydrocolloids have been broadly utilized as part of refreshments and juice, generally to work as rheological modifiers, suspension agents, and stabilizers. Cloudiness can be maintained amid prolonged storage by treating the juices with hydrocolloids. Hydrocolloids are expended in food suspensions types because of their useful properties, such as water-holding capacity, the direction of the rheological characters, and ionization in liquid solutions, prompting the stability of insoluble particles (Somogyi, 1996). Hydrocolloid communicated with protein in aqueous solution and impacted security properties in three distinctive ways. Low concentration of gum with net repulsive reaction prompted miscibility, while high concentration with net attractive prompted coacervation (complexation). Thermodynamic inconsistency created an isolated phase of protein and hydrocolloid-rich phases (Dickinson, 2003).

Fruit juices are clarified during processing in order to avoid undesirable turbidity, haze, and sediments in the final products. Visual perception of turbidity and haze in fruit juices is the result of light scattering caused by suspending substances. The immediate turbidity in freshly pressed fruit juices is generally considered to be a result of suspended pectin particles stemming from the plant cell walls, but other disrupted cell wall and cell materials may also contribute to juice turbidity (Binning and Possmann, 1993). In contrast, the development of turbidity during cold storage, usually referred to as haze formation, is assumed to be caused by interactions between haze-active proteins and polyphenols that form insoluble multimolecular structures (Siebert, 2006; Siebert et al., 1996).

Whatever the reason for clarification might be, the only protecting aspect is the use of a thermal treatment to inactivate the enzymes. The pectolytic enzymes were shown to be inactivated by heating the product from 90.5°C to 93.3°C. Such treatment would affect the organoleptic properties of the product, since it produces a cooked flavor. Therefore, any degrade of stability due to thermal treatment could adversely affect the quality of the final processed product (Chen et al., 1993). Many substances, known as complex polysaccharides, are capable of making water viscous. The most important functions performed by these materials are retention of water and regulation of viscosity. Such material could enhance the cloud stability of citrus juice concentrates. The most effective plant hydrocolloids include: plant exudates such as gum Arabic, seaweed extracts such as alginates, and seed gums such as guar gum. Pectin is the naturally occurring colloidal stabilizer that gives body to juices. When this colloid is degraded, the juices clarify and become watery. This leads to a settling of the colloidal suspended materials as well as the rapid separation of the coarse wares: insoluble solids called pulp (Baker and Cameron, 1999).

Tamarind (*Tamarindus indica* L.), a tropical fruit found in Africa and Asia, is highly valued for its pulp (Gunasena and Hughes, 2000). Tamarind fruit pulp has a sweet acidic taste due to a combination of high contents of tartaric acid and reducing sugars. The pulp is used for seasoning; in prepared foods; to flavor confections, curries, and sauces; and as a major ingredient in juices and other beverages (De Caluwe et al., 2009). The *T. indica* L. tree is underutilized because inadequate research efforts are directed at it. Tropical tree fruits are important in the supplementation and improvement of diets. The potential of indigenous tropical tree fruits has not been fully realized. Furthermore, low-cost processing techniques are essential for increased exploitation of tamarind. Tamarind is valued mostly for its fruit, especially the pulp, which is used for a wide variety of domestic and industrial purposes. Tamarind pulp is used as a raw material for the manufacture of several industrial products, such as tamarind juice concentrate, tamarind pulp powder, tartaric acid, pectin, tartrates, and alcohol (Kulkarni et al., 1993). The acidic pulp is used as a favorite ingredient in culinary preparations, such as curries, chutneys, sauces, ice cream, and sherbet in countries where the tree grows naturally. In India, the pulp is also eaten raw and sweetened with gar. The tropical forest and its adjoining savannah zone are rich in wild tree fruits, many of which are neither cultivated nor eaten in the zone. These lesser-known food plants are extremely rich in food value, while some have unusual tastes, not known among the cultivated species (Salih et al., 1991; Adeola and Aworh, 2007). In India, the pulp is also eaten raw and sweetened with sugar (Lotschert and Beese, 1994). Practically all parts of the plant are used in folk medicine, and it has numerous therapeutic applications in humans, including its usage as a digestive, tranquilizer, laxative, and expectorant (Komutarin, 2004). Tamarind pulp has a unique sour taste due to the natural occurrence of sugars and plant acids together. *T. indica* L. belongs to the Leguminosae family (Caesalpinioideae). It is originally from Africa and is grown in humid or arid, tropical and subtropical regions with an average annual temperature of 25°C; it requires great light intensity and is sensitive to cold. It is harvested by pulling the pod from its stalk. The first crop of fruits occurs between the 8th and the 12th year after planting. A mature tree may be capable of producing up to 175 kg (386 lb) of fruit per year. Extensive tamarind orchards in India produce 275,500 tons (250,000 MT) annually. Annual yields of 150 to 2000 kg per tree have been reported from well-managed groves (http://www.fao.org). The objective of the present research was to retard the occurrence of clarification in *T. indica* juice using two colloidal stabilizers (gum Arabic and sodium alginate). Also, the average appearance scores of *T. indica* L. juice were determined.

Materials and methods

Experimental design

The experiment was conducted using a 2 × 3 × 9 factorial arrangement in a completely randomized design with two replications, the factors being two stabilizers (sodium alginate and gum Arabic) at different concentrations of 0.1, 0.2, and 0.3% w/v and 1.0, 2.0, and 3.0% w/v, respectively, and nine different sampling times (0, 3, 6, 9, 12 15, 18, 21, and 24 days).

Study area

The study was conducted in the national research center and biotechnology lab of Africa City of Technology, Khartoum, Sudan.

Source of the samples

Tamarindus samples were selected by visual inspection. Good appearances without any defects, such as size, length, color, and nonbreaking of the shell, were used in this research. *T. indica* L. was purchased from South Kordofan (Sudan).

Extraction of Tamarindus indica *juice*

After being received, the seeds were lightly rinsed with distilled water to remove waste from seeds, then extracted with tap water for 30 min, at a ratio of 1:3, seed to tap water, respectively, at room temperature. Then the mixture was filtered, and 3% sugar and stabilizers were added to the juice. The homogenized juice was stored in plastic bottles in the refrigerator at 5°C.

Preparation of stabilizers and their concentrations

In this experiment, sodium alginate at concentrations of 0.1, 0.2, and 0.3% w/v and gum Arabic at concentrations of 1.0, 2.0, and 3.0% w/v were used as stabilizers. These colloidal stabilizers were added to *Tamarindus* juice and then bottled in a 200-mL plastic container. The bottles were shaken for 5 min. The control was also shaken for an equal period. Gum Arabic was heated in a solution at 75°C for 1 h to inactivate oxidases and peroxidases before being used in the study.

Storage of juice

Samples of *Tamarindus* juice, bottled in 200-mL plastic containers, were stored in the refrigerator at temperature of 5°C.

Sampling

Different samples of juice and control during storage were periodically (every 3, 6, 9, 12, 15, 18, 21, and 24 days) withdrawn from the bottles for the stability test after being shaken uniformly.

Determination of cloud stability in juice

Twenty mL of each juice sample were taken for stability tests. Samples were agitated uniformly and then centrifuged at 1700 rpm/15 min. The supernatant was measured at 540 nm against water as blank (Atkins et al., 1966).

Average score of appearance of Tamarindus indica *L. juice*

Average score of appearance of juice samples was determined at the end of storage using 10 members. Considering the total score of 10 points, a score below 4 was considered not acceptable.

Results and discussion

Effect of sodium alginate on cloud stability

The stability of *T. indica* L. juice treated with gum Arabic and sodium alginate at different concentrations was determined during storage at 5°C for 24 days by measuring the percentage of light transmittance. The results obtained are shown in Table 15.1.

From the results obtained in Table 15.1, it was observed that sodium alginate was able to inhibit the cloud separation for up to 15 days, whereas the optimum concentration of sodium alginate was found to be 0.3%. Samples of juice treated with 0.1 and 0.2% sodium alginate inhibited clarification of the prepared juice up to 9 days only when stored at 5°C. Results are in accordance with those stated by Mirhosseini et al. (2008), who reported that high absorbance was corresponded to the high level of cloudiness of emulsion. Therefore, the highest cloudiness that was observed in the juice with 0.3% (w/v) sodium alginate implied the highest amount of colloid.

Effect of gum Arabic on cloud stability

Results in Table 15.2 show that gum Arabic is usually used in high concentration. The effect of gum Arabic on the stabilization process of *T. indica* L. juice was found to be efficient. Results in Table 15.2 indicated that gum Arabic at a concentration of 3% was capable to retard clarification of the prepared juice up to 12 days when stored

Table 15.1 Effect of sodium alginate on cloud stability of *Tamarindus indica* L. juice

Treatment	Percentage of transmission during storage period (days)								
Additions	0	3	6	9	12	15	18	21	24
Control	0.8	1.0	2.5	3.9	4.2	6.3	9.0	12	14
Sodium alginate 0.1%	0.09	0.08	0.08	0.10	0.24	0.3	0.9	1.5	2.9
Sodium alginate 0.2%	0.09	0.07	0.08	0.10	1.97	3.90	5.22	7.53	8.71
Sodium alginate 0.3%	0.06	0.08	0.07	0.09	0.10	0.16	1.18	1.98	2.80

Table 15.2 Effect of gum Arabic on cloud stability of *Tamarindus indica* L. juice

Treatment	Percentage of transmission during storage period (days)								
Additions	0	3	6	9	12	15	18	21	24
Control	0.8	1.0	2.5	3.9	4.2	6.3	9.0	12	14
Gum Arabic 1.0%	0.09	0.09	0.18	2.18	3.10	3.35	4.50	5.6	6.7
Gum Arabic 2.0%	0.08	0.08	0.09	2.08	2.20	3.90	4.50	5.06	7.80
Gum Arabic 3.0%	0.08	0.07	0.09	0.17	0.19	2.07	2.1	3.4	6.5

at 5°C. Other concentrations (1 and 2%) of samples treated using gum Arabic inhibited cloud up to 6 days only. Therefore, results reveal that samples treated with 3% gum Arabic were the best to maintain juice stability, whereas the lowest appearance scores were observed in samples treated with 1 and 2% gum Arabic. *T. indica* juice is a widely consumed product due to its high nutritional value and desirable sensorial characteristics. The juice cloud, which is composed of finely divided particulates of pectin, cellulose, hemicellulose, proteins, and lipids in suspension, is considered a desirable characteristic, since it favorably affects turbidity, flavor, and the characteristic color of juice. Koksoy and Kilic (2004) and Paraskevopoulou et al. (2005) stated that in order to improve the cloudy stability of juices for prolonged periods, hydrocolloids have been added.

Average score of appearance of Tamarindus indica *L. juice*

The appearance scores of *T. indica* L. juice are presented in Table 15.3. On subjecting the products to panel evaluation, the scores were in favor of 0.3% sodium alginate and 3% gum Arabic. The appearance of *T. indica* L. juice depends mainly on the cloud stability.

Sensory evaluation of the juice after treatment with gum Arabic and sodium alginate was performed using average scores of appearance of juice. The average hedonic scale ratings of the judges were evaluated, and they suggest that the appearance of the juice after treatment was found to be significant. From the obtained results, it is clear that *T. indica* L. juice treated using 3% gum Arabic and 0.3% sodium alginate maintained stability and prevented juice clarification for 12 and 15 days, respectively, at 5°C. Similarly, Sharoba et al. (2012) found that papaya–apricot nectar containing gum Arabic had a high score in overall acceptability. The use of hydrocolloids increased cloudy juice yield and improved its quality. Both gum Arabic and sodium alginate

Table 15.3 Average score of appearance of *Tamarindus indica* L. juice

Stabilizers	Gum Arabic			Sodium alginate		
Concentrations %	1.0	2.0	3.0	0.1	0.2	0.3
Average score	4.0	5.5	8.2	5.7	6.0	9.2

stabilized the cloudy juice. As the concentration of both stabilizers in tamarind juice increased, the average score of appearance increased, too. Thus, the addition of food colloidal stabilizers to enhance the stability of local citrus beverages, prepared from our indigenous citrus trees, should be of primary importance to increase exploitation of tamarind.

It could be concluded that the application of various gums and stabilizers into different juices can stabilize the sensory attributes as well as nutritional quality of juices. More researches are required regarding the application of hydrocolloids in different juices as well as their effect on the nutritional attributes of juices.

References

Adeola, A.A., Aworh, O.C., 2007. Physico-chemical and nutritional properties of Nigerian Tamarind (*Tamarindus indica* L.) fruits. A paper presented at the 7th International Food Data Conference, Sao Paulo, Brazil.

Atkins, C., Rouse, A., El-Moore, 1966. Determination of cloud retention in frozen concentrated orange and grapefruit juices with various colorimeters. In: Proceedings of the Florida State Horticultural Societyp. 124, 68.

Azam, S., Hafiz, A., Muhammad, S., Hafiz, A., Rabia, R., 2013. Effect of various hydrocolloids on cloud stability and nutrition of carrot juice. J. Global Innov. Agric. Social Sci. 1 (1), 22–27, http://www.fao.org/Ag/agp/agpc/doc/gbase/data/Pf000172.HTM.

Baker, R., Cameron, R., 1999. Clouds of citrus juices and juice drinks. Food Technol. 53 (1), 64–69.

Binning, R., Possmann, P., 1993. Apple juice. In: Nagy, S., Chen, C.S., Shaw, P.E. (Eds.), Fruit Juice Processing Technology. AGSCIENCE, Inc, Auburndale, FL, USA, pp. 271–317.

Chen, C.S., Shaw, P.E., Parish, M.E., 1993. Orange and tangerine juices. In: Nagy, S., Chen, C.S., Shaw, P.E. (Eds.), Fruit Juice Processing Technology. AGSCIENCE, Inc, Auburndale, FL, USA., pp. 110–165.

De Caluwe, E., De Smedt, S., Assogbadjo, A.E., Samson, R., Sinsin, B., Van Damme, P., 2009. Ethnic differences in use value and use patterns of baobab (*Adansonia digitata* L.) in northern Benin. Afr. J. Ecol. 47, 433–440.

Dickinson, E., 2003. Hydrocolloids at interfaces and the influence on the properties of dispersed system. Food Hydrocoll. 17, 25–39.

Dondain, G., Phillips, G.O., 1999. The regulatory journey of gum Arabic. Food Ingredients J. Jpn. 179, 38–56.

FDA., 1974. Proposed affirmation of GRAS status for gum arabic. Fed. Regist. 39, 34203–34208.

Gunasena, H.P.M., Hughes, A., 2000. Tamarind. International Centre for Underutilized Crops, Southampton, United Kingdom, pp. 169.

Islam, A.M., Phillips, G.O., Sljivo, A., Snowden, M.J., Willians, P.A., 1997. A review of recent developments on the regulatory, structural and functional aspects of gum Arabic. Food Hydrocoll. 11, 493–505.

Koksoy, A., Kilic, M., 2004. Use of hydrocolloids in textural stabilization of a yoghurt drink, ayran. Food Hydrocoll. 18, 593–600.

Komutarin, T., 2004. Extract of the seed coat of *Tamarindus indica* inhibits nitric oxide production by murine macrophages in vitro and in vivo. Food Chem. Toxicol. 42 (4), 649–658.

Kulkarni, R., Gangaprasad, S., Swamy, G., 1993. *Tamarindus indica*: economically an important minor forest product. Minor Forest Prod. News 3, 6.

Lotschert, W., Beese, G., 1994. Tropical plants. Collins Photo Guide. Harper Collins Publishers, Hammersmith London, pp. 223.

Mirhosseini, H., Tan, C.P., Aghlara, A., Hamid, S.A., Yusof, S., Chern, B.H., 2008. Influence of pectin and CMC on physical stability turbidity loss rate, cloudiness and flavor release of orange beverage emulsion during storage. Carbohydr. Polym. 73, 83–91.

Mothe, C.G., Rao, M.A., 1999. Rheological behavior of aqueous dispersions of cashew gum and gum Arabic: effect of concentration and blending. Food Hydrocoll. 13, 501–506.

Paraskevopoulou, A., Boskou, D., Kiosseoglou, V., 2005. Stabilization of olive oil–lemon juice emulsion polysaccharides. Food Chem. 90, 627–634.

Salih, O.M., Nour, A.M., Harper, D.B., 1991. Chemical and nutritional composition of two famine food sources used in Sudan, Mukheit (*Boscia senegalensis*) and Maikah (*Dobera roxburghi*). J. Sci. Food Agric. 57, 367–377.

Sharoba, A.M., El-Desouky, A.I., Mahmoud, M.H., 2012. Effect of addition some hydrocolloids and sweeteners on flow behavior and sensory properties of papaya-apricot nectar blends. J. Food Process. Technol. 3, 170.

Siebert, K.J., 2006. Haze formation in beverages. LWT-Food Sci. Technol. 39, 987–994.

Siebert, K.J., Carrasco, A., Lynn, P.Y., 1996. Formation of protein–polyphenol haze in beverages. J. Agric. Food Chem. 44, 1997–2005.

Sinchaipanit, P., Kerr, W.L., Chamchan, R., 2013. Effect of sweeteners and hydrocolloids on quality attributes of reduced-calorie carrot juice. Available from: http://www.pak foods.

Somogyi, L.S., 1996. Direct food additives in fruit processing. In: Somogyi, L.P., Ramaswamy, H.S., Hui, Y.H. (Eds.), Biology, Principles, and Applications. American Food and Nutrition Center Davis, California.

Telesphore, M., He, Q., 2009. Optimization of processing parameters for cloudy passion fruit juice processing using pectolytic and amylolytic enzymes. Pak. J. Nutr. 8, 1806–1813.

Verbeken, D., Dierckx, S., Dewettinck, K., 2003. Exudate gums: occurrence, production and applications. Appl. Microb. Biotechnol. 63, 10–21.

Part III

Medical and Health Uses

Health Benefits of Gum Arabic and Medical Use

16

Abdelkareem A. Ahmed
University of Nyala, Nyala, Sudan

Introduction

The reports from the World Health Organization (WHO) indicated that approximately 80% of the world's population rely on traditional medicine for treatment of a variety of disorders. Several decades ago, the industrial countries have also witnessed an ascending trend in the consumption of complementary and alternative medicine (CAM), mainly herbal remedies (Mahomoodally, 2013; Mahomoodally and Muthoorah, 2014; Nunkoo and Mahomoodally, 2012). Herbal medicines consist of herbs, herbal substances, herbal mixtures, herbal extracts, and completed herbal products, which include parts of plants or different plant materials as active ingredients. Ninety percent of the Ethiopian population have a propensity to use medicinal plants remedies for their primary healthcare. Studies carried out in developed countries such as Canada and Germany also tend to reveal that at least 70% of the population of these countries have tried CAM at least once (Gurib-Fakim, 2006; Nunkoo and Mahomoodally, 2012). It is likely that the perceptive knowledge of medicinal plant remedies in traditional cultures has been developed through several steps of trial and error over many centuries and, along with the main cures, was guardedly transmitted verbally from generation to generation. Certainly, recent allopathic medication has its origin in this ancient medicine, and it is likely that a lot of vital new remedies will be developed and commercialized in the future from the African and Arabian biodiversity, since it has been up to till date by following the leads provided by traditional knowledge and experiences (Gurib-Fakim, 2006; Nunkoo and Mahomoodally, 2012; Suroowan and Mahomoodally, 2011; Wells, 2011).

The widespread utilization of traditional medicine both in Africa and the Middle East, composed mainly of medicinal plants, has been documented to be associated with economic and cultural origins. Therefore, WHO supports both African and Arabian member countries incorporating and promoting the traditional medical practices in their health systems. These medicinal plants contain mixtures of various phytochemicals that are known as intermediate metabolites that might serve independently, additively, or synergistically to improve potential health benefits. In fact, medicinal plants are different from pharmacological drugs, which usually have several chemical compounds functioning together synergistically and catalytically to generate a combined effect, which can surpass the activity of total constituents in the body. The combined actions of these compounds are likely to enhance the action of the main medicinal ingredients via slowing down or speeding up their integration into

Gum Arabic. http://dx.doi.org/10.1016/B978-0-12-812002-6.00016-6

the body. Secondary metabolites of medicinal plant origins might be likely to increase the stability of the phytochemicals or active compounds, diminishing the rate of undesired unfavorable side effects, and have an additive antagonistic or potentiating effect. It has been hypothesized that a vast diversity of chemical structures existing in these medicinal plants are not waste products; however, their particular action as secondary metabolites has been implicated in the relationship of the organism with the surrounding environment, for instance, attractants of pollinators, signal products, protective compounds against parasites and predators, or the resistance against microbial diseases and pests. A single plant may contain bitter constituents that enhance digestion and that may contain antiinflammatory compounds that reduce pain and swelling. These principals include phenolic compounds that can act as venotonics, antioxidants, antifungal and antibacterial tannins that serve as diuretic substances, and natural antibiotics that increase the elimination of waste, toxins, and alkaloids that promote good temper and give a sense of well-being (Mahomoodally, 2013; Suroowan and Mahomoodally, 2011). Although some may view the isolation of phytochemicals and their use as single chemical entities as a better alternative medicine that has resulted in the replacement of plant extract use, nowadays, a view that there may be some advantages of the medical use of crude and/or standardized extracts as opposed to isolated single compounds is gaining much momentum in the scientific community.

History of gum Arabic

Gum Arabic (GA) has been used for many years in traditional medicine and in daily applications for several diseases. The Egyptians used GA as a pain reliever base and as a glue. In the Middle East, Arabic doctors used GA to treat a wide variety of ailments. At present, GA is used broadly in the pharmaceutical and food industries, predominantly in the food preparation industry to provide body and texture to processed food products.

GA is a derivative of exudates of *Acacia seyal* or *Acacia senegal* trees. It is composed of a mix of complex polysaccharides in addition to oligosaccharides and glycoproteins (Goodrum et al., 2000); however, its composition was reported to vary depending on its source, type of soil, and climate (Figs. 16.1 and 16.2).

Sudan is considered the world's main country producing GA, followed by a number of other African countries. GA is easily dissolved in water to produce colloid solutions, which are characterized by low stickiness and viscosity. These properties allow GA to be used in various daily applications (Patel and Goyal, 2015). GA has been used as thickening agent, an emulsifier, and stabilizer of flavor in both the food industries and pharmaceutical preparations (Ali et al., 2009a). It also has been used in textile and cosmetics industries and pottery. The FAO/WHO Joint Expert Committee for Food Additives defined GA as dried exudations obtained from the stems or stalks of *A. senegal* or closely interrelated species of *Acacia* (Fig. 16.3).

Figure 16.1 Accacia tree.

Figure 16.2 Gum Arabic.

Chemistry of gum Arabic

The chemical composition of GA is documented to vary based on its source, the age of the trees, the place from which it was obtained, soil environment, and climatic conditions. GA is considered a branched chain, complex polysaccharide (Table 16.1), neutral or slightly acidic, found as mixed potassium, calcium, and magnesium salt in a polysaccharidic acid (Arabic acid) (Ali et al., 2009a).

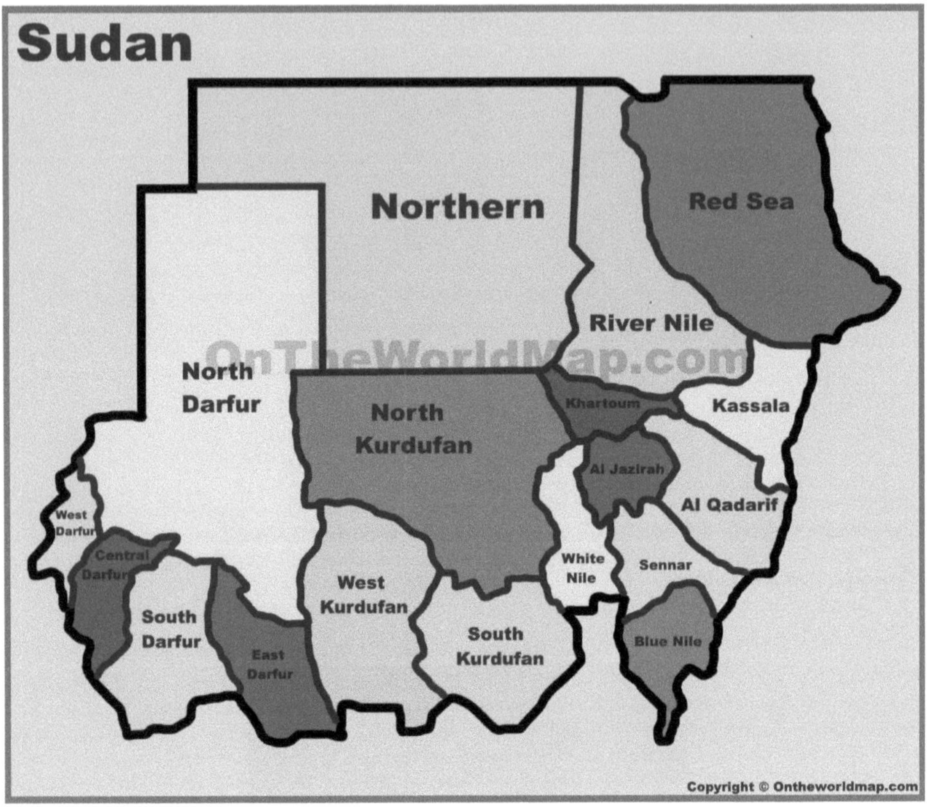

Figure 16.3 Sudan map.

Table 16.1 Chemical characteristics of gum from *Acacia senegal* and *Acacia seyal*

The percent of sugars	*Acacia senegal*	*Acacia seyal*
% Galactose	44.00	38.00
% Arabinose	27.00	46.00
% Rhamnose	13.00	4.00
% Glucuronic acid	14.50	6.50
4-O-methyl glucuronic acid	1.50	5.50
% Nitrogen	0.36	0.15
Specific rotation/degrees	30.00	51.00
Molecular mass (mw)	380,000	850,000

GA also contains proteins (Table 16.2), and the major amino acids present in it are an arabinogalactan protein complex. Arabinogalactan consists of serine, praline, and hydroxyproline, while in glycoprotein, aspartic acid is the most abundant (Ali et al., 2009b).

Table 16.2 **Amino acid content in gum Arabic obtained from *Acacia senegal***

Amino acid	(nmol/mg) GA	% Amino acid
Hydroxyproline	28.700	0.208
Proline	15.100	0.166
Aspartic acid	8.290	0.085
Phenylalanine	5.130	0.045
Isoleucine	2.300	0.037
Methionine	0.000	0.000
Hydroxyproline	28.700	0.208
Proline	15.100	0.166
Aspartic acid	8.290	0.085
Phenylalanine	5.130	0.045
Isoleucine	2.300	0.037
Methionine	0.000	0.000
Hydroxyproline	28.700	0.208
Proline	15.100	0.166
Aspartic acid	8.290	0.085
Phenylalanine	5.130	0.045
Isoleucine	2.300	0.037

Absorption and metabolism of gum Arabic

GA is mostly indigestible to both animals and humans, as it is not degraded in the small intestine; however, it is fermented in the large intestine, particularly in the colon, due to the enzymatic action of microorganisms (Phillips, 1998). Several studies have reported that GA can be fermented by intestinal bacteria to short-chain fatty acid, mainly propionate (Kishimoto et al., 2006).

Potential health benefits of gum Arabic

In North Africa and the Middle East, GA has been used as an oral hygiene material by different communities for centuries (Tyler and Robbers, 1977). Previous studies revealed that a high ingestion of dietary fiber (DF), including GA, is associated with beneficial effects on fats metabolism (Ali et al., 2009a; Slavin, 2003). DF was found to promote satiation and satiety, change glycemic index, influence gastric emptying, and aid in gut hormone secretion, and therefore, it helps to manage body weight (Chandalia et al., 2000). In western Sudan, particularly in Kordofan and Darfur, the people have used GA for several centuries with no restrictions. It is fermented in the colon of human and experimental animals such as rats and mice to provide short-chain fatty acids, which have a wide range of potential health benefits (Phillips and Phillips, 2011). These benefits include antiobesity effects, hypoglycemic effects, hyperlipidemic effects (Ahmed et al., 2015b; Mohamed et al., 2015), antidiabetic effects

(Ahmed et al., 2015a), prebiotic effects (Calame et al., 2008), and antioxidant (Ahmed et al., 2016a), in addition to a protective role against cardiac (Abd-Allah et al., 2002), hepatic, and nephritic toxicities (Al-Majed et al., 2003).

Antiobesity effect of gum Arabic

Obesity is a worldwide health concern associated with high mortality and morbidity. Therapeutic strategies such as synthetic drugs and surgery may cause dangerous complications and health consequences with high costs. The antiobese effects of DF have widely been accepted in a large body of literature. It has been reported that DF can prevent obesity through various mechanisms, such as decreasing caloric density of food (Schneeman, 1987) and reducing fats absorption in the small intestine (Vahouny et al., 1988). In addition, a soluble DF, for example, alginate, is found to have a strong capacity to limit lipid absorption (Brownlee et al., 2005). Moreover, a number of studies reported the association between DF intake and obesity (Brauchla et al., 2012), food intake (Kristensen and Jensen, 2011), and body weight (Adam et al., 2014).

GA is considered as a DF that is found to decrease the body fat deposition. Ingestion of GA was reported to reduce body mass index (BMI) and percentage of body fat among healthy adult females (Babiker et al., 2012). GA is found to have several antiobesity properties as a dietary supplement, both in humans and experimental animals. It serves as a DF, which helps reduction of body weight and fat deposition. The lowering of caloric density of the diet is the most possible mechanism involved (Schneeman, 1987). Moreover, our recent experiments propose that other complex mechanisms may be implicated (Ahmed et al., 2016b). The lowering of glucose and fat absorption is also another proposed mechanism; however, this mechanism was debated in earlier reports (Ushida, 2012). The improvement of fat utilization in adipose tissues also has been suggested by our recent experiments using a real-time polymerase chain reaction approach (Ahmed et al., 2016b). In this mechanism, GA suppressed diet-induced obesity by alternating the expression of mRNA levels of genes involved in lipid metabolism in mouse liver (Ahmed et al., 2016b). In another experiment, GA consumption in the form of drinking water decreased visceral adipose tissue (VAT) associated with downregulation of 11β-hydroxysteroid dehydrogenase type 1 in liver and muscle of mice (Ahmed et al., 2015b) (Fig. 16.4).

Various mechanisms are projected that the DF prevent obesity; for example, DF frequently has lower energy levels, which play a crucial role in reduction of the energy density (Slavin, 2005). DF needs to be eaten for longer, resulting in an increase of time needed for food chew and thus the feeling of satiety. The fiber makes up viscous solutions, which may delay the passage of food through the gastrointestinal tract (GIT) and contribute to an increase in satiety and a reduction in energy consumption (Pasman et al., 1997). It is also important to know that the effects of DF on body weight may be due to the influences of different gut hormones and enzymes that regulate energy intake, satiety, and/or pancreatic functions (Aleixandre and Miguel, 2008).

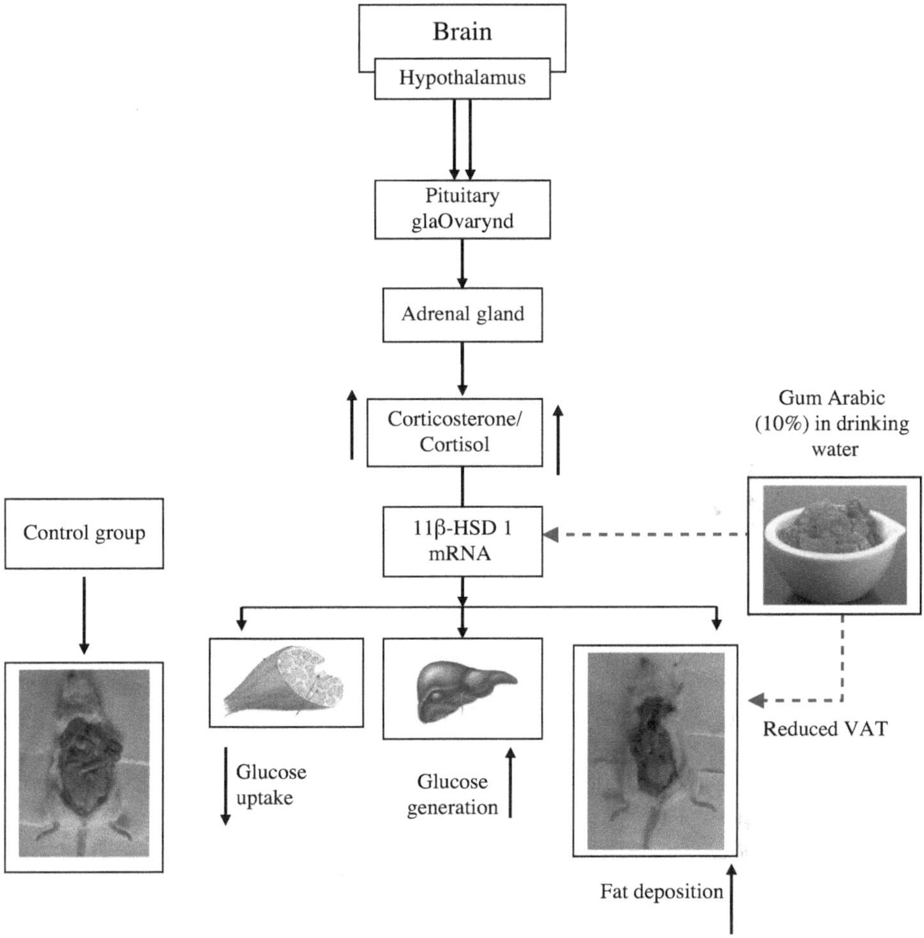

Figure 16.4 Effects of gum Arabic on visceral adipose tissue and plasma lipid profile in mice.

The effects of gum Arabic on food intake

Energy intake is balanced by food intake and energy expenditure, which is affected by several factors, including basal metabolism rate (Blundell et al., 2012), physical activity (Sallis and Hinckson, 2014), body weight, life span (Speakman, 2005), and thermogenesis (Rosen and Spiegelman, 2006). Disturbances of either central (Avena and Bocarsly, 2012) or peripheral (Izzo et al., 2009) signals lead to a state of obesity or anorexia. The increase in obesity, a predisposing factor for developing several diseases, including hypertension, diabetes, hyperlipidemia, cancer, and other metabolic disorders, has driven a major interest in the regulation of food intake, appetite, and fat deposition (Mercer et al., 2011). Energy homeostasis is balanced by a complex neuroendocrine system including appetite regulatory hypothalamic peptides (Bi et al., 2012), as well as adipocytes-derived peripheral signals, such as leptin (Patterson et al., 2012). These signals act in a reciprocal manner to integrate information about energy status, a system referred to as the hypothalamus–adipose tissue axis (Zhang et al., 2014).

The central nervous system, including the hypothalamus, is involved in the regulation of feeding behavior and energy balance in mammals and birds (Koch and Horvath, 2014; Liu et al., 2014). A variety of orexigenic and anorexigenic neuropeptides have been reported to play a critical role in food intake and energy balance (Oyama et al., 2010; Smitka et al., 2013). Orexigenic neuropeptides, including orexin (Martin-Fardon and Boutrel, 2012), agouti-related peptide (AGRP) (Ilnytska et al., 2011), and neuropeptide Y (NPY) (Pazos et al., 2014), are known to stimulate food intake. On the other hand, anorexigenic neuropeptides, including corticotropin-releasing hormone (Richard et al., 2000), proopiomelanocortin (POMC) (Wardlaw, 2011), glucagon-like peptide-1 (GLP-1) (Hayes et al., 2010), and adenosine monophosphate-activated protein kinase (AMPK) (Pimentel et al., 2013) are known to control food intake. The fat mass and obesity-associated gene (FTO) is found to associate with BMI and increased food intake without significant effects on metabolic rates or physical activity (Wardle et al., 2008). The role of hypothalamic expression of NPY (Bi et al., 2012), FTO (Tung et al., 2010), POMC (Conde-Sieira et al., 2010), AMP (Martinez de Morentin et al., 2012), and agouti (Ilnytska and Argyropoulos, 2008) on food intake and energy homeostasis is well documented.

The role of DF in treatment and prevention of obesity has been extensively studied in both humans and experimental animals (Howarth et al., 2001; Pereira and Ludwig, 2001). The consumption of GA seems to be a promising dietary strategy to treat or prevent obesity because of its several biological mechanisms (Chandalia et al., 2000), although GA's influence on energy intake and body weight regulation need more and deeper investigations. An increasing amount of scientific evidence indicates that GA consumption has significant effects on reduction of caloric intake with an increased individual feeling of satiety (Calame et al., 2011).

One study showed a strong positive association between BMI and consumption of GA (Babiker et al., 2012). A recent study used experimental mice as animal model; supplementation with GA in drinking water significantly decreased food intake. The reduction of body weight by GA may be due to the fact that a number of epidemiological studies proposed that a high intake of DF, including GA, is associated with beneficial effects on fat metabolism (Ali et al., 2009a; Slavin, 2003). In addition, the DF was reported to promote satiety and satiation, alter glycemic index, affect gastric emptying and therefore stomach hormone secretion, and reduce weight (Chandalia et al., 2000). Moreover, it is reported that GA inhibits intestinal glucose absorption via interaction with membrane abundance of sodium-glucose transport protein 1 in mice (Nasir et al., 2010a). In the same experiment, GA treatment increased hypothalamic POMC expression in mice. Keenan et al. (2006) results provide the first evidence that GA alters hypothalamic neuropeptide expression in mice. These authors reported GA potentially has three major properties as a part of the diet: metabolic energy dilution, a bulking and satiety effect, and fermentation to produce short-chain fatty acids and increase GLP-1 and PYY (Keenan et al., 2006). The bulking and satiety effect could be due to the high fiber content in GA diet. If the bulking makes the alterations in POMC, destroying vagal afferent nerves supposedly prevents the alterations, as distension signals from the stomach to the brain are vagal afferent nerve dependent (Phillips and Powley, 2000).

The effects of gum Arabic on serum lipid profiles

High intake of DF (commonly >25 g/day for women) is recommended due to the various associated health benefits. Confirmation from randomized controlled trial, observational reports, and experimental animal models showed that DF decreases serum lipid profile (Mumford et al., 2011). Recent reports suggested that GA decreased plasma triglyceride, total cholesterol, and low-density lipoprotein (LDL) concentrations in human (Mohamed et al., 2015) and mice (Ahmed et al., 2015b). GA supplementation also decreased plasma very low-density lipoprotein concentrations. In the same study, GA treatment increased HDL concentrations in mice (Ahmed et al., 2016b). Various mechanisms have been suggested to reveal the hypocholesterolemic effects of DF (Dvir et al., 2009). One possible explanation is that dietary increases the intestinal contents' viscosity, consequently interfering with nutrient absorption and micelle formation, which, in turn, decreases lipid absorption from the intestine (Lattimer and Haub, 2010). Another mechanism proposed that soluble DF disrupts enterohepatic circulation of bile acids, leading to increased bile acid excretion (Parnell and Reimer, 2010) and afterward reduces plasma cholesterol concentrations (Babio et al., 2010). Moreover, the viscosity of fermentable DF is found to contribute substantially to the lipid-lowering effects in rat (Brockman et al., 2014).

Gum Arabic and lipid metabolism gene

A number of key enzymes are known to play a critical role in lipid metabolism in lipid associate tissues (Zhao et al., 2010). Acetyl-CoA carboxylase (Liu et al., 1994), fatty acid synthase (Smith et al., 2003), and glucose-6-phosphate dehydrogenase (Young et al., 1964) are the key lipogenic enzymes that play vital roles in fatty acids biosynthesis through alteration of their activities. Peroxisome proliferator-activated receptor gamma has been also known as lipid metabolism regulator (Grindflek et al., 2000). On the other hand, monoacylglycerol lipase (Taschler et al., 2011), and·hormone-sensitive lipase (HSL) (Kraemer and Shen, 2002) are the key lipolytic enzymes that play a vital role in the lipolysis process. Ahmed et al. (2016b) found that the treatment of GA downregulated adipose triglyceride lipase·and super-conserved receptor expressed in brain 2 gene expression in the liver mice fed with a high-fat diet. In addition, the supplementation of GA was considerably upregulated tumor necrosis factor-α (TNF-α) and HSL in mice liver (Ahmed et al., 2016b). These authors indicated that GA has powerful effects to suppress obesity via alteration of lipid metabolic genes mRNA expression.

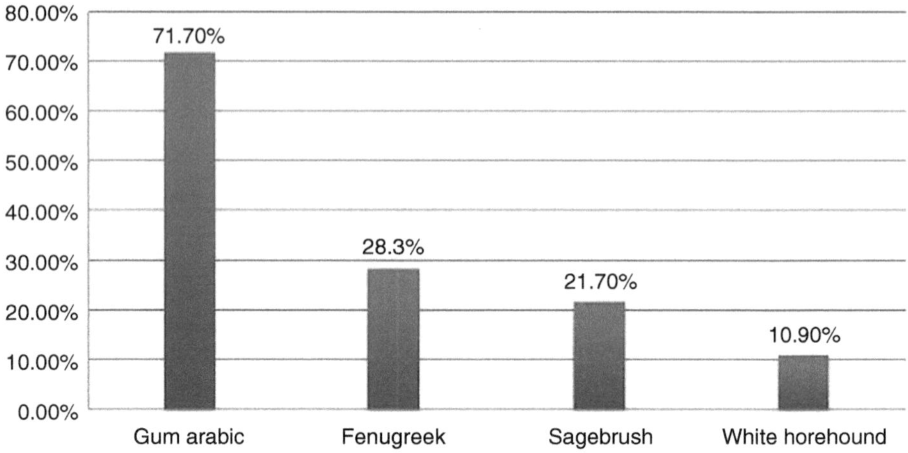

Figure 16.5 The uses of gum Arabic as one of medicinal plant in Tunisia.

Protective effect of gum Arabic for type 2 diabetes mellitus complications

Diabetes mellitus (DM) is a chronic and most common metabolic disease that has become the epidemic of the 21st century (Tabish, 2007). Approximately 422 million people were affected with DM in 2014 worldwide. The WHO expects that DM will be the seventh leading cause of death in 2030. The intake of DF has been highlighted in the recommendations of most nutritional and diabetes associations (Weickert and Pfeiffer, 2008). It is established that gel-forming and viscous properties of soluble DF inhibit intestinal absorption of macronutrient (Lattimer and Haub, 2010) and decrease postprandial glucose index (O'Keefe et al., 2008). The colonic fermentation of naturally occurring high-fiber foods may possibly also be mainly attributed to soluble DF, whereas no difference between soluble and insoluble DF consumption on the regulation of body weight has been observed. In prospective cohort studies, the principally insoluble cereal DF is frequently associated with reduced diabetes risk (InterAct, 2015), indicating that further, unidentified mechanisms are possible to be involved; thus, further studies in this direction are required to elucidate novel mechanisms or pathways. Recent studies indicated that DF consumption plays a vital role in enhancement of insulin sensitivity (Weickert, 2012) and modification of certain gut hormones secretion (Kim et al., 2016), and affects a variety of metabolic and inflammatory biomarkers (Sánchez et al., 2011), which are associated with metabolic syndrome (Dall'Alba et al., 2013).

In India, GA is one of the indigenous medicines, and it has many medicinal uses. In Tunisia, the use of GA as a hypoglycemic medicinal plant to treat diabetic patients was found to reach more than 70% compared to other medicinal plants (Othman et al., 2013) (Fig. 16.5).

A clinical trial used 40 participants with a daily supplement of powdered GA (10 g/day) for 16 weeks in healthy individuals, prediabetic patients, type 2 DM patients, and diabetic nephropathy patients. The supplementation of GA significantly decreased fasting blood glucose levels and glycosylated hemoglobin, together with significant reduction of blood uric acid and total protein concentrations (Omaima Nasir and Salim, 2016).

Antioxidant activity of gum Arabic

Oxidative stress is mainly induced by the disturbances of oxidants/antioxidants damage of biological macromolecules, including lipids, carbohydrates, proteins, and nucleic acids, resulting in the dysregulation of cellular homeostasis and thus the production of other reactive oxygen molecules that ultimately cause more oxidative damage (Birben et al., 2012). Activity of key antioxidant enzymes plays a critical role in the introduction of hyperglycemia-related tissue damage (Wu et al., 2014). The significance of oxidative stress and its association with the pathology of DM, along with its complications, have been comprehensively investigated (Afanas'ev, 2010; Wu et al., 2014). Several studies revealed that production of reactive oxygen species in diabetes might initiate the development of chronic diabetic lesions on many tissues for instance vessels (Son, 2012), retina (Zhao et al., 2012), kidneys (Bondeva and Wolf, 2014), and neurodegenerative diseases (Borza, 2014).

Catalase (CAT) (Tiedge et al., 1998), superoxide dismutase (SOD) (Gill and Tuteja, 2010), and glutathione peroxidase (GPx) (Tsai et al., 2012) are the most vital defense mechanisms against reactive oxygen metabolites, which are implicated in the oxidative damage (Sharma et al., 2012; Tiwari et al., 2013). Oxidative stress in DM is reported to cause several adverse effects on the cellular physiology (Tiwari et al., 2013). It found to decrease the level of reduced glutathione (GSH) concentrations (Dincer et al., 2002), reduce the activity of CAT (Goth, 2000), decrease the level of renal SOD (Wang et al., 2011), and increase the level of heat shock protein 70 (Nakhjavani et al., 2010). The reduction of antioxidant components has been reported to be the main factors in the onset of pathogenesis and diabetic complications (Nakhjavani et al., 2013). The treatment of alloxan was found to cause a significant decrease in antioxidant enzyme activities including SOD, CAT, and GPx. The reduction in antioxidant enzyme activities were associated with obvious increases in production of malondialdehyde that may reflect oxidative stress, and thus mirrors a decreased antioxidant defense potential (Fujita et al., 2009; Tiwari et al., 2013). The treatment of GA increased the activity of SOD, CAT, and GPx in the liver of diabetic rat. GA may act by either directly scavenging free radicals or reactive oxygen metabolites, due to the presence of various antioxidant compounds in GA (Ali et al., 2009a, 2013b; Kong et al., 2014) or via increasing the synthesis of antioxidant biomolecules. The mechanism of action through which GA improves the antioxidant capacity may be due to the fact that GA contains several types of amino acid residues such as lysine, tyrosine, and histidine, which are commonly considered as antioxidant biomolecules (Marcuse, 1960; Park et al., 2005). Moreover, the antioxidant prosperities of GA in biological systems require a more direct knowledge of the antioxidant capacity (Ali et al., 2009a; Liu et al., 2015).

Hepatoprotective effect of gum Arabic

Blood alanine transaminase (ALT) and aspartate transaminase (AST) activities are commonly used in clinical chemistry for the evaluation of liver damage. In our previous experiment (Ahmed et al., 2015a), we have used a diabetic rats model to examine the hepatoprotective effect of GA. The treatment of GA revealed significant increases in the activities of hepatic key enzymes, including ALT and AST (Table 16.3). Moreover, GA treatment significantly protected the liver of diabetic rats when compared with control and untreated diabetic rats (Ahmed et al., 2015a) (Fig. 16.6).

Another study (Singh, 2014) has reported that the treatment of animals with pod hydroalcoholic extract of *A. senegal* significantly decreased liver damage and symptoms of liver injury by restoration of architecture of liver tissue, which was indicated by lower levels of serum bilirubin when compared with the normal and silymarin-treated rat. Histological changes were also evident, as the liver was protected when compared with CCl4-induced liver damage rat. The treatment of pod hydroalcoholic extract of *A. senegal* showed significant hepatoprotective activity against CCl4-induced liver injury model.

Table 16.3 Effects of gum Arabic on liver function parameters in diabetic rat

Group	ALT (U/L)	AST (U/L)	Albumin (g/L)	Globulin (g/L)	Total protein (g/L)
Control	62.6 ± 3.0[a]	160.8 ± 15.7[a]	39.68 ± 2.1[a]	36.6 ± 2.8[a]	75.18 ± 4.1[a]
Diabetic	76 ± 5.7[c]	220.25 ± 9.3[b]	37.3 ± 0.50[a]	32.23 ± 0.60[a]	69.53 ± 1.1[a]
D + GA	54.6 ± 9.7[ab]	168.8 ± 24.4[a]	35.92 ± 1.2[a]	31.46 ± 1.5[a]	67.38 ± 1.1[a]

ALT, Alanine transaminase: *AST*, aspartate transaminase.
[a]Histology of the liver of the control group showed normal hepatocytes morphology with no visible fatty liver changes.
[b]However, liver of diabetic rats revealed marked degeneration in liver cells, with severe vacoulations.
[c]The treatment of GA significantly protected the liver of diabetic rats compared with control and untreated diabetic rats.

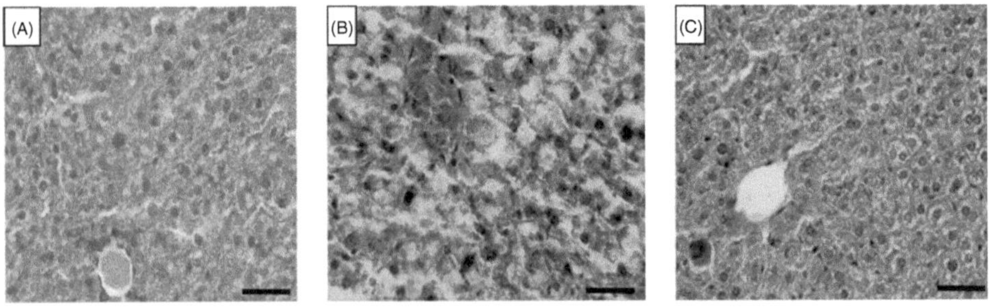

Figure 16.6 Hepatoprotective effect of gum Arabic in diabetic rat.

Nephroprotective effect of gum Arabic

The administration of GA in the form of drinking water (15% w/v) was found to reduce the concentrations of plasma urea and creatinine in adenine-induced chronic renal failure (CRF) in rats. It also decreased the clearance of creatinine and induced significant increases in the inflammatory mediator concentration. Further, the treatment with GA significantly ameliorated all adverse effects that inuced by adenine. The mechanism underlying the salutary effect of GA in adenine-induced CRF may associate with mitigation of the adenine-induced inflammation and generation of free radicals (Ali et al., 2013b).

Pretreatment of GA (7.5 g/kg/day per oral administration), starting 5 days before mercuric chloride injection, resulted in a complete reversal of Hg-induced increase in blood urea nitrogen, creatinine, thiobarbituric acid reactive substances, and total nitrate/nitrite to control values. Pretreatment of GA prevented Hg-induced degenerative changes of kidney tissues. This indicated that GA is an efficient cytoprotective agent against Hg-induced nephrotoxicity (Gado and Aldahmash, 2013). The protective effect of GA on renal function was also confirmed to cause significant reduction in blood creatinine and urea nitrogen concentrations in diabetic nephropathy patients (Omaima Nasir and Salim, 2016).

Recently, Musa et al. (2016) investigated the effect of GA on the diabetic rat. They found that GA significantly reduced serum urea and urinary protein. In addition, GA significantly decreased α-SAM and TGF-β1 gene expression in kidney when compared to streptozotocin (STZ)-treated rats. The treatment of GA also increased kidney cytokeratin-19 and E-cadherin gene expression when compared to the STZ-treated group. This indicates that GA may attenuate the development of nephropathy in type 1 diabetes rat.

Antiulcer activity of gum Arabic

In the past, low DF was suggested to patients with gastric, duodenal ulcers; however, in the late 1970s, there was evidence that a higher DF was beneficial in treatment of gastric or duodenal ulcers. Then the evidences began to accumulate. GA was, as DF, found to accelerate the healing of gastric ulcers at the different doses investigated. GA also produced a significant protection in the development of the gastric ulcers in rats (Abdulrahman and AL-Yahya, 2016; Ryan-Harshman and Aldoori, 2004). These findings may indicate that GA has beneficial effects in the protection and healing of gastric ulcers, thus indicating both gastric antisecretory and cytoprotective effects of GIT.

Effects of gum Arabic on wound healing

GA has been applied in pharmaceuticals as a demulcent. It is applied topically for wound healing, and it has been revealed to inhibit the growth of periodontic bacteria in addition to an early deposition of plaque (Samy et al., 2014).

Reproductive function improvement from gum Arabic

Positive beneficial health benefits of DF on long-lasting or chronic diseases have been observed. Yet the beneficial effect of DF intake on endogenous sexual hormones levels and other reproductive measures (for example, ovulation) in women has had inadequate reports (Gaskins et al., 2009). The effect of DF consumption on ovulation and reproductive hormone concentrations is of particular interest in reproductive-age women, known the role of these sex hormones on conception and maintenance of pregnancy (Bayram et al., 2000; Mason et al., 1984). The utilization of higher DF was reported to be associated with reduction of reproductive hormone levels and an improved risk of incidence of ovulation in young, healthy women (Gaskins et al., 2009). The significant association was also reported to have persisted and remained strong, whether consumption of DF was considered as a continuous variable or categorized according to groupings on the basis of dietary reference intake. The experiential links between high DF intake, increased risk of ovulation, and reduced reproductive hormone levels indicates the potential health benefits for reproductive health implications related to DF in young women that, up to that date, might not have been well recognized. DF was found to decrease the concentrations of estrogen mainly in older women. DF consumption was inversely associated with decreased estrogen levels and positively associated with the risk of an ovulation (Goldin et al., 1982) In addition, it was shown that DF binds to estrogen in the intestine, increasing its fecal excretion (Lewis et al., 1997; Shultz and Howie, 1986). Through the influence of DF on estrogen, DF consequently may influence other reproductive hormones because of its strict feedback loop mechanisms, which, in turn, dictate reproductive hormonal instability in the menstrual cycle (Liu and Yen, 1983; Micevych et al., 2007). However, DF is also found to decrease follicle-stimulating hormone (FSH) and luteinizing hormone (LH) levels, independent of estradiol (Gaskins et al., 2012). The biological mechanisms underlying the influence of DF on FSH and LH are still required to be investigated and need more research. The administration of GA considerably improved semen quality. GA may protect the testis through enhancement of antioxidant capacity, and therefore, it may be useful to meliorate the diabetic fertility complications. GA has been documented to have protective effects in a number of diseased conditions, for example, cardiotoxicity (Abd-Allah et al., 2002), nephrotoxicity (Al-Majed et al., 2002; Gado and Aldahmash, 2013), DM (Nasir, 2013), and inflammation (Ali et al., 2013a). In our previous experiment, we found that supplementation of GA protected the ovary from oxidative stress damage in mice fed with a high-fat diet. In another study, we found that the consumption of GA increased sperm quality in the diabetic rat (Ahmed et al., 2016a) (Figs. 16.7 and 16.8).

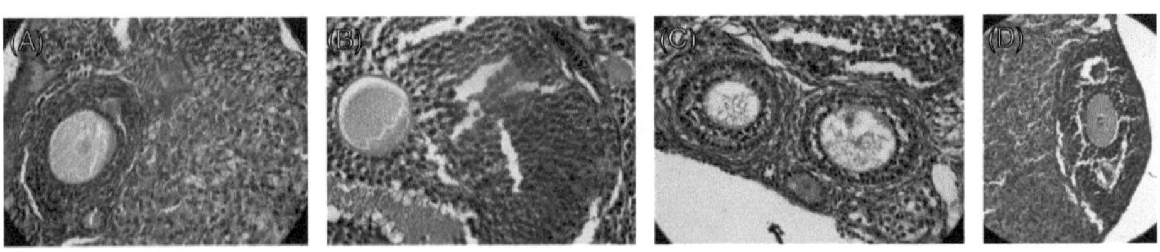

Figure 16.7 Protective effects of gum Arabic in the ovary of mice fed with high fat diet.

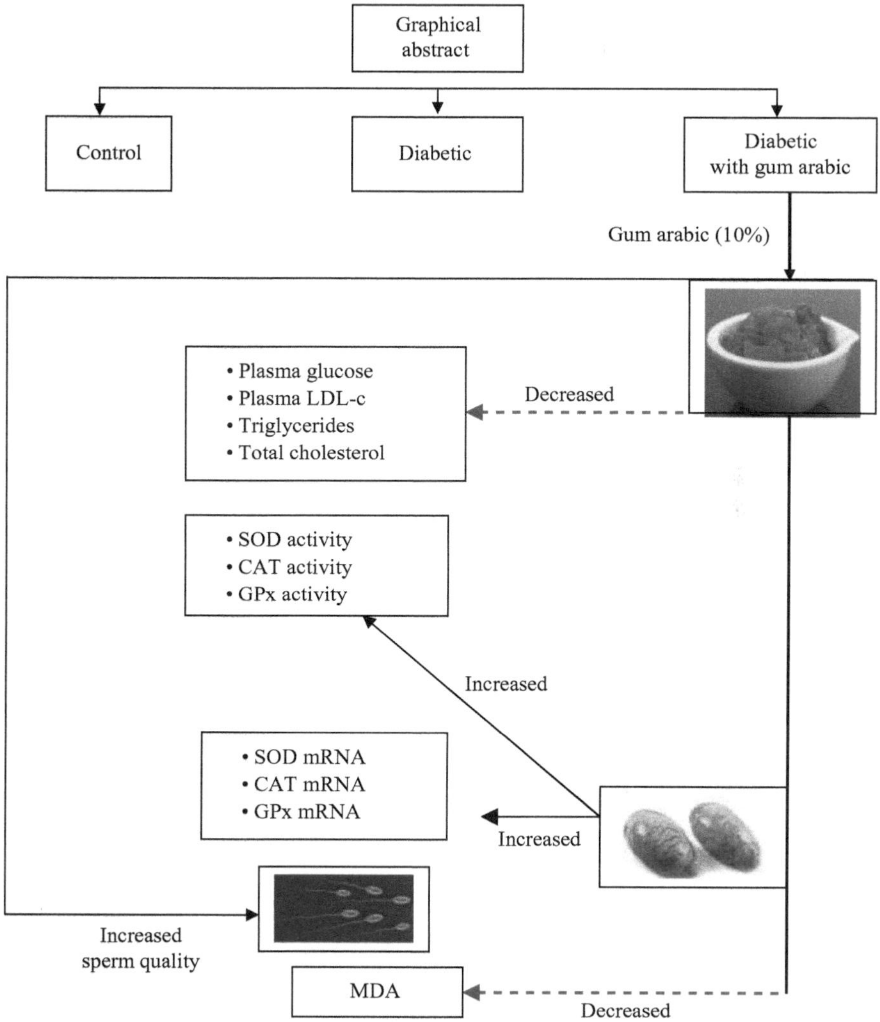

Figure 16.8 Effects of gum Arabic on semen quality in diabetic rat.

Immunomodulatory effect of gum Arabic

It is well known that the dendritic cells (DCs) are stimulated by microbial polysaccharides; accordingly, we hypothesized that the consumption of GA may correspondingly stimulate DCs. Since we do not currently have the research projected in such direction, we strongly recommend the researchers to conduct research on the effect of GA on immunomodulation both in vivo and in vitro. Up to date, only one study has investigated using mouse DCs, which were treated with either lipopolysaccharide (LPS) or GA to test expression of phagocytotic activity, maturation markers, cytokine production, and ability to motivate $CD4^+$ T cells. Both GA and LPS increased the percentage of $CD11c^+CD40^+$, $CD11c^+MHCII^+$, $CD11c^+CD86^+$, and $CD54^-$ expressing DCs, whereas they reduced their phagocytic activity. Likewise, GA and LPS stimulated the production of IL-6, IL-10, IL12p70, and TNF-α in a p38- and/or extracellular signal-regulated kinases (ERK)-dependent manner. The treatment of GA enhanced secretion of IL-10, whereas LPS was found to be more effective on the production of IL-6 and IL-12p70. In addition, GA and LPS stimulated DCs to enhance $CD4^+$ T cell proliferation. High levels of IL-6 and IL-10 were detected in the DCs that were treated with GA, while IL-12p70 and IFN-Γ production was comparable with GA or LPS-treated DCs. LPS transiently upregulated ERK1/2 and p38, whereas GA upregulated more sustained activation of ERK1/2. These findings disclose a powerful immunomodulatory effect of GA (Xuan et al., 2010).

Antibacterial activity of gum Arabic

Several types of gums are found to reduce diseases such as diarrhea, sore throat, and kidney, wound, and gum infections. Only scattered results are in writing assessing their antimicrobial potential, prompting underestimation and at some point, over desires from gums and gum items in therapeutics (Singh et al. 2015). Recently, it has been confirmed that GA bark extracts showed antimicrobial activity. It has found to serve against human pathogenic isolates, for example, *Streptococcus pneumoniae*, *Staphylococcus aureus*, *Escherichia coli*, *Shigella dysenteriae*, *Salmonella typhi*, *Klebsiella pneumoniae*, *Streptococcus pyogenes*, *Proteus vulgaris*, and *Pseudomonas aeruginosa* (Mahomoodally, 2013). Significant antibacterial potentiality of GA may possibly be due to the attendance of saponins and tannins in the plant. Moreover, it was reported that the GA extract was not toxic to humans following in vitro cytotoxicity evaluation.

Antimalarial effect of gum Arabic

GA exudates are considered a powerful immunomodulatory agent on antigen-presenting cells, which are involved in the activation of both innate and adaptive immune response. Digestion of GA was reported to produce short-chain volatile fatty acids, which have been known to enhance the expression of fetal hemoglobin (HbF) in red blood cell. GA is reported to increase the levels of HbF, which is known to hamper the intraerythrocytic growth of *Plasmodium* spp. and therefore provides slight protection against malaria infection. As indicated by the in vitro experiments, butyrate basically blunted parasitemia just at concentrations significantly higher (3 mM) than those experienced in vivo taken after GA ingestion ($<1\ \mu M$). As indicated by the in vivo analyses, the treatment of GA somewhat, yet essentially, diminished the parasitemia and fundamentally broadened the life expectancy of infected mice (Ballal et al., 2011). GA, to some extent, influences the parasitemia and survival of *Plasmodium*-infected mice. The mechanism underlying this health potentiality remain unknown. Further studies are required to clarify such mechanism.

Gum Arabic and sickle cell anemia

Increased percentage of HbF is revealed to decrease sickle-cell anemia (SCA) severity and lead to healthier survival. Based on in vitro and in vivo studies, butyrate was reported to increase HbF production. Butyrate consumption in clinical practice is disadvantaged, yet, by its short half-life. Serum butyrate concentrations might be enhanced by fermentation of GA by colonic microflora. The administration of GA was found to significantly increase HbF level, mean corpuscular volume, and hematocrit level. Moreover, 37% of SCA patients experienced slight side effects, which resolved within a week (Kaddam et al., 2015). Thus, GA may be useful to promote the production of HbF.

Prebiotic properties of gum Arabic

GA has prebiotic properties in healthy human volunteers in a dose-dependent manner. When normal human volunteers ingested different daily doses (5, 10, 20, and 40 g) of GA for up to 4 weeks, the genera of *Lactobacillus* and *Bifidobacterium* were taken as potentially beneficial bacteria, whereas *Bacteroides*, *Enterococci*, and *Clostridium difficile* were taken as potentially nonbeneficial. Compared with the negative control, the colons of *Lactobacillus* and *Bifidobacterium* 4 weeks after utilization were significantly higher for GA. The best dose was confirmed to be around 10 g. In addition, at this dose, the numbers of *Lactobacillus*, *Bifidobacterium*, and *Bacteroides* were significantly higher for GA than for inulin (polysaccharide). When compared with the negative control, the numbers of *Bifidobacterium* and *Lactobacillus* 4 weeks after consumption were significantly higher for GA, the optimal dose being around 10 g. Moreover, at this dose, the numbers of *Bifidobacterium*, *Lactobacillus*, and *Bacteroides* were significantly higher for GA than for inulin. No significant drawback was encountered during the study. It is concluded that GA establishes prebiotic efficacy, at least as good as inulin. The optimal daily dose was found to be 10 g. Thus, GA may establish prebiotic value, at least as well as insulin (Calame et al., 2008).

Gum Arabic electrolyte and water absorption

GA was reported to increase water and electrolyte movement, for example, water and sodium from the intestinal lumen to the bloodstream (Turvill et al., 2000). Another health benefit of GA is that its consumption does not modify the viscosity of the final solution (Turvill et al., 2000). Previous investigations have demonstrated that a soluble fiber, GA, upgrades water, electrolyte, and glucose absorption in animal models of diarrhea. The ways implicated in this impact have not been completely illustrated. This review inspected the possibility that paracellular transport is balanced by luminal GA, bringing about an upgraded rate of ingestion in the small digestive system. This speculation was tried by 3-hr jejunal perfusions on anesthetized rats with solutions containing 140 mM NaCl, 5 mM KCl, and 2 microCi/L (74 kBq) $3H_2O$, with either 2.5 g/L

GA (+GA) or in its absence (CTL), and one of the accompanying agents, fit for modifying paracellular transport: chenodeoxycholic acid at 0.5 mM (CDC), 2,4,6-triaminopyrimidine (TAP) at 20 mM, and protamine at 100 mg/L (PTM). Sodium, potassium, net water, and unidirectional water development were measured. The expansion of GA expanded sodium absorption in perfusions with CDC, TAP, or PTM only. Comparable impacts by GA on net water absorption rates were gotten in tissues permeabilized with CDC and PTM; however, GA added to TAP did not standardize the decrease created by TAP. In spite of the fact that PTM did not change net water absorption, expansion of GA to perfusates with PTM improved absorption values over those of CTL. GA reversed the solid negative impacts of CDC on potassium absorption, but was insufficient in such manner with TAP and PTM. The information gotten with those reagents that influence paracellular transport and the histological confirmation support the view that GA advances net absorption by this route in the small intestine of normal rats.

Gum Arabic and cancer

In the 1970s, several reports concluded that increased colorectal cancer prevalence was associated with low-fiber diet consumption. These assumptions were essentially based on differences in the rates of colorectal cancer among geographical regions and nations with low and high fiber intakes (Slavin, 2013). It has been found that GA effectively stabilizes gold nanoparticles in the body. The researchers have distinguished that over 98% of the gold salt was converted to GA labeled with gold nanoparticles. The resultant nanoparticles were more stable in biological fluids for at least 7 days (Ganta et al., 2009).

The treatment of GA changed several tumor-related gene expressions in mice. The administration of GA within 4 days significantly reduced the colonic mRNA levels of the angiogenetic factors, angiogenin 1, angiogenin 3, and angiogenin 4, by 78 ± 18%, 88 ± 15%, and 92 ± 13%, respectively. In addition, the treatment of GA-modified mRNA levels of Vav3 oncogene, CD38 antigen, interleukin18, aquaporin4, sulfatase1, ubiquitinD, gamma (+)-amino acid transporter, and chemokine ligand 5 in mice. When performed in western blotting, GA treatment likewise decreased angiogenin protein expression. Furthermore, using immunohistochemistry, GA reduced β-catenin expression. The treatment of chemical cancerogenesis to the mice significantly increased multiple colonic tumors within 12 weeks. However, the treatment of GA (10% wt/wt) in the form of drinking water significantly decreased the number of tumors by 70% (Nasir et al., 2010b). The findings revealed a powerful anticarcinogenic effect of GA. The mechanism of action through which GA modifies cancer-related genes' mRNA expression is still unclear. Further studies are required to reveal the mechanism of action.

Gum Arabic and cosmetics

GA has a nontoxic nature, antiallergic properties, and is free from causing dermatological allergic reactions, therefore, it has been used in cosmetics preparations. In protective creams and lotions, GA adds a smoothing feeling to the skin and serves as a protective coating. It also serves as an adhesive substance in preparation of facial masks. GA also acts as a binder in face powder compact, whereas in protective creams, it is used effectively as a stabilizer (Smolinske and Susan, 1992).

Gum Arabic and inks

Botanical research discovered a possible linkage between inks and medical treatments through the creation of charms in which ink was washed from the paper and drunk or used as a body wash. Since the GA has potential health benefits, it could be involved in this traditional medicine efficiently. Thus, using GA in indigenous medicine is a good idea of our scholars toward the treatment of various diseases (Biddle, 2011).

Toxicological effects of gum Arabic

GA has no carcinogenic or teratogenic properties. In rats, GA has no hematological or histopathological toxicity when administered at doses as high as 5 g/kg/day for 13 weeks.

It has been suggested that the renoprotective effect of GA is possibly through inhibition of the production of oxygen-free radicals that cause peroxidation. However, this has yet to be proven. It has, however, been shown that AG has no effect on the concentration of some free-radical scavengers (reduced glutathione, ascorbic acid, lipid peroxidation, and SOD) on the kidneys and liver of healthy rats. Occasionally, the use of GA can have some adverse side effects on health. These side effects include gas formation, bloating, or even diarrhea. Due to the individual variation, certain persons may progress the development of allergic reactions, skin reactions, or asthma. Conjunctivitis and rhinitis may also appear in the form of allergic reactions to GA.

Gum Arabic interactions with other medication

It has been reported that GA has the threat of interactions with other medication. GA hinders the absorption of some antibiotics, such as amoxicillin, and therefore proves fruitless in the patient who taking the medicine. GA also interferes with mineral absorption, for instance, iron, and this mainly has significant consequences in pregnant women who require supplements of iron during the period of their pregnancy. GA should be taken with extreme caution in patients who are on medication for respiratory diseases because when GA is used in combination, it may trigger an allergic reaction.

In conclusion, many in vivo and in vitro studies have confirmed that GA has strong antioxidant properties, and it reduces the adverse effects in chronic kidney disease and decreases diabetic complications in type 2 diabetes patients. GA contains soluble DF that reduces the total cholesterol, HDL, and LDL cholesterol concentrations in the blood, which, in turn, has an inhibitory effect on coronary heart disease. GA has antimicrobial potentiality and thus inhibits the growth of various pathogenic bacteria in vitro and in vivo due to the fact that GA is digested in the small intestine, and thus, it reduces the pH in the intestine. GA has potent impact on minerals, and it increases the absorption of potassium and sodium from the diet, which is beneficial to health. GA reduces excretion of urea nitrogen and increases the excretion of nitrogen in the feces. GA also reduces production of urea, which leads to fewer overloads on kidney. GA improves reproductive function, both in male and female, via improvement to semen quality and ovary function. GA has an immunomodulatory property via activating CD4, and this property is due to the presence of polysaccharides in GA. GA has antimalarial activity through increases of hemoglobin concentration. GA increases hemoglobin and thus has a positive impact on sickle-cell anemia. GA increases water and electrolyte absorption. GA has antiobesity potentiality through decreases in food intake, body weight, and plasma lipids profiles. GA decreases obesity via decreases of VAT and modifies lipid metabolism gene mRNA.

List of abbreviations

ACC	Acetyl-CoA carboxylase
AGRP	Agouti-related peptide
ALT	Alanine transaminase
AMP	Adenosine monophosphate
AMPK	Adenosine monophosphate-activated protein kinase
AST	Aspartate transaminase
ATGL	Adipose triglyceride lipase
BMI	Body mass index
CAM	Complementary and alternative medicine
CAT	Catalase
CCl4	Carbon tetrachloride
CD4	Cluster of differentiation 4
CNS	Central nervous system
CRF	Chronic renal failure
CRH	Corticotropin-releasing hormone
DCs	Dendritic cells
DF	Dietary fiber
DM	Diabetes mellitus
DRI	Dietary reference intake
ERK1/2	Extracellular signal-regulated kinases 1/2
FAO	Food and Agriculture Organization
FAS	Fatty acid synthase
FTO	Fat mass and obesity associated protein
G-6-PDH	Glucose-6-phosphate dehydrogenase
GA	Gum Arabic
GIT	Gastrointestinal tract
GLP-1	Glucagon-like peptide-1
GPx	Glutathione peroxidase
HbF	Fetal hemoglobin

HDL	High-density lipoprotein
HSL	Hormone-sensitive lipase
HSP70	Heat shock protein 70
IL	Interleukin
LDL	Low-density lipoprotein
LPS	Lipopolysaccharide
MCV	Mean corpuscular volume
MGL	Monoacylglycerol lipase
NPY	Neuropeptide Y
ORX	Orexin
PCR	Polymerase chain reaction
POMC	Pro-opiomelanocortin
PPAR-g	Peroxisome proliferator-activated receptor gamma
PYY	Peptide YY
RBC	Red blood cell
ROS	Reactive oxygen species
SCA	Sickle cell anemia
SGLT1	Sodium-glucose transport protein 1
SOD	Superoxide dismutase
SREB2	Super-conserved receptor expressed in brain 2
STZ	Streptozotocin
TGF-β1	Transforming growth factor beta 1
TNF-α	Tumor necrosis factor alpha
VAT	Visceral adipose tissue
VLDL	Very low-density lipoprotein
WHO	World Health Organization
α-SAM	Alpha smooth muscle actin

References

Abd-Allah, A.R., Al-Majed, A.A., Mostafa, A.M., Al-Shabanah, O.A., Din, A.G., Nagi, M.N., 2002. Protective effect of arabic gum against cardiotoxicity induced by doxorubicin in mice: a possible mechanism of protection. J. Biochem. Mol. Toxicol. 16, 254–259.

Abdulrahman, A.I., AL-Yahya, M.A., 2016. Antiulcer activity of gum Arabic and its interaction with antiulcer effect of ranitidine in rats. Biomed. Res. 4, 1102–1106.

Adam, C.L., Williams, P.A., Dalby, M.J., Garden, K., Thomson, L.M., Richardson, A.J., Gratz, S.W., Ross, A.W., 2014. Different types of soluble fermentable dietary fibre decrease food intake, body weight gain and adiposity in young adult male rats. Nutr. Metab. (Lond) 11, 36.

Afanas'ev, I., 2010. Signaling of reactive oxygen and nitrogen species in Diabetes mellitus. Oxid. Med. Cell. Longevity 3, 361–373.

Ahmed, A.A., Fedail, J.S., Musa, H.H., Kamboh, A.A., Sifaldin, A.Z., Musa, T.H., 2015a. Gum Arabic extracts protect against hepatic oxidative stress in alloxan induced diabetes in rats. Pathophysiology 22, 189–194.

Ahmed, A.A., Musa, H.H., Fedail, J.S., Sifaldin, A.Z., Musa, T.H., 2015b. Gum Arabic decreased visceral adipose tissue associated with downregulation of 11β-hydroxysteroid dehydrogenase type I in liver and muscle of mice. Bioact. Carbohydr. Dietary Fibre 6, 31–36.

Ahmed, A.A., Fedail, J.S., Musa, H.H., Musa, T.H., Sifaldin, A.Z., 2016a. Gum Arabic supplementation improved antioxidant status and alters expression of oxidative stress gene in ovary of mice fed high fat diet. Middle East Fertil. Soc. J. 21, 101–108.

Ahmed, A.A., Musa, H.H., Fedail, J.S., Sifaldin, A.Z., Musa, T.H., 2016b. Gum Arabic suppressed diet-induced obesity by alteration the expression of mRNA levels of genes involved in lipid metabolism in mouse liver. Bioact. Carbohydr. Dietary Fibre 7, 15–20.

Aleixandre, A., Miguel, M., 2008. Dietary fiber in the prevention and treatment of metabolic syndrome: a review. Crit. Rev. Food Sci. Nutr. 48, 905–912.

Ali, B.H., Ziada, A., Blunden, G., 2009a. Biological effects of gum Arabic: a review of some recent research. Food Chem. Toxicol. 47, 1–8.

Ali, B.H., Ziada, A., Blunden, G., 2009b. Biological effects of gum Arabic: a review of some recent research. Food Chem. Toxicol. 47, 1–8.

Ali, B.H., Al-Husseni, I., Beegam, S., Al-Shukaili, A., Nemmar, A., Schierling, S., Queisser, N., Schupp, N., 2013a. Effect of gum Arabic on oxidative stress and inflammation in adenine-induced chronic renal failure in rats. PloS One 8, e55242.

Ali, B.H., Al-Husseni, I., Beegam, S., Al-Shukaili, A., Nemmar, A., Schierling, S., Queisser, N., Schupp, N., 2013b. Effect of gum Arabic on oxidative stress and inflammation in adenine–induced chronic renal failure in rats. PLoS One 8, e55242.

Al-Majed, A.A., Mostafa, A.M., Al-Rikabi, A.C., Al-Shabanah, O.A., 2002. Protective effects of oral arabic gum administration on gentamicin-induced nephrotoxicity in rats. Pharmacol. Res. 46, 445–451.

Al-Majed, A.A., Abd-Allah, A.R., Al-Rikabi, A.C., Al-Shabanah, O.A., Mostafa, A.M., 2003. Effect of oral administration of Arabic gum on cisplatin-induced nephrotoxicity in rats. J. Biochem. Mol. Toxicol. 17, 146–153.

Avena, N.M., Bocarsly, M.E., 2012. Dysregulation of brain reward systems in eating disorders: neurochemical information from animal models of binge eating, bulimia nervosa, and anorexia nervosa. Neuropharmacology 63, 87–96.

Babiker, R., Merghani, T.H., Elmusharaf, K., Badi, R.M., Lang, F., Saeed, A.M., 2012. Effects of gum Arabic ingestion on body mass index and body fat percentage in healthy adult females: two-arm randomized, placebo controlled, double-blind trial. Nutr. J. 11, 111.

Babio, N., Balanza, R., Basulto, J., Bullo, M., Salas-Salvado, J., 2010. Dietary fibre: influence on body weight, glycemic control and plasma cholesterol profile. Nutr. Hosp 25, 327–340.

Ballal, A., Bobbala, D., Qadri, S.M., Foller, M., Kempe, D., Nasir, O., Saeed, A., Lang, F., 2011. Anti-malarial effect of gum Arabic. Malar. J. 10, 139.

Bayram, N., van Wely, M., Vandekerckhove, P., Lilford, R., van Der Veen, F., 2000. Pulsatile luteinising hormone releasing hormone for ovulation induction in subfertility associated with polycystic ovary syndrome. Cochrane Database Syst. Rev., CD000412.

Bi, S., Kim, Y.J., Zheng, F., 2012. Dorsomedial hypothalamic NPY and energy balance control. Neuropeptides 46, 309–314.

Biddle, M., 2011. Inks in the Islamic manuscripts of Northern Nigeria old recipes, modern analysis and medicine. J. Islamic Manuscripts 2, 1–35.

Birben, E., Sahiner, U.M., Sackesen, C., Erzurum, S., Kalayci, O., 2012. Oxidative stress and antioxidant defense. World Allergy Organ J. 5, 9–19.

Blundell, J.E., Caudwell, P., Gibbons, C., Hopkins, M., Naslund, E., King, N., Finlayson, G., 2012. Role of resting metabolic rate and energy expenditure in hunger and appetite control: a new formulation. Dis. Model. Mech. 5, 608–613.

Bondeva, T., Wolf, G., 2014. Reactive oxygen species in diabetic nephropathy: friend or foe? Nephrol. Dial. Transplant. 29, 1998–2003.

Borza, L.R., 2014. A review on the cause-effect relationship between oxidative stress and toxic proteins in the pathogenesis of neurodegenerative diseases. Rev. Med. Chir. Soc. Med. Nat. Iasi 118, 19–27.

Brauchla, M., Juan, W., Story, J., Kranz, S., 2012. Sources of dietary fiber and the association of fiber intake with childhood obesity risk (in 2–18 year olds) and diabetes risk of adolescents 12–18 year olds: NHANES 2003–2006. J. Nutr. Metab. 2012, 736258.

Brockman, D.A., Chen, X., Gallaher, D.D., 2014. High-viscosity dietary fibers reduce adiposity and decrease hepatic steatosis in rats fed a high-fat diet. J. Nutr. 144, 1415–1422.

Brownlee, I.A., Allen, A., Pearson, J.P., Dettmar, P.W., Havler, M.E., Atherton, M.R., Onsøyen, E., 2005. Alginate as a source of dietary fiber. Crit. Rev. Food Sci. Nutr. 45, 497–510.

Calame, W., Weseler, A.R., Viebke, C., Flynn, C., Siemensma, A.D., 2008. Gum Arabic establishes prebiotic functionality in healthy human volunteers in a dose-dependent manner. Br. J. Nutr. 100, 1269–1275.

Calame, W., Thomassen, F., Hull, S., Viebke, C., Siemensma, A.D., 2011. Evaluation of satiety enhancement, including compensation, by blends of gum Arabic. A methodological approach. Appetite 57, 358–364.

Chandalia, M., Garg, A., Lutjohann, D., von Bergmann, K., Grundy, S.M., Brinkley, L.J., 2000. Beneficial effects of high dietary fiber intake in patients with type 2 diabetes mellitus. N. Engl. J. Med. 342, 1392–1398.

Conde-Sieira, M., Agulleiro, M.J., Aguilar, A.J., Miguez, J.M., Cerda-Reverter, J.M., Soengas, J.L., 2010. Effect of different glycaemic conditions on gene expression of neuropeptides involved in control of food intake in rainbow trout; interaction with stress. J. Exp. Biol. 213, 3858–3865.

Dall'Alba, V., Silva, F.M., Antonio, J.P., Steemburgo, T., Royer, C.P., Almeida, J.C., Gross, J.L., Azevedo, M.J., 2013. Improvement of the metabolic syndrome profile by soluble fibre – guar gum – in patients with type 2 diabetes: a randomised clinical trial. Br. J. Nutr. 110, 1601–1610.

Dincer, Y., Akcay, T., Alademir, Z., Ilkova, H., 2002. Assessment of DNA base oxidation and glutathione level in patients with type 2 diabetes. Mutat. Res. 505, 75–81.

Dvir, I., Stark, A.H., Chayoth, R., Madar, Z., Arad, S.M., 2009. Hypocholesterolemic effects of nutraceuticals produced from the red microalga *Porphyridium* sp. in rats. Nutrients 1, 156–167.

Fujita, H., Fujishima, H., Chida, S., Takahashi, K., Qi, Z., Kanetsuna, Y., Breyer, M.D., Harris, R.C., Yamada, Y., Takahashi, T., 2009. Reduction of renal superoxide dismutase in progressive diabetic nephropathy. J. Am. Soc. Nephrol. 20, 1303–1313.

Gado, A.M., Aldahmash, B.A., 2013. Antioxidant effect of Arabic gum against mercuric chloride-induced nephrotoxicity. Drug Des. Dev. Ther. 7, 1245–1252.

Ganta, S., Chadwick, S., Deshpande, D., Amiji, M., 2009. Role of eco-friendly strategies in the development of biomedical nanotechnology. Int. J. Green Nanotechnol. Biomed. 1, B9–B23.

Gaskins, A.J., Mumford, S.L., Zhang, C., Wactawski-Wende, J., Hovey, K.M., Whitcomb, B.W., Howards, P.P., Perkins, N.J., Yeung, E., Schisterman, E.F., BioCycle Study, G., 2009. Effect of daily fiber intake on reproductive function: the BioCycle Study. Am. J. Clin. Nutr. 90, 1061–1069.

Gaskins, A.J., Mumford, S.L., Wactawski-Wende, J., Schisterman, E.F., 2012. Effect of daily fiber intake on luteinizing hormone levels in reproductive-aged women. Eur. J. Nutr. 51, 249–253.

Gill, S.S., Tuteja, N., 2010. Reactive oxygen species and antioxidant machinery in abiotic stress tolerance in crop plants. Plant Physiol. Biochem. 48, 909–930.

Goldin, B.R., Adlercreutz, H., Gorbach, S.L., Warram, J.H., Dwyer, J.T., Swenson, L., Woods, M.N., 1982. Estrogen excretion patterns and plasma levels in vegetarian and omnivorous women. N. Engl. J. Med. 307, 1542–1547.

Goodrum, L.J., Patel, A., Leykam, J.F., Kieliszewski, M.J., 2000. Gum Arabic glycoprotein contains glycomodules of both extensin and arabinogalactan-glycoproteins. Phytochemistry 54, 99–106.

Goth, L., 2000. Lipid and carbohydrate metabolism in acatalasemia. Clin. Chem. 46, 564–566.

Grindflek, E., Sundvold, H., Lien, S., Rothschild, M.F., 2000. Rapid communication: physical and genetic mapping of the Peroxisome Proliferator Activated Receptor gamma (PPAR-gamma) gene to porcine chromosome 13. J. Anim. Sci. 78, 1391–1392.

Gurib-Fakim, A., 2006. Medicinal plants: traditions of yesterday and drugs of tomorrow. Mol. Aspects Med. 27, 1–93.

Hayes, M.R., De Jonghe, B.C., Kanoski, S.E., 2010. Role of the glucagon-like-peptide-1 receptor in the control of energy balance. Physiol. Behav. 100, 503–510.

Howarth, N.C., Saltzman, E., Roberts, S.B., 2001. Dietary fiber and weight regulation. Nutr. Rev. 59, 129–139.

Ilnytska, O., Argyropoulos, G., 2008. The role of the Agouti-Related Protein in energy balance regulation. Cell Mol. Life Sci. 65, 2721–2731.

Ilnytska, O., Stutz, A.M., Park-York, M., York, D.A., Ribnicky, D.M., Zuberi, A., Cefalu, W.T., Argyropoulos, G., 2011. Molecular mechanisms for activation of the agouti-related protein and stimulation of appetite. Diabetes 60, 97–106.

InterAct, C., 2015. Dietary fibre and incidence of type 2 diabetes in eight European countries: the EPIC-InterAct Study and a meta-analysis of prospective studies. Diabetologia 58, 1394–1408.

Izzo, A.A., Piscitelli, F., Capasso, R., Aviello, G., Romano, B., Borrelli, F., Petrosino, S., Di Marzo, V., 2009. Peripheral endocannabinoid dysregulation in obesity: relation to intestinal motility and energy processing induced by food deprivation and re-feeding. Br. J. Pharmacol. 158, 451–461.

Kaddam, L., FdleAlmula, I., Eisawi, O.A., Abdelrazig, H.A., Elnimeiri, M., Lang, F., Saeed, A.M., 2015. Gum Arabic as fetal hemoglobin inducing agent in sickle cell anemia; in vivo study. BMC Hematol. 15, 19.

Keenan, M.J., Zhou, J., McCutcheon, K.L., Raggio, A.M., Bateman, H.G., Todd, E., Jones, C.K., Tulley, R.T., Melton, S., Martin, R.J., Hegsted, M., 2006. Effects of resistant starch, a non-digestible fermentable fiber, on reducing body fat. Obesity (Silver Spring) 14, 1523–1534.

Kim, E.K., Oh, T.J., Kim, L.-K., Cho, Y.M., 2016. Improving effect of the acute administration of dietary fiber-enriched cereals on blood glucose levels and gut hormone secretion. J. Korean Med. Sci. 31, 222–230.

Kishimoto, A., Ushida, K., Phillips, G.O., Ogasawara, T., Sasaki, Y., 2006. Identification of intestinal bacteria responsible for fermentation of gum Arabic in pig model. Curr. Microbiol. 53, 173–177.

Koch, M., Horvath, T.L., 2014. Molecular and cellular regulation of hypothalamic melanocortin neurons controlling food intake and energy metabolism. Mol. Psychiatry 19, 752–761.

Kong, H., Yang, J., Zhang, Y., Fang, Y., Nishinari, K., Phillips, G.O., 2014. Synthesis and antioxidant properties of gum Arabic-stabilized selenium nanoparticles. Int. J. Biol. Macromol. 65, 155–162.

Kraemer, F.B., Shen, W.J., 2002. Hormone-sensitive lipase: control of intracellular tri-(di-)acylglycerol and cholesteryl ester hydrolysis. J. Lipid Res. 43, 1585–1594.

Kristensen, M., Jensen, M.G., 2011. Dietary fibres in the regulation of appetite and food intake. Importance of viscosity. Appetite 56, 65–70.

Lattimer, J.M., Haub, M.D., 2010. Effects of dietary fiber and its components on metabolic health. Nutrients 2, 1266–1289.

Lewis, S.J., Heaton, K.W., Oakey, R.E., McGarrigle, H.H., 1997. Lower serum oestrogen concentrations associated with faster intestinal transit. Br. J. Cancer 76, 395–400.

Liu, J.H., Yen, S.S., 1983. Induction of midcycle gonadotropin surge by ovarian steroids in women: a critical evaluation. J. Clin. Endocrinol. Metab. 57, 797–802.

Liu, C.Y., Grant, A.L., Kim, K.H., Mills, S.E., 1994. Porcine somatotropin decreases acetyl-CoA carboxylase gene expression in porcine adipose tissue. Domest. Anim. Endocrinol. 11, 125–132.

Liu, L., Song, Z., Jiao, H., Lin, H., 2014. Glucocorticoids increase NPY gene expression via hypothalamic AMPK signaling in broiler chicks. Endocrinology 155, 2190–2198.

Liu, Y., Hou, Z., Yang, J., Gao, Y., 2015. Effects of antioxidants on the stability of beta-Carotene in O/W emulsions stabilized by gum Arabic. J. Food Sci. Technol. 52, 3300–3311.

Mahomoodally, M.F., 2013. Traditional medicines in Africa: an appraisal of ten potent african medicinal plants. Evid. Based Complement. Altern. Med. 2013, 617459.

Mahomoodally, M.F., Muthoorah, L.D., 2014. An ethnopharmacological survey of natural remedies used by the Chinese community in Mauritius. Asian Pac. J. Trop. Biomed. 4, S387–S399.

Marcuse, R., 1960. Antioxidative effect of amino-acids. Nature 186, 886–887.

Martinez de Morentin, P.B., Whittle, A.J., Ferno, J., Nogueiras, R., Dieguez, C., Vidal-Puig, A., Lopez, M., 2012. Nicotine induces negative energy balance through hypothalamic AMP-activated protein kinase. Diabetes 61, 807–817.

Martin-Fardon, R., Boutrel, B., 2012. Orexin/hypocretin (Orx/Hcrt) transmission and drug-seeking behavior: is the paraventricular nucleus of the thalamus (PVT) part of the drug seeking circuitry? Front. Behav. Neurosci. 6, 75.

Mason, P., Adams, J., Morris, D.V., Tucker, M., Price, J., Voulgaris, Z., Van der Spuy, Z.M., Sutherland, I., Chambers, G.R., White, S., 1984. Induction of ovulation with pulsatile luteinising hormone releasing hormone. Br. Med. J. (Clin. Res. Ed.) 288, 181–185.

Mercer, R.E., Chee, M.J., Colmers, W.F., 2011. The role of NPY in hypothalamic mediated food intake. Front. Neuroendocrinol. 32, 398–415.

Micevych, P.E., Chaban, V., Ogi, J., Dewing, P., Lu, J.K., Sinchak, K., 2007. Estradiol stimulates progesterone synthesis in hypothalamic astrocyte cultures. Endocrinology 148, 782–789.

Mohamed, R.E., Gadour, M.O., Adam, I., 2015. The lowering effect of gum Arabic on hyperlipidemia in Sudanese patients. Front. Physiol. 6, 160.

Mumford, S.L., Schisterman, E.F., Siega-Riz, A.M., Gaskins, A.J., Wactawski-Wende, J., VanderWeele, T.J., 2011. Effect of dietary fiber intake on lipoprotein cholesterol levels independent of estradiol in healthy premenopausal women. Am. J. Epidemiol. 173, 145–156.

Musa, H.H., Ahmed, A.A., Fedail, J.S., Musa, T.H., Sifaldin, A.Z., 2016. Gum Arabic attenuates the development of nephropathy in Type 1 diabetes rat. Gums and stabilisers for the food industry 18: hydrocolloid functionality for affordable and sustainable global food solutions. R. Soc. Chem., 245–255.

Nakhjavani, M., Morteza, A., Khajeali, L., Esteghamati, A., Khalilzadeh, O., Asgarani, F., Outeiro, T.F., 2010. Increased serum HSP70 levels are associated with the duration of diabetes. Cell Stress Chaperones 15, 959–964.

Nakhjavani, M., Morteza, A., Nargesi, A.A., Mostafavi, E., Esteghamati, A., 2013. Appearance of leptin–HSP70 correlation, in type 2 diabetes. Meta Gene 1, 1–7.

Nasir, O., 2013. Renal and extrarenal effects of gum Arabic (*Acacia senegal*)–what can be learned from animal experiments? Kidney Blood Press. Res. 37, 269–279.

Nasir, O., Artunc, F., Wang, K., Rexhepaj, R., Foller, M., Ebrahim, A., Kempe, D.S., Biswas, R., Bhandaru, M., Walter, M., Mohebbi, N., Wagner, C.A., Saeed, A.M., Lang, F., 2010a. Downregulation of mouse intestinal Na(+)-coupled glucose transporter SGLT1 by gum Arabic (*Acacia senegal*). Cellular physiology and biochemistry. Int. J. Exp. Cell. Physiol. Biochem. Pharmacol. 25, 203–210.

Nasir, O., Wang, K., Foller, M., Bhandaru, M., Sandulache, D., Artunc, F., Ackermann, T.F., Ebrahim, A., Palmada, M., Klingel, K., Saeed, A.M., Lang, F., 2010b. Downregulation of angiogenin transcript levels and inhibition of colonic carcinoma by gum Arabic (*Acacia senegal*). Nutr. Cancer 62, 802–810.

Nunkoo, D.H., Mahomoodally, M.F., 2012. Ethnopharmacological survey of native remedies commonly used against infectious diseases in the tropical island of Mauritius. J. Ethnopharmacol. 143, 548–564.

O'Keefe, J.H., Gheewala, N.M., O'Keefe, J.O., 2008. Dietary strategies for improving post-prandial glucose, lipids, inflammation, and cardiovascular health. J. Am. Coll. Cardiol. 51, 249–255.

Omaima Nasir, S.B., Salim, Abdel-Moneim M., 2016. Protective effect of gum Arabic supplementation for type 2 diabetes mellitus and its complications. Int. J. Multidiscip. Curr. Res. 4, 294.

Othman, R.B., Ibrahim, H., Mankai, A., Abid, N., Othmani, N., Jenhani, N., Tertek, H., Trabelsi, N., Trimesh, A., Mami, F.B., 2013. Use of hypoglycemic plants by Tunisian diabetic patients. Alexandria J. Med. 49, 261–264.

Oyama, L.M., do Nascimento, C.M., Carnier, J., de Piano, A., Tock, L., Sanches Pde, L., Gomes, F.A., Tufik, S., de Mello, M.T., Damaso, A.R., 2010. The role of anorexigenic and orexigenic neuropeptides and peripheral signals on quartiles of weight loss in obese adolescents. Neuropeptides 44, 467–474.

Park, E.Y., Murakami, H., Matsumura, Y., 2005. Effects of the addition of amino acids and peptides on lipid oxidation in a powdery model system. J. Agric. Food Chem. 53, 8334–8341.

Parnell, J.A., Reimer, R.A., 2010. Effect of prebiotic fibre supplementation on hepatic gene expression and serum lipids: a dose-response study in JCR:LA-cp rats. Br. J. Nutr. 103, 1577–1584.

Pasman, W.J., Saris, W.H., Wauters, M.A., Westerterp-Plantenga, M.S., 1997. Effect of one week of fibre supplementation on hunger and satiety ratings and energy intake. Appetite 29, 77–87.

Patel, S., Goyal, A., 2015. Applications of natural polymer gum Arabic: a review. Int. J. Food Prop. 18, 986–998.

Patterson, C.M., Villanueva, E.C., Greenwald-Yarnell, M., Rajala, M., Gonzalez, I.E., Saini, N., Jones, J., Myers, Jr., M.G., 2012. Leptin action via LepR-b Tyr1077 contributes to the control of energy balance and female reproduction. Mol. Metab. 1, 61–69.

Pazos, P., Lima, L., Dieguez, C., Garcia, M.C., 2014. Energy balance regulating neuropeptides are expressed through pregnancy and regulated by interleukin-6 deficiency in mouse placenta. Int. J. Endocrinol. 2014, 537603.

Pereira, M.A., Ludwig, D.S., 2001. Dietary fiber and body-weight regulation. Observations and mechanisms. Pediatr. Clin. North Am. 48, 969–980.

Phillips, G.O., 1998. Acacia gum (gum Arabic): a nutritional fibre; metabolism and calorific value. Food Addit. Contam. 15, 251–264.

Phillips, A.O., Phillips, G.O., 2011. Biofunctional behaviour and health benefits of a specific gum Arabic. Food Hydrocoll. 25, 165–169.

Phillips, R.J., Powley, T.L., 2000. Tension and stretch receptors in gastrointestinal smooth muscle: re-evaluating vagal mechanoreceptor electrophysiology. Brain Res. Brain Res. Rev. 34, 1–26.

Pimentel, G.D., Ropelle, E.R., Rocha, G.Z., Carvalheira, J.B., 2013. The role of neuronal AMPK as a mediator of nutritional regulation of food intake and energy homeostasis. Metabolism 62, 171–178.

Richard, D., Huang, Q., Timofeeva, E., 2000. The corticotropin-releasing hormone system in the regulation of energy balance in obesity. Int. J. Obesity Relat. Metab. Disord. 24 (Suppl. 2), S36–S39.

Rosen, E.D., Spiegelman, B.M., 2006. Adipocytes as regulators of energy balance and glucose homeostasis. Nature 444, 847–853.

Ryan-Harshman, M., Aldoori, W., 2004. How diet and lifestyle affect duodenal ulcers. Review of the evidence. Can. Family Physician 50, 727–732.

Sallis, J.F., Hinckson, E.A., 2014. Reversing the obesity epidemic in young people: building up the physical activity side of energy balance. Lancet Diabetes Endocrinol. 2, 190–191.

Samy, W.M., Ghoneim, A.I., Elgindy, N.A., 2014. Novel microstructured sildenafil dosage forms as wound healing promoters. Expert Opin. Drug Deliv. 11, 1525–1536.

Sánchez, D., Quiñones, M., Moulay, L., Muguerza, B., Miguel, M., Aleixandre, A., 2011. Soluble fiber-enriched diets improve inflammation and oxidative stress biomarkers in Zucker fatty rats. Pharmacol. Res. 64, 31–35.

Schneeman, B.O., 1987. Dietary fiber: comments on interpreting recent research. J. Am. Diet. Assoc. 87, 1163.

Sharma, P., Jha, A.B., Dubey, R.S., Pessarakli, M., 2012. Reactive oxygen species, oxidative damage, and antioxidative defense mechanism in plants under stressful conditions. J. Bot. 2012, 26.

Shultz, T.D., Howie, B.J., 1986. In vitro binding of steroid hormones by natural and purified fibers. Nutr. Cancer 8, 141–147.

Singh, R.P.M.S.H.C.S.B.J., 2014. Hepatoprotective activity of *Acacia senegal* Pod against carbon tetrachloride-induced hepatotoxicity in rats. Int. J. Pharm. Sci. Rev. Res. 1, 165–168.

Slavin, J., 2003. Why whole grains are protective: biological mechanisms. Proc. Nutr. Soc. 62, 129–134.

Slavin, J.L., 2005. Dietary fiber and body weight. Nutrition 21, 411–418.

Slavin, J., 2013. Fiber and prebiotics: mechanisms and health benefits. Nutrients 5, 1417–1435.

Smith, S., Witkowski, A., Joshi, A.K., 2003. Structural and functional organization of the animal fatty acid synthase. Prog. Lipid Res. 42, 289–317.

Smitka, K., Papezova, H., Vondra, K., Hill, M., Hainer, V., Nedvidkova, J., 2013. The role of "mixed" orexigenic and anorexigenic signals and autoantibodies reacting with appetite-regulating neuropeptides and peptides of the adipose tissue-gut-brain axis: relevance to food intake and nutritional status in patients with anorexia nervosa and bulimia nervosa. Int. J. Endocrinol. 2013, 483145.

Smolinske and Susan, C., 1992. Handbook of Food, Drug, and Cosmetic Excipients. 7.

Son, S.M., 2012. Reactive oxygen and nitrogen species in pathogenesis of vascular complications of diabetes. Diabetes Metab. J. 36, 190–198.

Speakman, J.R., 2005. Body size, energy metabolism and lifespan. J. Exp. Biol. 208, 1717–1730.

Suroowan, S., Mahomoodally, F., 2011. Complementary and alternative medicine use among Mauritian women. Complement. Ther. Clin. Pract. 19, 36–43.

Tabish, S.A., 2007. Is diabetes becoming the biggest epidemic of the twenty-first century? Int. J. Health Sci. (Qassim) 1, V–V10.

Taschler, U., Radner, F.P., Heier, C., Schreiber, R., Schweiger, M., Schoiswohl, G., Preiss-Landl, K., Jaeger, D., Reiter, B., Koefeler, H.C., Wojciechowski, J., Theussl, C., Penninger, J.M., Lass, A., Haemmerle, G., Zechner, R., Zimmermann, R., 2011. Monoglyceride lipase deficiency in mice impairs lipolysis and attenuates diet-induced insulin resistance. J. Biol. Chem. 286, 17467–17477.

Tiedge, M., Lortz, S., Munday, R., Lenzen, S., 1998. Complementary action of antioxidant enzymes in the protection of bioengineered insulin-producing RINm5F cells against the toxicity of reactive oxygen species. Diabetes 47, 1578–1585.

Tiwari, B.K., Pandey, K.B., Abidi, A.B., Rizvi, S.I., 2013. Markers of oxidative stress during diabetes mellitus. J. Biomarkers 2013, 8.

Tsai, C.-J., Hsieh, C.-J., Tung, S.-C., Kuo, M.-C., Shen, F.-C., 2012. Acute blood glucose fluctuations can decrease blood glutathione and adiponectin levels in patients with type 2 diabetes. Diabetes Res. Clin. Pract. 98, 257–263.

Tung, Y.-C.L., Ayuso, E., Shan, X., Bosch, F., O'Rahilly, S., Coll, A.P., Yeo, G.S.H., 2010. Hypothalamic-specific manipulation of the ortholog of the human obesity gene affects food intake in rats. PLoS One 5, e8771.

Turvill, J.L., Wapnir, R.A., Wingertzahn, M.A., Teichberg, S., Farthing, M.J., 2000. Cholera toxin-induced secretion in rats is reduced by a soluble fiber, gum Arabic. Dig. Dis. Sci. 45, 946–951.

Tyler, V.B., Robbers, L., 1977. Pharmacognosy. Lea & Febiger, Philadelphia, pp. 64–68.

Ushida, K., 2012. Gum Arabic and its anti-obese effect. Gum Arabic. R. Soc. Chem., 285–290.

Vahouny, G.V., Satchithanandam, S., Chen, I., Tepper, S.A., Kritchevsky, D., Lightfoot, F.G., Cassidy, M.M., 1988. Dietary fiber and intestinal adaptation: effects on lipid absorption and lymphatic transport in the rat. Am. J. Clin. Nutr. 47, 201–206.

Wang, C., Li, S., Shang, D.J., Wang, X.L., You, Z.L., Li, H.B., 2011. Antihyperglycemic and neuroprotective effects of one novel Cu-Zn SOD mimetic. Bioorg. Med. Chem. Lett. 21, 4320–4324.

Wardlaw, S.L., 2011. Hypothalamic proopiomelanocortin processing and the regulation of energy balance. Eur. J. Pharmacol. 660, 213–219.

Wardle, J., Carnell, S., Haworth, C.M., Farooqi, I.S., O'Rahilly, S., Plomin, R., 2008. Obesity associated genetic variation in FTO is associated with diminished satiety. J. Clin. Endocrinol. Metab. 93, 3640–3643.

Weickert, M.O., 2012. What dietary modification best improves insulin sensitivity and why? Clin. Endocrinol. (Oxf) 77, 508–512.

Weickert, M.O., Pfeiffer, A.F., 2008. Metabolic effects of dietary fiber consumption and prevention of diabetes. J. Nutr. 138, 439–442.

Wells, T.N.C., 2011. Natural products as starting points for future anti-malarial therapies: going back to our roots? Malar. J. 10, S3–S13.

Wu, Y., Tang, L., Chen, B., 2014. Oxidative stress: implications for the development of diabetic retinopathy and antioxidant therapeutic perspectives. Oxid. Med. Cell. Longevity 2014, 752387.

Xuan, N.T., Shumilina, E., Nasir, O., Bobbala, D., Gotz, F., Lang, F., 2010. Stimulation of mouse dendritic cells by Gum Arabic. Cellular physiology and biochemistry. Int. J. Exp. Cell. Physiol. Biochem. Pharmacol. 25, 641–648.

Young, J.W., Shrago, E., Lardy, H.A., 1964. Metabolic control of enzymes involved in lipogenesis and gluconeogenesis. Biochemistry 3, 1687–1692.

Zhang, W., Cline, M.A., Gilbert, E.R., 2014. Hypothalamus-adipose tissue crosstalk: neuropeptide Y and the regulation of energy metabolism. Nutr. Metab. (Lond) 11, 27.

Zhao, S., Wang, J., Song, X., Zhang, X., Ge, C., Gao, S., 2010. Impact of dietary protein on lipid metabolism-related gene expression in porcine adipose tissue. Nutr. Metab. (Lond) 7, 6.

Zhao, Y., Yang, K., Wang, F., Liang, Y., Peng, Y., Shen, R., Wong, T., Wang, N., 2012. Associations between metabolic syndrome and syndrome components and retinal microvascular signs in a rural Chinese population: the Handan Eye Study. Graefes Arch. Clin. Exp. Ophthalmol. 1250, 1755–1763.

Gum Arabic Beneficial Effects, Clinical Applications, and Future Prospective

17

Lamis Kaddam, Imad Fdl-Elmula**, Amal Saeed†*
*AlNeelain University, Khartoum, Sudan; **Assafa Academy, Khartoum, Sudan;
†University of Khartoum, Khartoum, Sudan

Introduction

Gum Arabic (GA) is an edible, dried gummy exudate from the trees of *A. senegal* and *Acacia seyal*. GA is in nonviscous, soluble fiber. It is defined as a dried exudation obtained from the stem of *A. senegal* (Ali et al., 2009). GA has wide manufacturing usages as a stabilizer, thickening agent, and emulsifier, mainly in the food and drug production (e.g., in soft drink syrup, gummy candies, and marshmallows). The American Food and Drug Administration recognized GA as one of the safest nutritional fibers (Ali et al., 2009; Babiker et al., 2012). Both human and animals cannot digest GA, so colonic intestinal bacteria ferments GA, which results in the formation of various degradation products, such as short-chain fatty acids (SCFA) (Ballal et al., 2011). GA consumption increases serum short-chain fatty acid concentration, mainly butyrate and propionate (Ali et al., 2009).

The aim of this chapter is to review the recent literature regarding different beneficial effects of GA and new scientific researches and discoveries.

Gum Arabic chemical structure

GA is obtained from *A. senegal* or *A. seyal*. It is a high-molecular-weight polysaccharide containing galactose, rhamnose, and glucuronic acid and arabinose residues (Al-Majed et al., 2003; Patel and Goyal, 2015); also, calcium, potassium and magnesium are contained in GA (Calame et al., 2008). GA also contains 1.5%–2.6% protein and 0.22%–0.39% nitrogen (Ali et al., 2009). Hydroxyproline, serine, proline, and aspartic acid are the main amino acids (Ali et al., 2009). It is highly water-soluble (Codipilly et al., 2006), but insoluble in alcohol. GA is considered to react differently in various applications than other gums according to its low-viscosity effect (Calame et al., 2011). GA is principally indigestible to both humans and animals. GA is gaining its name from Arabian traders who presented it to Europe and played a vital role in its popularity (Patel and Goyal, 2015).

Gum Arabic. http://dx.doi.org/10.1016/B978-0-12-812002-6.00017-8

Health benefits of gum Arabic

GA has a capability to positively modulate the physiological status of human subjects (Khalid et al., 2014). It is claimed to have many biological and beneficial effects that had been investigated in many studies, which will be listed in the next few pages.

Gum Arabic as prebiotic

Prebiotics are nondigestible food elements that initiate the development of bifidogenic and lactic acid bacteria in the gastrointestinal tract (Slavin, 2013). GA is known as a nondigestible food component that positively selectively stimulates the growth and/or activity of one or an exact number of colonic bacteria without encouraging the growth of unwanted bacteria, and thus contributes to host well-being and fitness (Calame et al., 2008). The metabolism of GA is due to bacterial mass found in the cecum. GA is known to augment the endurance of probiotic bacteria under physiological conditions in the gut (Calame et al., 2008; Khalid et al., 2014). *Lactobacilli* and *Bifidobacteria* are the most common target species of probiotics (Slavin, 2013). In vitro study showed that dry spread of GA improved the existence of *Lactobacillus paracasei* that lead to more efficient probiotic delivery (Desmond et al., 2002). In a study conducted in 2008 in healthy volunteers, researchers found that GA has a prebiotic effect (Calame et al., 2008), since GA significantly increases *Bifidobacteria*, *Lactobacteria*, and *Bacteroides* in the gut (Calame et al., 2008). Acacia exudate was shown to produce a superior proliferation in *Bifidobacteria* and *Lactobacilli* than an equivalent dose of inulin, and resulted in less gastrointestinal symptoms, such as gas and bloating (Slavin, 2013).

Gum Arabic and butyrate

Butyrate is an SCFA representing the most important end product of colonic bacterial aerobic fermentation of fiber and starch (Matsumoto et al., 2006). Butyrate is considered a crucial nutrient responsible for metabolic activity and growth of colonocytes, and may work as an elementary protective element against colonic disorders (Slavin, 2013). GA is not degraded in the small intestine, but fermented in the large intestine (Khalid et al., 2014) by microorganisms to SCFA, particularly propionate, acetate, and butyrate (Ali et al., 2009; Ushida et al., 2011). Dietary fiber has an impact on bacterial mass and enzyme activity; the cecum's bacteria count may rise due to more dietary fiber content without altering the sorts of bacteria. The relative amounts of acetate and butyrate produced depend on the amount of GA fed (Eastwood and Morris, 1992). GA increases butyrate serum level (Ali et al., 2009; Ballal et al., 2011; Matsumoto et al., 2006; Younes et al., 2006). SCFAs are water-soluble and are absorbed into the bloodstream (Slavin, 2013). Serum butyrate concentration amplified following oral ingestion of GA in healthy subjects (Ali et al., 2009; Matsumoto et al., 2006).

GA's biologically significant effects have been accredited to butyrate, including modulation of cell proliferation, apoptosis, and regulation of angiogenesis and inflammation (Smith et al., 1998).

Gum Arabic effect on renal functions

GA has a protective role against nephrotoxicity; it clearly raises creatinine clearance and modulates electrolyte excretion, suggesting favorable actions in renal insufficiency (Ali et al., 2013c). It had been used for a long time as traditional medicine for chronic renal failure (CRF) (Ali et al., 2009; 2010a; Nasir et al., 2010). The renal effects of GA are, however, ill-defined (Khalid et al., 2014). GA increases fecal nitrogen excretion and lowers serum urea nitrogen concentration in CRF patients (Al-Mosawi, 2007; Ali et al., 2009, 2010a; Khalid et al., 2014; Matsumoto et al., 2006), and it reduced the need for dialysis from three to two times per week (Al-Majed et al., 2003; Ali et al., 2013a). Study from Central Sudan showed supplementation with 50 g/day of GA led to an improvement of biochemical profile of patients undergoing regular hemodialysis (Ali et al., 2008; Khalid et al., 2014). Regarding animal models, GA was shown experimentally to be beneficial in rats with adenine-induced CRF, but not in rats with surgically-induced CRF (Ali et al., 2010b). GA mechanism of action involves opposing apoptosis induced by adenine (Ali et al., 2013b). In diabetic mice model, GA treatment reduced glucosuria without noticeably influencing the plasma glucose concentration (Nasir et al., 2012). In the same study, GA treatment diminished proteinuria, lowers plasma phosphate level, and increases the glomerular filtration rate (Nasir et al., 2012).

Gum Arabic effect on gastrointestinal tract

GA is capable of modifying intestinal absorption (Rehman et al., 2001). GA ingestion leads to increase in net water, sodium, and glucose absorption in the gastrointestinal tract in animal model, showing proabsorptive properties (Codipilly et al., 2006; Rehman et al., 2001). GA also enhanced zinc (Ibrahim et al., 2004) and Ca^{2+} absorption when orally dissolved in isotonic solutions to animals. In other studies, GA blunted intestinal absorption of Na^{+} and water in healthy mice and enhanced calcium and magnesium uptake, but not in rat model of chronic osmotic diarrhea (Nasir et al., 2010). Direct administration of GA to perfused jejunal segments did not alter intestinal glucose uptake (Nasir, 2012). Mechanism of action either GA modifies nitric oxide (NO) release, acting both as a scavenger of NO (Rehman et al., 2001, 2004) and as a competitive inhibitor of NO synthase (NOS) (Codipilly et al., 2006; Rehman et al., 2004). NO scavenging effect was found to be more significant than NOS inhibition (Rehman et al., 2004). This showed the potential effectiveness of GA in the cases of diarrhea or malabsorption caused by infectious or functional origins (Codipilly et al., 2006; Rehman et al., 2001). In a study conducted in Indonesia, GA supplementation to ORS sachet (oral hydration solution) significantly reduce the period of diarrhea, increasing the effectiveness of ORS (Salwan et al., 2007). A study conducted in South Korea revealed that composite yogurt enriched with acacia gum and *Bifidobacterium lactis* has greater therapeutic effects in patients with irritable bowel syndrome than standard yogurt, and it alleviates constipation among the test group (Min et al., 2012).

Gum Arabic effect on lipid profile and blood pressure

GA ingestion decreases both systolic and diastolic blood pressure in healthy (Doi et al., 2006) and adenine-induced CRF rats (Ali et al., 2010b). GA supplementation decreased arterial blood pressure and proteinuria in diabetic mice (Nasir et al., 2012). Other studies demonstrated the same effect on humans (Al-Mosawi, 2002; Delzenne and Cani, 2005; Patel and Goyal, 2015).

Reports on the effects of GA on lipid metabolism in clinical trials and animal experiments are inconsistent (Ali et al., 2009; Ushida et al., 2011). However, in several studies, GA ingestion reduced plasma cholesterol and low-density lipoprotein concentrations in humans, but the fall was modest (Mee and Gee, 1997). Other studies showed no outcome of the GA on lipid profile (Jensen et al., 1993).

Gum Arabic effects on blood glucose concentration and body mass

GA daily intake results in a substantial fall in body mass index (BMI) and body fat proportion among healthy adult females (Babiker et al., 2012). The influence of GA on obesity in humans may probably be in part due to its impact on satiety (Calame et al., 2011; da Silva et al., 2013; Nasir, 2012). Researchers from the Netherlands found that GA digestion is able to decrease the caloric intake significantly 3 h after consumption (Calame et al., 2011). In animal models, Nasir et al. (2008) revealed that GA significantly blunted the increase in body weight, fasting plasma glucose, and fasting insulin concentrations during high-fat diets (Nasir, 2014; Nasir et al., 2010). Mechanism of action was downregulation of SGLT1 protein expression (Nasir et al., 2010). Intake of GA for 6 months significantly affected the development of the visceral adipose tissue in female mice (Ushida et al., 2011). A double randomized trial was conducted by Sudanese researchers among healthy female volunteers; it showed that GA ingestion for 6 weeks causes substantial reduction in BMI and body fat percentage (Babiker et al., 2012), and suggested the effect could be exploited in the treatment of obesity.

Gum Arabic immune-modulatory and antiinflammatory properties

There are convincing data to suggest that the consumption of prebiotic fibers can modulate immune parameters in gut associated lymphoid tissues (GALT), secondary lymphoid tissues, and peripheral circulation (Schley and Field, 2002). GA beneficial effects may be mediated immunomodulatory, antiinflammatory, and antioxidant actions (Ali et al., 2010b). GA modulates immunity in mice (Ali et al., 2013a). It was found to increase antiinflammatory cytokine IL10 and decrease tumor necrotic factor alpha (TNF-α) and C-reactive protein (CRP) (Ali et al., 2013a,c), demonstrating antiinflammatory effects (Ali et al., 2013c; Xuan et al., 2010). GA intake exhibited decreases in TNF-α expression in the adipose tissues (Ushida et al., 2011). Moreover, GA was found to block the hepatic macrophage function (Ali et al., 2009; Gamal el-din et al., 2003). In Arab folk medicine, GA is used to treat inflammation of the intestinal mucosa and as an external soothing agent (Al-Yahya et al., 2009; Ali et al., 2013; Gamal el-din et al., 2003) GA exerts local antiinflammatory effects on the small intestine by modulating nuclear factor-κB (Wapnir et al., 2008).

GA had the ability to modify the maturation of dendritic cells (DCs) (Xuan et al., 2010). In vitro studies showed that GA has a particularly strong effect on IL-10 and IL-6 formation, which are essential regulators of the immune response (Xuan et al., 2010). GA stimulates the formation of the IL-6, IL-10, and IL-12p70 as well as TNF-α (Xuan et al., 2010). GA-stimulated DCs improved $CD4^+$ T cell proliferation (Xuan et al., 2010). Stimulation of DC by GA resulted in an increased activity of extracellular signal-regulated kinase 1/2 (ERK1/2) evidenced by GA stimulated ERK1/2 phosphorylation on western blotting (Xuan et al., 2010).

Gum Arabic as antioxidant agent

GA has been supposed to act as an antioxidant and cytoprotective agent (Al-Yahya et al., 2009; Ali et al., 2013c, 2014b), and it can protect against experimental hepatic, renal, and cardiac toxicities in rats (Ali et al., 2009). All of these were based on the assumption that GA has strong antioxidant properties (Ali et al., 2010b, 2013c, 2014b); it may enhance the activity of superoxide dismutase in kidney (Ali et al., 2010a). The amino acids tyrosine, histidine, and methionine appear to be responsible for the antioxidant capacity of GA against reactive oxygen species (ROS) (Rodrigues et al., 2012). In induced CRF in rats, GA administration decreased the superoxide production to control levels and raised the level of glutathione (GSH) and total antioxidant capacity (TAC) (Ali et al., 2013a). Al-Yahya et al. (2009) discovered that GA offers protection against cyclophosphamide-induced urinary bladder cytotoxicity in a rat model by neutralizing ROS and mitigating oxidative stress significantly. GA was addressed as a potent superoxide scavenger in doxorubicin-induced cardiotoxicity mice model (Abd-Allah et al., 2002). GA reduced MDA renal level on cisplatin-induced nephrotoxicity rat model (Al-Majed et al., 2003). In animal experiments, GA protects mice from acetaminophen-induced liver injury, proved by the substantial fall in serum alanine transaminase, aspartate transaminase, and hepatic lipid peroxides (Gamal el-din et al., 2003). Kaddam et al. (2017) studied the effect of GA among sickle cell anemia patients. Their work revealed that GA has potent antioxidative effects, as demonstrated by its ability to increase TAC and decrease oxidative stress markers in humans. These observations suggested that GA's potent antioxidant property can be utilized in various clinical conditions and diseases characterized by increased lipid peroxidation and tissue injury.

Gum Arabic hematological effect

One of the recently discovered novel effects of GA its ability to increase fetal hemoglobin level among Sudanese sickle cell patients. GA was found to increase the mean corpuscular volume and decrease mean corpuscular hemoglobin concentration (Kaddam et al., 2015). GA was also found to have mild strength anticoagulant effect. In a two-arm controlled trial, GA found to increase prothrombin time significantly within the physiological range (Abdalla et al., 2014).

Gum Arabic anticancer properties

One of very impressive and promising effects of GA is its ability to reduce angiogenin transcript level expression and improve survival in colon cancer (Nasir et al., 2010). It seems that the main mediator of GA's action is its ability to increase the serum butyrate level (Nasir et al., 2010; Smith et al., 1998).

Gum Arabic and erythropoietin

Erythropoietin (EPO) is a hematopoietic cytokine that regulates red blood cell production. It is produced by peritubular interstitial cells in the renal cortex and outer medulla in response to hypoxia (Wen et al., 1993). It is important to maintain high levels of EPO activity when treating anemic patients (Way et al., 2005). The relation between GA administration and EPO is not well understood because of lack of literature. Ali et al. (2014a) studied the effect of GA administration on the renal EPO level in the CRF animal model. In two separate studies, GA treatment increased EPO level and significantly ameliorated anemia caused by adenine administration (Ali et al., 2013c, 2014a). In animal model of colon cancer and malaria, GA amended the anemic effect of chemical carcinogenesis (Nasir et al., 2010) and malaria parasitemia (Ballal et al., 2011). This may suggest a relationship between GA and EPO level. However, more scientific research is required to investigate the effect of GA on EPO level.

In conclusion, GA has been utilized in the treatment of CRF. Animal experiments indeed reveal several favorable renal effects of GA, including decrease of plasma phosphate concentration, lowering of blood pressure, and reduction of proteinuria. Moreover, GA exerts several extra renal effects with therapeutic potential, such as slowing of intestinal glucose transport, which could be valuable in the prophylaxis and treatment of obesity and diabetes as well as decrease of angiogenin and β-catenin expression, which may inhibit growth of large-bowel carcinoma and inflammatory disorders. However, most of these studies were conducted among animals or in vitro studies. Further clinical trials are warranted to test whether similar beneficial effects could be accomplished in human disease.

List of abbreviations

BMI Body mass index
CRF Chronic renal failure
GA Gum Arabic
HDL High-density lipoproteins
LDL Low-density lipoproteins
SCFA Short-chain fatty acid
TNF-α Tumor necrotic factor alpha

References

Abdalla, M.E., Badi, R.M., Ahmed, A.A., Saeed, A.M., 2014. Human in vivo study: the effect of gum Arabic on coagulation cascade. Int. Res. J. Pharm. Appl. Sci. (IRJPAS) 4 (6), 5–8.

Abd-Allah, A.R., Al-Majed, A.A., Mostafa, A.M., Al-Shabanah, O.A., Din, A.G.E., Nagi, M.N., 2002. Protective effect of arabic gum against cardiotoxicity induced by doxorubicin in mice: a possible mechanism of protection. J. Biochem. Mol. Toxicol. 16 (5), 254–259.

Al-Majed, A.A., Abd-Allah, A.R., Al-Rikabi, A.C., Al-Shabanah, O.A., Mostafa, A.M., 2003. Effect of oral administration of Arabic gum on cisplatin-induced nephrotoxicity in rats. J. Biochem. Mol. Toxicol. 17 (3), 146–153.

Al-Mosawi, A.J., 2002. The challenge of chronic renal failure in the developing world: possible use of acacia gum. Pediatr. Nephrol. 17 (5), 390–391.

Al Mosawi, A.J., 2007. The use of acacia gum in end stage renal failure. J. Trop. Pediatr. 53, 362–365.

Al-Yahya, A.A., Al-Majed, A.A., Gado, A.M., Daba, M.H., Al-Shabanah, O.A., Abd-Allah, A.R., 2009. *Acacia senegal* gum exudate offers protection against cyclophosphamide-induced urinary bladder cytotoxicity. Oxid. Med. Cell. Longev. 2 (4), 207–213.

Ali, B.H., Al Zaabi, M., Ramkumar, A., Yasin, J., Nemmar, A., 2014a. Anemia in adenine-induced chronic renal failure and the influence of treatment with gum acacia thereon. Physiol. Res. 63 (3), 351–358.

Ali, B.H., Al-Husseni, I., Beegam, S., Al-Shukaili, A., Nemmar, A., Schierling, S., Queisser, N., Schupp, N., 2013a. Effect of gum Arabic on oxidative stress and inflammation in adenine-induced chronic renal failure in rats. PLoS One 8 (2), e55242.

Ali, B.H., Al-Salam, S., Al, H., Kayed, I., Al-Masroori, R.R., Al-Harthi, T., Al, Z.M., Nemmar, A., 2010a. Effects of gum Arabic in rats with adenine-induced chronic renal failure. Exp. Biol. Med. (Maywood) 235 (3), 373–382.

Ali, B.H., Al-Salam, S., Al Za'abi, M., Al Balushi, K.A., Ramkumar, A., Waly, M.I., Yasin, J., Adham, S.A., Nemmar, A., 2014b. Does swimming exercise affect experimental chronic kidney disease in rats treated with gum acacia? PLoS One 9 (7), e102528.

Ali, B.H., Al-Salam, S., Al Za'abi, M., Waly, M.I., Ramkumar, A., Beegam, S., Al-Lawati, I., Adham, S.A., Nemmar, A., 2013b. New model for adenine-induced chronic renal failure in mice, and the effect of gum acacia treatment thereon: comparison with rats. J. Pharmacol. Toxicol. Methods 68 (3), 384–393.

Ali, A.A., Ali, K.E., Fadlalla, A.E., Khalid, K.E., 2008. The effects of gum Arabic oral treatment on the metabolic profile of chronic renal failure patients under regular haemodialysis in Central Sudan. Nat. Prod. Res. 22 (1), 12–21.

Ali, B.H., Beegam, S., Al Lawati, I., Waly, M.I., Nemmar, A., 2013c. Comparative efficacy of three brands of gum Arabic on adenine induced chronic renal failure in rats. Physiol. Res. 62 (1), 47–56.

Ali, B.H., Ziada, A., Al, H., I Beegam, S., Nemmar, A., 2010b. Motor and behavioral changes in rats with adenine-induced chronic renal failure: influence of acacia gum treatment. Exp. Biol. Med. (Maywood) 236 (1), 107–112.

Ali, B.H., Ziada, A., Blunden, G., 2009. Biological effects of gum Arabic: a review of some recent research. Food. Chem. Toxicol. 47 (1), 1–8.

Babiker, R., Merghani, T.H., Elmusharaf, K., Badi, R.M., Lang, F., Saeed, A.M., 2012. Effects of gum Arabic ingestion on body mass index and body fat percentage in healthy adult females: two-arm randomized, placebo controlled, double-blind trial. Nutr. J. 11 (1), 111.

Ballal, A., Bobbala, D., Qadri, S.M., Foller, M., Kempe, D., Nasir, O., Saeed, A., Lang, F., 2011. Anti-malarial effect of gum Arabic. Malar. J. 10, 139.

Calame, W., Thomassen, F., Hull, S., Viebke, C., Siemensma, A.D., 2011. Evaluation of satiety enhancement, including compensation, by blends of gum Arabic. A methodological approach. Appetite 57 (2), 358–364.

Calame, W., Weseler, A.R., Viebke, C., Flynn, C., Siemensma, A.D., 2008. Gum arabic establishes prebiotic functionality in healthy human volunteers in a dose-dependent manner. Br. J. Nutr. 100 (6), 1269–1275.

Codipilly, C.N., Teichberg, S., Wapnir, R.A., 2006. Enhancement of absorption by gum Arabic in a model of gastrointestinal dysfunction. J. Am. Coll. Nutr. 25, 307–312.

da Silva, S.T., dos Santos, C.A., Bressan, J., 2013. Intestinal microbiota; relevance to obesity and modulation by prebiotics and probiotics. Nutr. Hosp. 28 (4), 1039–1048.

Delzenne, N.M., Cani, P.D., 2005. A place for dietary fibre in the management of the metabolic syndrome. Curr. Opin. Clin. Nutr. Metab. Care 8 (6), 636–640.

Desmond, C., Ross, R.P., O'Callaghan, E., Fitzgerald, G., Stanton, C., 2002. Improved survival of *Lactobacillus paracasei* NFBC 338 in spray-dried powders containing gum acacia. J. Appl. Microbiol. 93 (6), 1003–1011.

Doi, Y., Ichihara, T., Hagiwara, A., Imai, N., Tamano, S., Orikoshi, H., Ogasawara, K., Sasaki, Y., Nakamura, M., Shirai, T., 2006. A ninety-day oral toxicity study of a new type of processed gum Arabic, from Acacia tree (*Acacia senegal*) exudates, in F344 rats. Food Chem. Toxicol. 44 (4), 560–566.

Eastwood, M.A., Morris, E.R., 1992. Physical properties of dietary fiber that influence physiological function: a model for polymers along the gastrointestinal tract. Am. J. Clin. Nutr. 55 (2), 436–442.

Gamal el-din, A.M., Mostafa, A.M., Al-Shabanah, O.A., Al-Bekairi, A.M., Nagi, M.N., 2003. Protective effect of arabic gum against acetaminophen-induced hepatotoxicity in mice. Pharmacol. Res. 48 (6), 631–635.

Ibrahim, M.A., Kohn, N., Wapnir, R.A., 2004. Proabsorptive effect of gum Arabic in isotonic solutions orally administered to rats: effect on zinc and other solutes. J. Nutr. Biochem. 15 (3), 185–189.

Jensen, C.D., Spiller, G.A., Gates, J.E., Miller, A.F., Whittam, J.H., 1993. The effect of acacia gum and a water-soluble dietary fiber mixture on blood lipids in humans. J. Am. Coll. Nutr. 12 (2), 147–154.

Kaddam, L., FdleAlmula, I., Eisawi, O.A., Abdelrazig, H.A., Elnimeiri, M., Lang, F., Saeed, A.M., 2015. Gum Arabic as fetal hemoglobin inducing agent in sickle cell anemia; in vivo study. BMC Hematol. 15 (19). doi: 10.1186/s12878-015-0040-6.

Kaddam, L., Fadl-Elmula, I., Eisawi, O.A., Abdelrazig, H.A., Salih, M.A., Lang, F., Saeed, A.M., 2017. Gum Arabic as novel anti-oxidant agent in sickle cell anemia, phase II trial. BMC Hematol. 17 (4), 1–16.

Khalid, S.A., Musa, A.M., Saeed, A.M., Abugroun, E.A., Ahmed, E.O.S., Ghalib, M.B., Elnima, E.I., Alkarib, S.Y., Abdelsalam, T.M., Abdelgader, A., 2014. Manipulating dietary fibre: gum Arabic making friends of the colon and the kidney. Bioact. Carbohydr. Dietary Fibre 3 (2), 71–76.

Matsumoto, N., Riley, S., Fraser, D., Al-Assaf, S., Ishimura, E., Wolever, T., Phillips, G.O., Phillips, A.O., 2006. Butyrate modulates TGF-β1 generation and function: potential renal benefit for Acacia(sen) SUPERGUM (gum Arabic)? Kidney Int. 69 (2), 257–265.

Mee, K.A., Gee, D.L., 1997. Apple fiber and gum Arabic lowers total and low-density lipoprotein cholesterol levels in men with mild hypercholesterolemia. J. Am. Diet. Assoc. 97 (4), 422–424.

Min, Y.W., Park, S.U., Jang, Y.S., Kim, Y.H., Rhee, P.L., Ko, S.H., Joo, N., Kim, S.I., Kim, C.H., Chang, D.K., 2012. Effect of composite yogurt enriched with acacia fiber and *Bifidobacterium lactis*. World J. Gastroenterol. 18 (33), 4563–4569.

Nasir, O., 2012. Renal and extrarenal effects of gum Arabic (*Acacia senegal*)—what can be learned from animal experiments? Kidney Blood Press. Res. 37 (4–5), 269–279.

Nasir, O., 2014. Effect of gum Arabic (*Acacia Senegal*) on glucose metabolism and body weight gain in mice. J. Biol. Agric. Healthcare 4 (9), 34–41.

Nasir, O., Artunc, F., Saeed, A., Kambal, M.A., Kalbacher, H., Sandulache, D., Boini, K.M., Jahovic, N., Lang, F., 2008. Effects of gum Arabic (*Acacia senegal*) on water and electrolyte balance in healthy mice. J. Ren. Nutr. 18 (2), 230–238.

Nasir, O., Artunc, F., Wang, K., Rexhepaj, R., Foller, M., Ebrahim, A., Kempe, D.S., Biswas, R., Bhandaru, M., Walter, M., Mohebbi, N., Wagner, C.A., Saeed, A.M., Lang, F., 2010. Downregulation of mouse intestinal Na^+-coupled glucose transporter SGLT1 by gum Arabic (*Acacia Senegal*). Cell. Physiol. Biochem. 25 (2–3), 203–210.

Nasir, O., Umbach, A.T., Rexhepaj, R., Ackermann, T.F., Bhandaru, M., Ebrahim, A., Artunc, F., Kempe, D.S., Puchchakayala, G., Siraskar, B., Foller, M., Saeed, A., Lang, F., 2012. Effects of gum Arabic (*Acacia senegal*) on renal function in diabetic mice. Kidney Blood Press. Res. 35 (5), 365–372.

Patel, S., Goyal, A., 2015. Applications of natural polymer gum Arabic: a review. Int. J. Food Prop. 18 (5), 986–998.

Rehman, K.U., Codipilly, C.N., Wapnir, R.A., 2004. Modulation of small intestinal nitric oxide synthase by gum Arabic. Exp. Biol. Med. 229 (9), 895–901.

Rehman, K., Wingertzahn, M.A., Harper, R.G., Wapnir, R.A., 2001. Proabsorptive action of gum Arabic: regulation of nitric oxide metabolism in the basolateral potassium channel of the small intestine. J. Pediatr. Gastroenterol. Nutr. 32 (5), 529–533.

Rodrigues, E., Mariutti, L.R., Faria, A.F., Mercadante, A.Z., 2012. Microcapsules containing antioxidant molecules as scavengers of reactive oxygen and nitrogen species. Food Chem. 134 (2), 704–711.

Salwan, H., Isnada, A.B., Ismail, R., Bahar, E., 2007. Benefits of gum Arabic supplementation to oral rehydration solution in managing acute diarrhea. Paediatr. Indones. 47 (6), 265.

Schley, P.D., Field, C.J., 2002. The immune-enhancing effects of dietary fibres and prebiotics. Br. J. Nutr. 87 (Suppl 2), S221–S230.

Slavin, J., 2013. Fiber and prebiotics: mechanisms and health benefits. Nutrients 5 (4), 1417–1435.

Smith, J.G., Yokoyama, W.H., German, J.B., 1998. Butyric acid from the diet: actions at the level of gene expression. Crit. Rev. Food Sci. Nutr. 38 (4), 259–297.

Ushida, K., Hatanaka, H., Inoue, R., Tsukahara, T., Phillips, G.O., 2011. Effect of long term ingestion of gum Arabic on the adipose tissues of female mice. Food Hydrocoll. 25 (5), 1344–1349.

Wapnir, R.A., Sherry, B., Codipilly, C.N., Goodwin, L.O., Vancurova, I., 2008. Modulation of rat intestinal nuclear factor NF-kappaB by gum arabic. Rat small intestine by gum arabic. Dig. Dis. Sci. 53, 80–87.

Way, J.C., Lauder, S., Brunkhorst, B., Kong, S.M., Qi, A., Webster, G., Campbell, I., McKenzie, S., Lan, Y., Marelli, B., Nguyen, L.A., Degon, S., Lo, K.M., Gillies, S.D., 2005. Improvement of Fc-erythropoietin structure and pharmacokinetics by modification at a disulfide bond. Protein Eng. Des. Sel. 18 (3), 111–118.

Wen, D., Boissel, J.P., Tracy, T.E., Gruninger, R.H., Mulcahy, L.S., Czelusniak, J., Goodman, M., Bunn, H.F., 1993. Erythropoietin structure–function relationships: high degree of sequence homology among mammals. Blood 82 (5), 1507–1516, Available from: PM: 8364201.

Xuan, N.T., Shumilina, E., Nasir, O., Bobbala, D., Goetz, F., Lang, F., 2010. Stimulation of mouse dendritic cells by gum Arabic. Cell. Physiol. Biochem. 25 (6), 641–648.

Younes, H., Egret, N., Hadj-Abdelkader, M., Rémésy, C., Demigné, C., Gueret, C., Deteix, P., Alphonse, J.C., 2006. Fermentable carbohydrate supplementation alters nitrogen excretion in chronic renal failure. J. Ren. Nutr. 16, 67–74.

Use of Acacia Gum in the Treatment of Skin Lesions of Two Children With Kwashiorkor

18

Insaf Abdel Karim Elhag Ali
King Faisal University, Al-Ahsa, Saudi Arabia

List of abbreviations

FAO	Food and Agricultural Organization
GA	Gum Arabic
JECFA	Joint Expert Committee for Food Additives
WHO	World Health Organization

Introduction

Acacia gum, or gum Arabic (GA), is defined by the Food and Agricultural Organization/World Health Organization/Joint Expert Committee for Food Additives as "a dried exudation obtained from the stems of *Acacia senegal* or closely related species of *Acacia* (family Leguminosae) (FAO, 1999)." The acacia tree (*A. senegal*) is a thorny, scraggly tree that grows to heights of about 15 ft. It grows most prolifically in regions of Africa, in particular in the Republic of the Sudan. During times of drought, the bark of the tree splits, exuding a sap that dries in small droplets, or "tears" (Dobelis, 1986).

GA is a complex mixture of polysaccharides, protein, and arabinogalactan protein species. It has been shown to be highly heterogeneous and is found in nature as mixed calcium, magnesium, potassium, and sodium salts of a polysaccharidic acid (Arabic acid). However, other heavy elements, such as Zn, Al, Cd, Cu, Cr, Pb, and Co, may also be present, but in very small quantities (Anderson et al., 1983; Osman et al., 1993).

History and origins of gum Arabic

GA is certainly the most ancient and the most well-known of all gum types. The term "gum Arabic" was coined by European merchants who imported it from Arab ports such as Jeddah and Alexandria. This gum was supposedly also used to fix pigments into hieroglyphic paintings. According to a Sudanese researcher, the word *mana* (manna) mentioned in the Koran (*Surah Al-Baqarah*) as the best food available to man is, in fact, a direct reference to GA. The word *mana* seemingly also refers to GA in the

Gum Arabic. http://dx.doi.org/10.1016/B978-0-12-812002-6.00018-X

Torah, where it is described as an essential food and designated by Moses to the Israelis as God-given bread. There are close to 900 acacia species capable of producing gum. These are primarily located in tropical climates, with about 130 of them located specifically on the African continent (ITC, 2008).

Uses of acacia gum

The usage of GA dates back to 5000 years ago. GA has been regarded as therapeutic for chronic kidney diseases in Middle Eastern countries (Nasir et al., 2012).

In folk medicine, GA has been reported to be used internally for the treatment of inflammation of the intestinal mucosa, and externally to cover inflamed surfaces. It has been reported that GA has nitric oxide scavenging properties (Rehman et al., 2001). Despite the fact that GA is widely used as a vehicle for drugs in experimental, physiological, and pharmacological experiments and is assumed to be an "inert" substance, some recent reports have claimed that GA possesses antioxidant, nephron-protectant, and other effects (Ali et al., 2008).

Clinically, it has been tried in patients with chronic renal failure, and it was claimed that it helps reduce urea and creatinine plasma concentrations and reduces the need for dialysis from three to two times per week (Suliman et al., 2000). GA has been used in pharmaceuticals as a demulcent, and soothes irritated mucous membranes. Consequently, it is used widely in topical preparations to promote wound healing and as a component of cough and some gastrointestinal preparations. In addition, it has been shown to inhibit the growth of periodontitis bacteria and the early deposition of plaque (Clark et al., 1993). Nishi et al. (2007) investigated the adjuvant role of oxidized GA in amphotericin B therapy against fungal pathogens: *Candida albicans* and *Cryptococcus neoformans*, and *Leishmania* caused by *Leishmania donovani*. The drug conjugate was stable, nonhemolytic, and nontoxic to the internal organs of the animal and showed good antifungal and antileishmanial activity in vitro (Nishi et al., 2007).

Ballal et al. (2011) conducted both in vitro and in vivo studies to test the inhibitory effect of GA on malaria. The human erythrocytes were in vitro infected with *Plasmodium falciparum*. Mice were in vivo infected with *Plasmodium berghei* ANKA by intraperitoneal injection of parasitized murine erythrocytes. The oral administration of 10% GA, 10 days prior to the infection, showed an attenuated level of parasites in blood. The GA degradation product butyrate is attributed to the reduced vigor of malaria (Ballal et al., 2011).

Severe malnutrition, manifested by wasting, edema, or both, occurs almost exclusively in children. Marasmus is defined as severe wasting; marasmic kwashiorkor, as severe wasting in the presence of edema; and kwashiorkor, as malnutrition with edema. The word "kwashiorkor" originated from a local language in Ghana; which implies "the disease that the young child developed when displaced from his mother by another child or pregnancy (Millward and Jackson, 2004)." Kwashiorkor is a form of severe protein–energy malnutrition characterized by edema, irritability, ulcerating dermatoses, and an enlarged liver with fatty infiltrates. Sufficient calorie intake, but with insufficient protein consumption, distinguishes it from marasmus. Kwashiorkor cases occur in areas of famine or poor food supply (Krebs et al., 2009).

A characteristic dermatosis is often seen in young children suffering from kwashiorkor; the skin lesions usually first occur in areas subjected to friction or pressure, for example, the groin, behind the knees, on the buttocks, and at the elbows; in advanced cases, the dermatosis may be almost anywhere on the body: trunk, limbs, and head. Darker pigmented patches form, and these may peel or desquamate, rather like old, sunbaked, blistered paint. This has led to the terms "peeling paint" or "flaky paint" dermatosis. Underneath these flakes are atrophic, depigmented areas that may resemble a healing burn (Latham, 1991).

Steps of treating dermatosis in severe malnutrition include:

- applying a barrier cream or ointment to raw areas;
- zinc and castor oil ointment;
- nystatin ointment or cream for diaper areas; and
- if severe (+++), bath for 10–15 min per day in 1% potassium permanganate solution (WHO, 1999).

Acacia has also been known to have its beneficial effects in treating various skin diseases, including the pores in the skin and the leading acne. Acne is usually common in teenagers. The pores on the skin usually get infected, which further lead to acne breakouts. Acacia honey contains a large concentration of fructose sugar in it, and it does not crystallize. This acacia honey actually possesses antimicrobial properties in it, which, in turn, can help in reducing acne, especially when combined with other medications (https://www.epainassist.com/articles/health-benefits-of-acacia-and-its-side-effects). *A. senegal* can be applied externally for skin inflammations and to treat minor cuts, wounds, and sores; it helps relieve the pain and prevent infection.

A. senegal is a very good oil-in-water emulsifier, which is in normal use in the food and pharmaceutical businesses. Restoratively, GA is utilized widely in pharmaceutical arrangements and is a nourishment-added substance endorsed as toxicologically safe by the Codex Alimentarius. It has been utilized as demulcent, skin defensive agent, and pharmaceutical guides, for example, emulsifier and stabilizer of suspensions and added substances for strong formulations. It is, some of the time, used to treat bacterial and fungal diseases of the skin and mouth. It has been accounted for to soothe the mucous films of the digestive organs and to treat inflamed skin (Aliyu, 2006; Okoro et al., 2011). The demulcent, emollient gum is utilized inside against irritation of intestinal mucosa and remotely to cover inflamed surfaces, such as burns, sore nipples, and nodular leprosy. Moreover, it has likewise been reported to be utilized as antitussive, expectorant, astringent, catarrh, and against colds, coughs, diarrhea, gonorrhea, discharge, sore throat, typhoid, and urinary tract ailments (Okoro et al., 2011). The gum of *A. senegal* has been pharmaceutically utilized, chiefly in the fabrication of emulsions and in making pills and troches, as a demulcent for inflammations of the throat or stomach, and as a masking agent for acrid-tasting substances, for example, capsicum, and furthermore as a film-framing operator in peel-off masks. GA is additionally utilized broadly as an ingredient in foods like confections and soft drinks, as the gum has the properties of glue but is safe to eat. GA is broadly utilized as a part of natural items as a common contrasting option to substance folios and is utilized as a part of business emulsification for the generation of drinks and flavor concentrates (Aliyu, 2006; Brendler et al., 2010; Mahomoodally, 2013; Okoro et al., 2011).

We investigated the topical use of GA powder in treating skin lesions of two cases of kwashiorkor, documenting the effect of the gum powder on healing of these skin lesions by taking photos before and after using the gum powder.

The first case (case A), was a 9-month-old boy diagnosed as having kwashiorkor; he was edematous, had severe skin lesions with desquamation, and "peeling paint" ulcers. The patient was very ill, septicemic, and treated with broad-spectrum antibiotics, topical gentian violet, and adequate nutrition for 10 days without improvement. After taking verbal consent from the mother, the skin lesion was treated by applying the acacia gum powder over the whole body anteriorly, waiting until the powder was absorbed, and then applied posteriorly. We observed that within 5–7 days, there was complete resolution of the skin lesions (Fig. 18.1), reduction of the edema, and improvement in the general condition of the infant.

In Fig. 18.1, the first column shows the lesion before using acacia gum powder, and the other two columns demonstrate the same lesions after using the acacia gum powder for 5 and 7 days.

The second case (case B) is a 2-year-old boy with kwashiorkor; he had lower limb edema and desquamated and hypopigmented skin changes in his four limbs. After topical use of acacia gum powder for 5 days, the skin color became homogeneous, as shown in Fig. 18.2.

From this figure, the first column shows skin desquamation and hypopigmentation of the malnourished infant before using acacia gum powder, and the other two columns demonstrate the same lesions after using the acacia gum powder for 5 days.

Discussion

In undamaged skin, the epidermis (surface layer) and dermis (deeper layer) form a protective barrier against the external environment. When the barrier is broken, an orchestrated cascade of biochemical events set into motion to repair the damage. This process is divided into predictable phases: blood clotting (hemostasis), inflammation, tissue growth (proliferation), and tissue remodeling (maturation) (Orgill and Blanco, 2009; Rieger et al., 2014). Wound healing is a complex and dynamic process of replacing devitalized and missing cellular structures and tissue layers (Gilmore, 1991). Malnutrition or nutritional deficiencies have a recognizable impact on wound healing post trauma or surgical intervention. Nutrients including proteins, carbohydrates, arginine, glutamine, polyunsaturated fatty acids, vitamin A, vitamin C, vitamin E, magnesium, copper, zinc, and iron all play significant roles in wound healing (Arnold and Barbul, 2006). Fats and carbohydrates provide the majority of energy required for wound healing. Glucose is the most prominent source of fuel, and it is used to create cellular adenosine triphosphate, providing energy for angiogenesis and the deposition of new tissues (Guo and DiPietro, 2010).

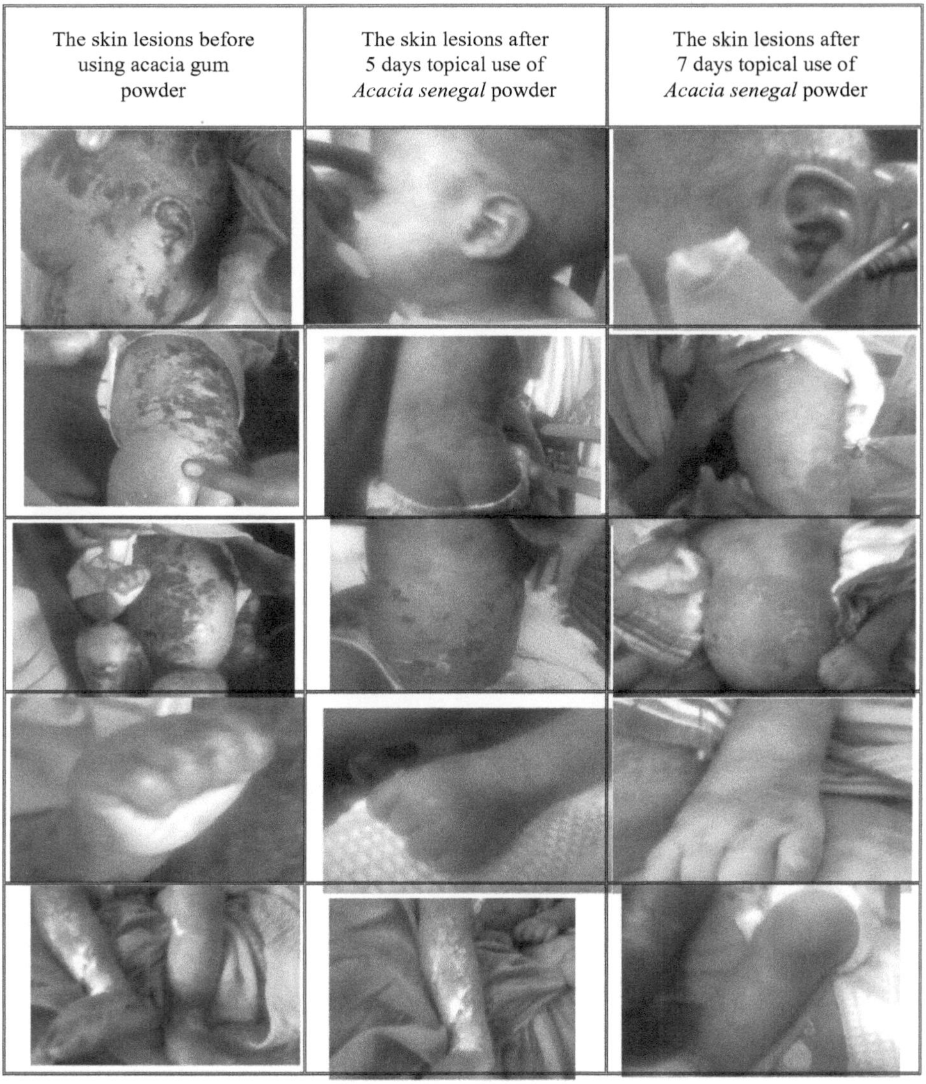

Figure 18.1 Effect of applying the acacia gum powder over skin lesions of a case of kwashiorkor (case A).

In the literature, there is one study done to compare efficacy of *A. senegal* barrier for stoma care in children after doing colostomy. It compared the protective effects of *A. senegal* barrier with zinc sulfate ointment in a total of 60 infants (30 as case study and 30 as control). Results showed that there was a statistically significant difference in peristomal skin inflammation rate in groups; infants who had *A. senegal* barrier showed lower and less sever inflammation rates ($p = 0.05$). Compared with zinc oxide, they found a lower rate of dermatitis in the acacia group (Hosseinpour et al., 2012).

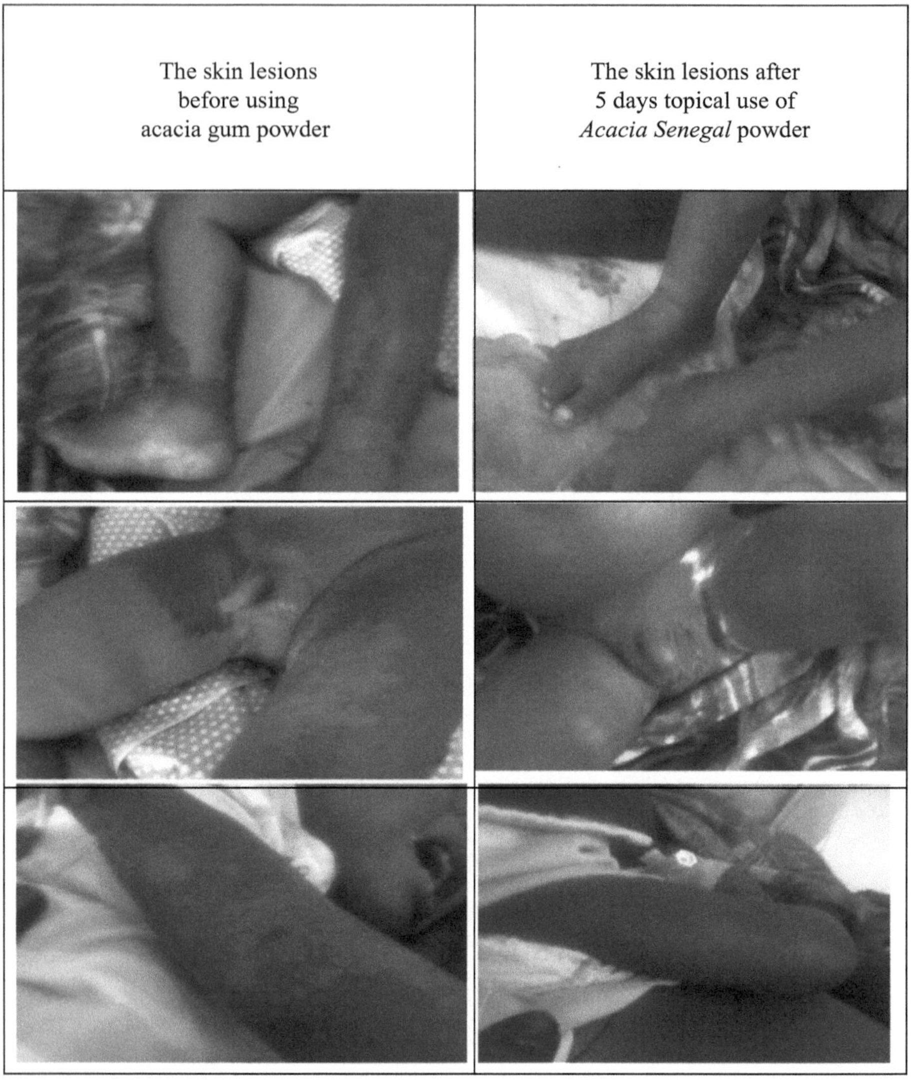

Figure 18.2 Effect of topical use of acacia gum powder on skin lesions of a case of kwashiorkor (case B).

It is mentioned that the gum flows naturally from the bark of the *A. senegal* tree in the form of a thick and rather frothy liquid, and speedily concretes in the sun into tears; sometimes, the discharge is promoted by wounding the trunk. During times of drought, the bark of the tree splits, exuding a sap that dries in small droplets or tears (Dobelis, 1986).

When we used acacia gum in treating inflamed skin in two malnourished infants, there was rapid and dramatic skin repair. As exudation of GA from the bark of the

A. senegal tree is a natural response that helps in healing of the bark when it is wounded by hot weather and drought or infected by microorganisms, we recommend to do further studies on the constituents of the GA and the mechanism by which it can affect the healing process of skin lesions.

References

Ali, A., Ali, E., Fadlalla, E., Khalid, E., 2008. The effects of G.A. oral treatment on the metabolic profile of chronic renal failure patients under regular haemodialysis in Central Sudan. Nat. Prod. Res. 22, 12–21.

Aliyu, B.S., 2006. Common Ethnomedicinal Plants of the Semiarid Regions of West Africa. Triumph Publishing, Kano, Nigeria.

Anderson, D., Bridgeman, M., Farouhar, J., McNab, C., 1983. The chemical characterization of the test article used in toxicological studies of gum arabic (*Acacia senegal* (L.) Willd). Int. Tree Crops J. 2 (3–4), 245–254.

Arnold, M., Barbul, A., 2006. Nutrition and wound healing. Plast. Reconstr. Surg. 117 (7), 42–58.

Ballal, A., Bobbala, D., Qadri, S., Foller, M., Kempe, D., Nasir, O., Lang, F., 2011. Anti-malarial effect of gum Arabic. Malar. J. 10, 139.

Brendler, T., Eloff, L.N., Gurib-Fakim, A., Phillips, L.D., 2010. African Herbal Pharmacopeia. AAMPS Publishing, Mauritius.

Clark, T., Gazi, I., Cox, W., Eley, M., Tinsley, F., 1993. The effects of Acacia arabica gum on the in vitro growth and protease activities of periodontopathic bacteria. J. Clin. Periodontol. 20 (4), 238–243.

Dobelis, N., 1986. Magic and Medicine of Plants. Reader's Digest Association, Pleasantville, NY.

FAO, 1999. Compendium of Food Additive Specifications Addendum 7. Food and Nutrition Paper, No. 52. Add. 7. Joint FAO/WHO Expert Committee on Food Additives 53rd Session, 1st–10th June, Rome, Italy.

Gilmore, M., 2010. Phases of wound healing, dimensions in Oncology. Nursing 5 (3), 32–34.

Guo, S., DiPietro, L., 2010. Factors affecting wound healing. J. Dent. Res. 89 (3), 219–229.

Hosseinpour, M., Fazeli, A., Agabeigi, M., 2012. Efficacy of *Acacia senegal* for stoma care in children with colostomy. Eur. J. Pediatr. Surg. 22 (3), 234–237.

ITC, 2008. Gum Arabic. Bulletin, Market News Service (MNS), Quarterly Edition.

Krebs, N., Primak, L., Hambridge, K., 2009. Normal childhood nutrition & its disorders. Current Paediatric Diagnosis & Treatment. McGraw-Hill Education, Blacklick, OH, USA.

Latham, M., 1991. The dermatosis of kwashiorkor in young children. Semin. Dermatol. 10 (4), 270–272.

Mahomoodally, M.F., 2013. Traditional medicines in Africa: an appraisal of ten potent African medicinal plants. Evid Based Complement. Alternat. Med. 2013, 617459.

Millward, D., Jackson, A., 2004. Protein/energy ratios of current diets in developed and developing countries compared with a safe protein/energy ratio: implications for recommended protein and amino acid intakes. Public Health Nutr. 7 (2), 387–405.

Nasir, O., Umbach, A., Rexhepaj, R., Ackermann, F., Bhandaru, M., Ebrahim, A., Lang, F., 2012. Effects of gum Arabic (*Acacia senegal*) on renal function in diabetic mice. Kidney Blood Press. Res. 35 (5), 365–372.

Nishi, K., Antony, M., Mohanan, P., Anilkumar, T., Loiseau, P., Jayakrishnan, A., 2007. Amphotericin B-gum Arabic conjugates: synthesis, toxicity, bioavailability, and activities against Leishmania and fungi. Pharm. Res. 24, 971–980.

Okoro, S.O., Kawo, A.H., Arzai, A.H., 2011. Phytochemical screening, antibacterial and toxicological activities of *Acacia senegal* extracts. Bayero J. Pure Appl. Sci. 5 (1), 163–170.

Orgill, D., Blanco, C., 2009. The pathophysiologic basis for wound healing and cutaneous regeneration. Biomaterials for Treating Skin Loss, Sawston, Cambridge, United Kingdom, p. 25.

Osman, M., Williams, P., Menzies, A., Phillips, G., Baldwin, T., 1993. The molecular characterisation of the polysaccharide gum from *Acacia senegal*. Carbohydr. Res. 246 (1), 303–318.

Rehman, K., Wingertzahn, A., Harper, G., Wapnir, A., 2001. Proabsorptive action of gum Arabic: regulation of nitric oxide metabolism in the basolateral potassium channel of the small intestine. J. Pediatr. Gastroenterol. Nutr. 35, 529–533.

Rieger, S., Zhao, H., Martin, P., Abe, K., Lisse, T., 2014. The role of nuclear hormone receptors in cutaneous wound repair. Cell Biochem. Funct. 33 (1), 1–13.

Suliman, S.M., Hamdouk, M.I., Elfak, M.B., 2000. Gum Arabic fibre as a supplement to low protein diet in chronic renal failure patients. Sudan Association of Physicians, 17th Conference, March 21–23, Friendship Hall, Khartoum, Sudan.

World Health Organization (WHO), 1999. Management of Severe Malnutrition: A Manual for Physicians and Other Senior Health Workers., Geneva.

Extraction and Utilization of Sudanese Acacia *Senegale* Seeds (Gum Arabic) Oil for Biodiesel Production in Sudan

19

Mai Ahmed Hamed, Elfatih Ahmed Hassan**, Abdelaziz Emad Atabani*[†]
*Africa City of Technology, Khartoum, Sudan; **Sudan University of Science and Technology, Khartoum, Sudan; [†]Erciyes University, Kayseri, Turkey

Introduction

Utilization of natural resources has become a top priority in developing countries due to the obvious advantages when it comes to the environment, sustainability, and viability issues (Gylfason et al., 1999; Kazmierczyk, 2005; Koenders and Deutscher, 2009). In recent years, the world had been keenly putting great effort into producing energy from nonrenewable resources, such as plastic and rubber, as well as renewable resources, such as plants (Elisabeth and Päivi, 2003; Kumar and Sharma, 2011).

Biodiesel has been considered as a viable alternative to fossil fuel, taking into consideration the environmental concern and sustainability issue. Well-established vegetable oils, such as palm oil, soybean oil, and Jatropha oil, were utilized in biodiesel production. Exploration of other potential vegetable oil sources continues where plants such as *Acacia* and *Balanites aegyptiaca* are subjected to some studies (Elisabeth and Päivi, 2003; Kumar and Sharma, 2011; Muhamad et al., 2013).

Maslin et al. (2003) reported that *Acacia*, as the cosmopolitan genus of the family Leguminosae, containing more than 1350 species, dominates in poor savannah and semideserts zones of the world in the Gum Arabic Belt (Fig. 19.1), which extends across central Sudan, for the most part between latitudes of 10°N and 14°N. In this belt, *A. senegal* is an indigenous species. The upper-east Al Fao, Gadarif, Kassala, and southeast of Blue Nile state are the only exceptions for the existence of *A. senegal* outside the gum belt mentioned above (Fig. 19.2).

The gum belt extends over an area of 520,000 km^2, constituting one-fifth of the area of the country. The existence of *A. senegal* diminishes at latitude 13°45′N, especially in Kordofan or Darfur (Quezel and Santa, 1963). The gum belt plays an important role in combating desertification over the unlimited locale of the Sudano–Sahelian zone. *A. senegal* gives an assortment of significant monetary and natural capacities, for example, gum Arabic for animal feed, fuelwood, shade, and additionally, numerous backhanded advantages associated with the tree. The extensive root system of the tree decreases soil disintegration and overflow. Being a leguminous plant, it fixes nitrogen and improves the soil fertility, supporting grasses and harvest development in

Gum Arabic. http://dx.doi.org/10.1016/B978-0-12-812002-6.00019-1

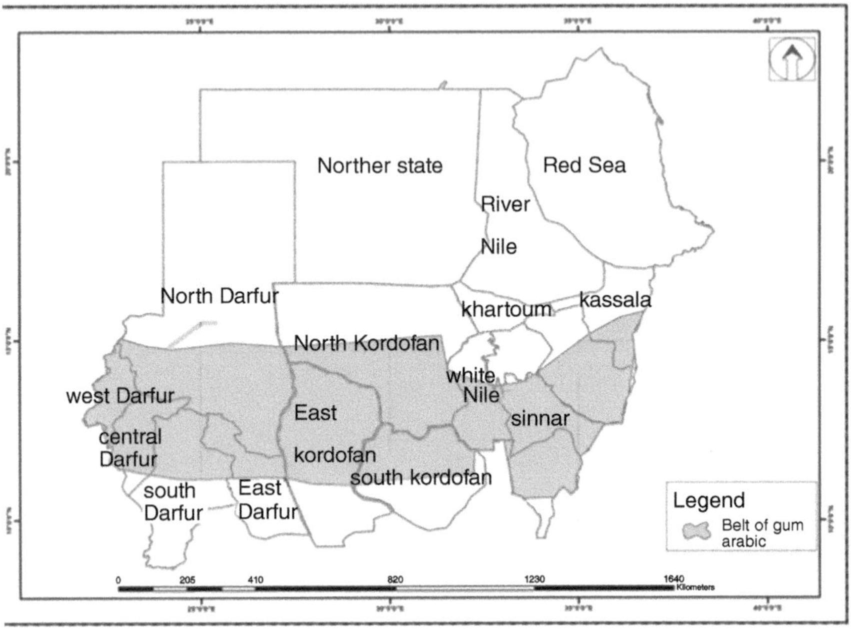

Arabic

Figure 19.1 Gum belt in Sudan.

Figure 19.2 *Acacia senegal* seeds.

these geographically vulnerable areas (Ayman, 2011; Elamin, 1989). The *A. senegal* tree botanically belongs to the family Leguminosae; subfamily, Mimosoideae; genus, *Acacia*; species, *senegal*, and variety *senegal*, locally named hashab (Elamin, 1990; Glicksman, 1969).

The accessibility of any feedstock for biodiesel creation relies upon the local atmosphere, land areas, nearby soil conditions, and horticultural practices (Atabani et al., 2013). The fundamental goal of this work is to obtain biodiesel from Sudanese resources that are not utilized as food. The main aim is to produce biofuel from the seeds of *A. senegal* var. *senegal*. Seeds are not utilized as eatable oil sources, food ingredients, or food additives.

Results and discussion

Physicochemical properties of acacia seeds and crude oils

Table 19.1 shows physical and chemical properties of *A. senegal* seed oil. The extraction process revealed that *A. senegal* seed had an oil content of 20% on dry weight basis of the whole seed.

Determination of biodiesel chemical and physical properties

The physical and chemical properties of the produced biodiesel from acacia seed oil are shown in Table 19.2. The following section will discuss some of the obtained results and compare them with each other. Some of the obtained results are discussed below and compared with published data for biodiesel.

Kinematic viscosity

Viscosity is the most critical property of any fuel, as it shows the resistance of a material to shear or stream. Along these lines, it influences the operation of the fuel infusion gear and shower atomization, especially at low temperatures when increment

Table 19.1 Physical and chemical properties of *Acacia senegal* seed oil

Test	Property
Density (g/cm^3)	0.915
Refractive index (40°C)	1.4628
Color (red/yellow)	1.5/29.04
Moisture	0.8%
Free fatty acid	3.8%
Saponification (mL/g)	848.4
Peroxide value (mL/g)	6.0
Iodine value (mL/g)	89.144

Table 19.2 **Physicochemical properties of biodiesel from *A. senegal***

Property	Biodiesel from acacia seed oil	Equipment	Manufacturer	ASTM D6751	ASTM D6751 limit
Kinematic viscosity	4.8378 mm^2/s	SVM 3000 – automatic	(Anton Paar, United Kingdom)	D 445	1.9–6.0
Flash point	198	Pensky Martens flash point, automatic, NPM 440	(Normalab, France)	D 93	130 min
Cloud and pour point	18 and 14	Cloud and pour point tester: automatic, NTE 450	(Normalab, France)	D 2500 and D 97	Report
Cold filter plugging point (CFPP) N/S	11	Cold filter plugging point: automatic, NTL 450	(Normalab, France)	D 6371	N/S[a]
Density at 15°C	0.8834	SVM3000: automatic	(Anton Paar, United Kingdom)	D 1298	N/S[a]
Density at 20°C	0.8808	SVM30010: automatic	(Anton Paar, United Kingdom)	D 1 298	N/S[a]

[a]N/S: ¼ not specified in ASTM D6751 test method.

in thickness influences the flow of the fuel. After conversion of the oil of *A. senegal* into *A. senegal* methyl ester, it's kinamatic viscosity was 4.8378 mm^2/s which is by all account, great as compared to published references (Atabani et al., 2012; Atabani et al., 2013; Sanford et al., 2009).

Flash point

Flash point may be defined as the temperature at which the fuel ignites when exposed to a spark or flame (Atabani et al., 2013). The methyl ester of the *A. senegal* obtained in this study was found to possess a flash point of 198°C, complying with the limits of the ASTM specification (D6751), which specify a value of 130°C as minimum.

Cloud, pour, and cold filter plugging points

The cloud point is the temperature at which wax crystals first get to be observed when the fuel is cooled. The pour point of a liquid is the temperature at which it becomes semisolid, loses its flow characteristics, and is adequate to gel the fuel; in this manner, it is the most reduced temperature at which the fuel can flow.

The cold filter plugging point (CFPP) is defined as the lowest temperature at which a specified volume of a diesel type of a fuel still passes through a standardized filtration system in a given time when cooled under certain conditions. It is expressed in degrees Celsius. The test is meant for indicating the lowest temperature at which the fuel will give trouble-free flow in certain fuel system.

Table 19.2 demonstrates the obtained results of CFPP at 18°C, 14°C, and 11°C, separately. These outcomes were as great as those of the previous studies (Atabani et al. 2012; Atabani et al., 2013; Sanford et al., 2009). Table 19.2 shows properties of biodiesel from *A. senegal.*

Infrared spectroscopy

Fig. 19.3 shows the Fourier-transform infrared spectroscopy (FTIR) spectrum of the biodiesel obtained from *A. senegal* seed oil.

The IR spectrum shows saturated (—C—H) stretching peaks found at 2923.81 and 2854.27 cm^{-1}, respectively, (—C—H) bending vibration of methylene group at 1436.53 cm^{-1}, (—C—H) bending vibration of methyl group at 1363.05 cm^{-1}, a peak of (C=O) at 1743.09 cm^{-1}, and a (C—O) peak at 1197.38 cm^{-1}. These peaks indicate presence of esters.

Gas chromatography–mass spectrometry analysis of the biodiesel

The result of gas chromatography–mass spectroscopy (GC–MS) is shown in Fig. 19.4. Acacia seed oil contains saturated and unsaturated fatty acids in range C_{16}–C_{24}; unsaturated fatty acids dominate the composition of the oil by nearly about 75%. Oleic acid had the highest percentage of the oil, nearly 32.98%, followed by palmitic acid at 15.73% and stearic acid at 6.7%. Table 19.3 shows components of biodiesel identified by GC–MS.

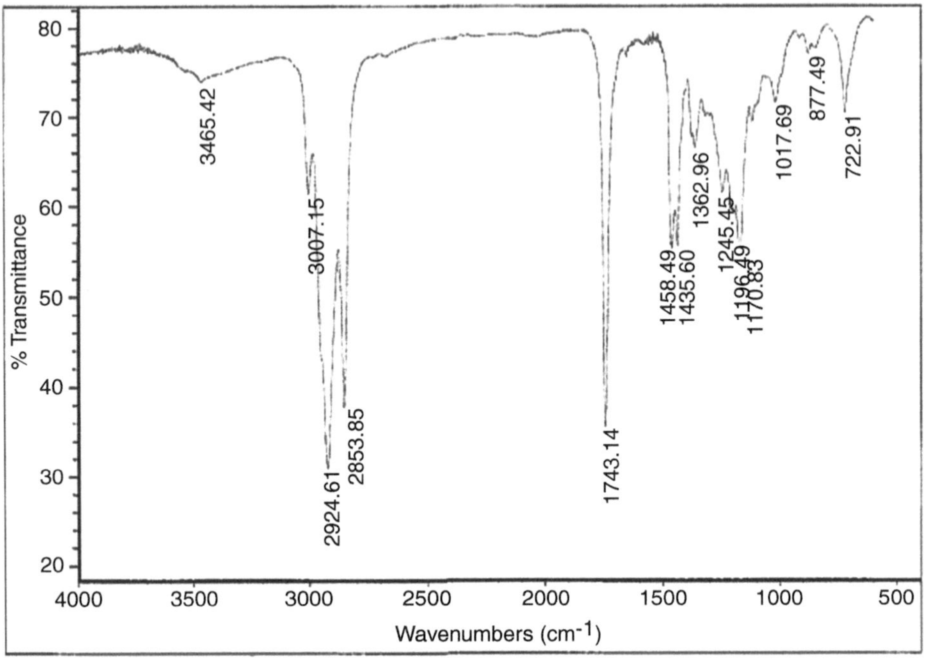

Figure 19.3 FT.IR spectrum of the biodiesel obtained from *A. senegal* seeds oil.

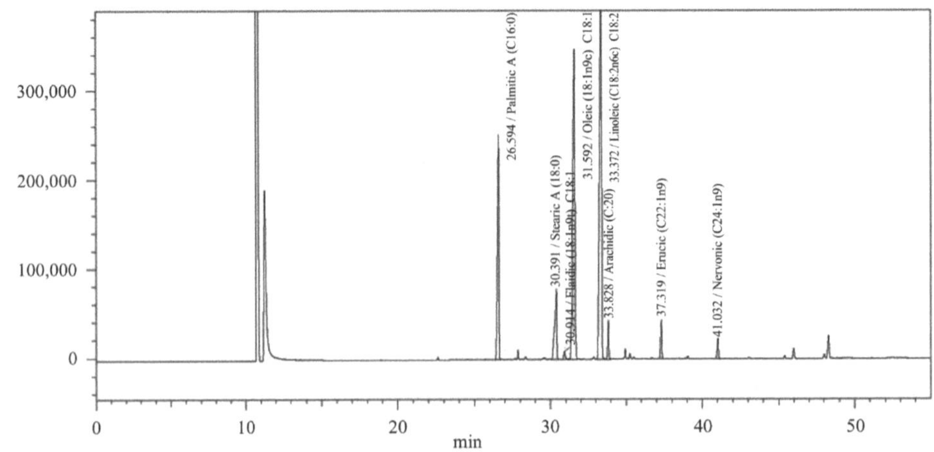

Figure 19.4 Shows a chromatogram of biodiesel showing the major fatty acid components.

Table 19.3 Components of biodiesel identified by GC–MS

Number	Compound name	Retention time	Area	Area %
1	Palmitic (C16:0)	26.594	1,703,779	15.7289
2	Stearic (C18:0)	30.391	728,443	6.7248
3	Elaidic (C18:1n9c) C1	30.914	61,327	0.5662
4	Oleic (C18:1n9c) C18	31.592	3,572,546	32.9809
5	Linoleic (C18:2n6c) C	33.372	4,200,376	38.7769
6	Arachidic (C20)	33.828	203,883	1.8822
7	Erucic (C22:1n9)	37.319	236,050	2.1792
8	Nervonic (C24:1n9)	41.032	125,757	1.1610
Total			10,832,161	100.0000

Table 19.4 Comparison between biodiesel from *A. senegal* seed oil and biodiesel from Jatropha

Test	Biodiesel from *A. senegal* seed oil	Biodiesel from Jatropha
Kinematic viscosity	4.8378 mm^2/s	4.9476 mm^2/s
Flash point	198	186.5°C
Pour point	14	10
Cloud point	18	10
Cold filter plugging point (CFPP)	11	10
Density	0.8834	0.8642

Table 19.3 shows components of *A. senegal* biodiesel identified by GC–MS. Table 19.4 shows the comparison between biodiesel from *A. senegal* seed oil and biodiesel from Jatropha. The qualities of the biodiesel obtained from oil of *A. senegal* seeds is comparable with the biodiesel obtained from Jatropha butter (Atabani et al., 2013).

Conclusion

Acacia seed oil is not suitable as an edible oil, as its fatty acid content exceeds the acceptable specifications limits. It possesses high free fatty acid content compared to known edible oils. Hence, it is more suitable for biofuel production. When the transesterification reaction of acacia seed oil was done, it was clean, yielding biodiesel with high qualities. GC–MS analysis showed that the biodiesel product was methyl ester and free from other undesired products. All chemical and physical properties of biodiesel from *A. senegal* seed oil were good compared to other biodiesel produced from Jatropha.

References

Atabani, A.E., Mahlia, T.M., Masjuki, H.H., Badruddin, I.A., Yussof, H.W., Chong, W.I., 2013. A comparative evaluation of physical and chemical properties of biodiesel synthesized from edible and non-edible oils and study on the effect of biodiesel blending. Energy 58, 296–304.

Atabani, A.E., Silitonga, A.S., Anjum, B.I., Mahlia, T.M.I., Masjuki, H.H., Mekhilef, S., 2012. A comprehensive review on biodiesel as an alternative energy resource and its characteristics. Renew. Sust. Energy Rev. 16 (4), 2070–2093.

Ayman, A.E., 2011. PhD thesis, Sudan university of Science and Technology, Khartoum, Sudan.

Elamin, H.M., 1990. Trees and shrubs of the Sudan, first ed. Ithaca press Exeter, UK.

Elamin, H.M., 1989. Sudan Acacias, the publishing Section Information Department, Forest Research Institute, Khartoum. Sudan.

Elisabeth, G., Päivi, L., 2003. The Location of Natural Resources Database. European Consortium for Political Research, Edinburgh, UK.

Glicksman, M., 1969. Gum Technology in the Food Industry (Food Science & Technological Monograph), first ed. Academic Press, New York, USA.

Gylfason, T., Tryggvi, T.H., Gylfi, Z.A., 1999. Mixed blessing: natural resources and economic growth. Macroecon. Dyn. 3, 204–255.

Kazmierczyk, P., 2005. Sustainable use and management of natural resources. European Environ. Agency, report 9, ISSN 1725–9177.

Koenders, B., Deutscher, E., 2009. Natural Resources and Pro-Poor Growth: The Economics and Politics. The Organisation for Economic Co-operation and Development, OECD Publications, ISBN 978-92-64-04182. http://www.oecd.org/dataoecd/56/63/42341879.pdf.com.

Kumar, A., Sharma, S., 2011. Potential non-edible oil resources as biodiesel feedstock: an Indian perspective. Renew. Sustain. Energy Rev. 15, 1791–1800.

Maslin, B.R., Miller, J.T., Seigler, D.S., 2003. Overview of the generic status of Acacia (Leguminosae: Mimosoideae). Aust. Syst. Bot. 16, 1–18.

Muhamad, S.A.W., Gimbun, J.C.K., Cheng, J.C.K., Nurdin, S., 2013. Enhancement of biodiesel yield from high FFA Malaysian rubber seed oil with sodium methoxide treated limestone. J. Eng. Technol. 4, 2180–3811.

Quezel, P., Santa, S., 1963. Nouvelle Flore de l'Algerie et des Regions Desertiques et Meridionales, Editions Tomes, CNRS, France, Paris.

Sanford, S.D., White, J.M., Shah, P.S., Wee, C., Valverde, M.A., Meier, G.R., 2009. Feedstock and Biodiesel Characteristics. Renewable Energy Group, USA, http://biodiesel.org/reports/20091117_gen-398.pdf.

Gum Arabic Dietary Fiber

20

*Abdalbasit A. Mariod**, **, †

*University of Jeddah, Alkamil, Kingdom of Saudi Arabia; **Sudan University of Science & Technology, Shambat, Sudan; †Indigenous knowledge and heritage center, Ghibaish College of Science & Technology, Ghibaish, Sudan

Introduction

Gum Arabic (GA), or acacia gum, is a natural forest product and one of the most well-known among all natural gums. It is a remarkably useful natural product, and an edible biopolymer and food hydrocolloid collected as dried exudate from the stems and branches of *Acacia senegal* (Vulgaris) and *Acacia seyal* (Gummiferae) trees (Leguminosae family). It is predominantly produced in Africa across sub-Saharan countries extending from Senegal, Mali, and Nigeria to Ethiopia and north of Kenya. Sudan is considered the first producer of GA worldwide, producing nearly 80% of the global total production of *Acacia* in the GA Belt zone of 520,000 km^2. The annual production of GA in Sudan is between 25,000 and 95,000 tons, averaging 50,000 tons per year (Williams and Phillips, 2000). No official statistics are available for the current annual production of GA in Sudan; however, several indicators suggest estimates in the range of 100,000–150,000 tons per annum. GA produced in Sudan is an organic, free of genetically modified organisms, and sustainable ingredient with considerable health benefits, and it meets clean label and natural requirements.

GA from *A. senegal*, known as hard gum, has a color of orange–brown, and it breaks in glass fraction, whereas *A. seyal* is known as flaky gum. In Sudan, *A. senegal* (known as hashab) produces high-quality gum, accounting for about 90% of Sudan's gum production, whereas *A. seyal* (known as talha) accounts for less than 10% of the total GA production, and is comparatively lower in quality and price.

The gum consists of a mixture of arabinogalactan (AG) oligosaccharides, polysaccharides, and glycoproteins (GPs), utilized in food processing under the E number E414.The world food industry consumes about 80% of the total industrial uses of GA. The gum is exceptionally dissolvable in water and structural solutions over an extensive variety of concentrations without being plainly viscous. GA is also used in several nonfood industries. GA codes include: (1) Harmonized System: 130120, (2) E number: E4141, (3) Chemical Administration Service: 9000-01-5, and (4) Codex Alimentarius SIN No. 414, INS No. 414.

Gum Arabic. http://dx.doi.org/10.1016/B978-0-12-812002-6.00020-8

Table 20.1 Chemical composition of *Acacia senegal* and *Acacia seyal*

Component	*Acacia senegal*	*Acacia seyal*	References
Dry matter (mg/g)	889.0	893.0	Lopez-Torrez et al. (2015)
Total sugar (%)	95.6	91.7	Gashua et al. (2015)
(mg/g dry matter)	940.0	950.0	Lopez et al. (2015)
Galactose (%)	38.9	44.2	Gashua et al. (2015)
(%) by GC–MS	35.8	36.9	Lopez et al. (2015)
Arabinose (%)	25.7	32.5	Gashua et al. (2015)
(%) by GC–MS	30.3	47.6	Lopez-Torrez et al. (2015)
Glucuronic acid (%)	21.5	13.0	Gashua et al. (2015)
(%) by GC–MS	17.4	6.7	Lopez-Torrez et al. (2015)
Rhamnose (%)	9.5	2.0	Gashua et al. (2015)
(%) by GC–MS	15.5	3.0	Lopez-Torrez et al. (2015)
4-O-Me-Glucuronic acid (%) by GC–MS	1.0	5.3	Lopez-Torrez et al. (2015)
Proteins (%)	2.09	0.99	Gashua et al. (2015)
(mg/g dry matter)	27.0	10.0	Lopez-Torrez et al. (2015)
Minerals (mg/g dry matter)	33.0	40.0	Lopez-Torrez et al. (2015)

GC–MS, Gas chromatography–mass spectrometry.

Table 20.2 Physicochemical properties of *Acacia senegal* and *Acacia seyal*

Component	*Acacia senegal*	*Acacia seyal*	References
Mw	6.4×10^5	1.14×10^6	Gashua et al. (2015)
Mn	3.52×10^5	6.84×10^5	Gashua et al. (2015)
Mw/Mn	1.9	1.7	Gashua et al. (2015)
R_g (%)	18.5	24.0	Gashua et al. (2015)
R_h	15.2	17.3	Gashua et al. (2015)
R_g/R_h	2.09	0.99	Gashua et al. (2015)

Gum Arabic chemical composition and physical properties

GA (acacia gum) is a complex carbohydrate with a molecular weight in the range of 2.5×10^5 to 1×10^6 Dalton, containing a small amount (about 2%–3%) of proteinaceous compound. On the total mass basis, GA is predominantly constituted of about 89% AG, 10% arabinogalactan–protein, and 1% GP (Tables 20.1 and 20.2).

Importance and sources of dietary fiber

Dietary fiber is the plant material in the diet that is impervious to enzymatic absorption, which incorporates cellulose, noncellulosic polysaccharides, for example, hemicellulose, pectic substances, gums, adhesives, and a noncarbohydrate part lignin. The diets rich in fiber, for example, grains, nuts, fruits, and vegetables positively affect health since their utilization has been identified with reduced incidence of several diseases. Dietary fiber can be utilized as a part of different useful functional foods, such as bakery, beverages, and meat items. Dissolvable and insoluble fibers make up the two fundamental classifications of dietary fiber. Cellulose, hemicellulose, and lignin are not solvent in water, while pectin, gums, and adhesives become sticky in water (Dhingra et al., 2012).

The significance of food fibers has prompted the advancement of a vast and potential market for fiber-rich items and ingredients, and as of late, there is a pattern to discover new sources of dietary fiber that can be utilized in food processing (Chau and Huang, 2003).

Dietary fiber is actually present in oats, fruits, vegetables, and nuts. The diets with a high substance of fiber have been found to positively affect health. Preparing the foods experience different physical, chemical, enzymatic and thermal treatments, which influence the composition of total fiber. Joining of fiber can change the consistency, texture, rheological behavior, and sensory properties of the finished products. The expansion of fiber in breakfast oats, bread, treats, cakes, yogurt, beverages, and meat items has been accounted for with great outcomes. Studies on changes in fiber amid different unit operations, extraction, and characterization of fiber from nonfood sources and improvement of fiber-enhanced items at temperate cost require quick consideration (Dhingra et al., 2012).

Gum Arabic dietary fiber

Dietary fibers affect digestion procedures and, in this way, metabolic reactions, while others enhance transit and stool yield. What's more, some of them seem, by all accounts, to be fundamental for mucosal barrier integrity and intestinal biological community to adjust by specifically empowering the development or potential movement of microscopic organisms, for example, *Bifidobacteria* or *Lactobacilli*, which are considered to profit the host. Regardless, a couple of fibers are immediately fermented into short-chain fatty acids (SCFA) and associated with a verbalized increment in gas production, a condition that may actuate undesirable indications, such as: bloating, flatus, stomach cramps, and diarrhea. In this manner, dietary fibers have impacts that advance well-being; these adverse occasions may restrict their utilization (Cherbut et al., 2003).

GA contains over 80% fiber. Because of its low consistency and its high solvency in water, it is easy to supplement foodstuff like drinks, dairy items, snacks, candy, and meat items with high sums (up to half) of GA. It is bland and unscented, thus giving no off-flavor. GA is a plant exudate of acacia trees. It is a 100% natural item, just physically cleaned and roller dried, yet not chemically adjusted. Dissolvable dietary fibers help diminish the total cholesterol and the low-density lipoprotein cholesterol, which affects coronal heart infections. Twenty-five grams of GA every day lessen total cholesterol essentially and thus have a positive impact on the avoidance of coronal heart diseases (McLean-Ross et al., 1983).

Prebiotic gum Arabic

Cherbut et al. (2003) compared the impacts of expanding doses of gum (GUM) to those of sucrose (SUC) on stool yield, grouping the primary bacterial populations in stools, event, and seriousness of intestinal symptoms (tooting, bloating, stomach cramps, diarrhea). They announced that the ingestion of GUM, 10 and 15 g/day for 10 days, expanded total lactic acid, creating bacteria and *Bifidobacteria* counts in the stools without influencing complete anaerobe and aerobe counts. The greatness of this particular impact was most noteworthy in subjects with a low starting fecal concentration of *Bifidobacteria*. Cherbut et al. (2003) added that the fecal digestibility of GUM was around 95%, and its caloric value was evaluated to be in the range of 5.5 and 7.7 kJ/g. What's more, stool weight expanded 30% in view of the more prominent fecal water. The digestive tolerance of GUM was high and not statistically different from that of SUC up to a day-by-day dosage of 30 g. Subsequently, GUM is an extremely all-around endured dietary fiber with bifidogenic properties accepted to benefit intestinal health (Cherbut et al., 2003).

Calame et al. (2008) showed that gum Arabic bears prebiotic efficacy within a dose range *similar to* or *lower than inulin* as established *via* the *quantitative* development of *bacteria* in *stool samples*. They did their investigation upon utilization by humans for up to 4 weeks. Human volunteers took different day-by-day dosages (5, 10, 20, and 40 g) of GA in water for up to 4 weeks. Every day, utilization of water was taken as the negative control and that of 10 g inulin as the positive control. At 0, 1, 2, and 4 weeks, measurement of bacterial numbers in feces samples was performed by means of constant polymerase chain reaction (PCR) procedures. Contrasted with the negative control, the quantities of *Bifidobacteria* and *Lactobacilli* 4 weeks after utilization were essentially higher for GA. Additionally, at this dose, the quantities of *Bifidobacteria*, *Lactobacilli*, and *Bacteroides* were fundamentally higher for GA than for inulin (Calame et al., 2008).

Cherbut et al. (2003) proposed that GUM is a prebiotic fiber, as it could specifically raise the extents of lactic acid microscopic organisms and *Bifidobacteria* in healthy subjects. It was matured gradually and delivered short-chain fatty acids, and its fecal edibility was around 95%. As on account of other solvent dietary fibers, it increased stool yield by increasing the water substance of stools. Likewise, its intestinal resilience gave off an impression of being great, and high day-by-day dosages could be consumed without inciting any antagonistic intestinal events.

Food applications

Nassar et al. (2008) proposed that 15% of orange peel and mash could be joined as an ingredient in making biscuits, as they are reasonable sources of dietary fiber with related bioactive compounds (flavonoids, carotenoids). The addition of dietary fiber to bakery items additionally enhances their nutritional quality, since it makes it conceivable to diminish the fatty substance by utilizing dietary fiber as a substitute for fat without loss of quality. Sharif et al. (2009) reasoned that the supplanting of wheat flour with defatted rice bran could be utilized without antagonistically influencing physical and sensory qualities of cookies. Rice bran supplementation fundamentally enhanced the dietary fiber, mineral, and protein substance of the cookies, and additionally, the cost of production was likewise diminished with proportionate increment of supplementation. Ice creams and solidified yogurts have higher fat levels, which have their specific functionalities. The addition of fiber components, for example, alginates, guar gums, and cellulose gels replace fat, as well as serve to give viscosity, enhance emulsion, foam, freeze/thaw strength, control dissolving properties, decrease syneresis, advances arrangement of smaller ice crystals, and encourage extrusion (Alexander, 1997). Guar gum, pectin, and inulin are additionally added amid cheddar in order to diminish its fat percentage without losing its organoleptic attributes, for example, texture and flavor. In the event of drinks and beverages, the addition of dietary fiber builds their viscosity and strength, the soluble fiber being the most utilized on the grounds that it is more dispersible in water than insoluble fiber. A few cases of solvent fibers are those from parts of grains, multiple natural products (Bollinger, 2001), pectin, β-glucans, and cellulose beet-root fiber (Nelson, 2001). Oat fiber can be joined into milkshakes, instant types of breakfast beverages, fruits and vegetable juices, iced tea, sports beverages, cappuccino, and wine.

Regulatory aspects of gum Arabic dietary fiber

As a result of its physical properties, GA hinders glucose ingestion, expands stool mass, and traps bile acids, and can possibly usefully change the physiological status of human subjects. Along these lines, since acacia gum fits in with the chemical definition and can reach the internal organ without absorption in the small digestive system, it can be classified as a nondigestible carbohydrate or dietary fiber. It fermented by intestinal microscopic organisms to SCFA in the internal organs. Acacia gum has been appearing to be a bifidogenic dietary fiber (Cherbut et al., 2003). It can specifically raise the extents of lactic acid microscopic organisms and *Bifidobacteria* in healthy subjects. It is fermented gradually and delivers short-chain fats, and the fecal digestibility is around 95%. It likewise builds stool yield by enlarging the water substance of stools. Likewise, its intestinal resilience is fantastic, and high day-by-day dosages can be devoured with no unfriendly intestinal occasions. All of these properties would support a potential role for dietary supplementation with GA in renal disease by changing the proteolytic to a saccharolytic fermentation pattern of colonic bacteria, as outlined above (Khalid et al., 2014).

The renal effects of GA are, however, ill-defined. A reduction of serum urea due to increased bacterial nitrogen excretion in the feces was reported almost 20 years ago in a cohort of patients with chronic renal disease, providing an additional approach to lowering serum urea nitrogen. Further evidence favoring a positive effect in renal disease was also provided by the work performed over 10 years ago in Khartoum, which suggested a beneficial effect on blood biochemistry following dietary supplementation with 50 g/day in patients with chronic kidney disease, predominantly undergoing renal replacement therapy (Suliman et al., 2000). A similar finding has more recently been reported from the Central Sudan in a cohort of patients undergoing regular hemodialysis in which supplementation with 50 g/day of GA led to an improvement of their biochemical profile (Ali et al., 2008). In view of the relatively poor status of the health-care system in the developing world, this approach may offer a potential supplementary therapy, which, when combined with dialysis, reduces its cost burden and improve patient symptomatology and well-being. Work undertaken in experimental animal models of progressive renal disease suggests that GA may also have a direct effect in modifying renal disease prior to the need of renal replacement therapy. Specifically, GA supplementation (7.5 g/100 mL in drinking water) in a rat model of acute nephrotoxic (gentamicin) renal failure led to a reduction in renal injury independently of its action of fecal bacterial ammonia metabolism (Ali et al., 2003). This direct renoprotective effect of GA is supported by observations in models of cisplatin-induced renal injury and a model of ischemic acute renal injury (Mahmoud et al., 2012). The experimental evidence in models of chronic renal injury is less clear, with the studies providing both positive and negative findings in two different rodent models of progressive renal injury (Ali et al., 2004, 2010).

References

Alexander, R.J., 1997. Moving toward low-calorie dairy products. Food Prod. Des. 7 (1), 74–98.

Ali, A.A., Ali, K.E., Fadlalla, A.E., Khalid, K.E., 2008. The effects of gum Arabic oral treatment on the metabolic profile of chronic renal failure patients under regular haemodialysis in Central Sudan. Nat. Prod. Res. 22 (1), 12–21.

Ali, B.H., Alqarawi, A.A., Ahmed, I.H., 2004. Does treatment with gum Arabic affect experimental chronic renal failure in rats? Fundam. Clin. Pharmacol. 18 (3), 327–329.

Ali, B.H., Al-Qarawi, A.A., Haroun, E.M., Mousa, H.M., 2003. The effect of treatment with gum Arabic on gentamicin nephrotoxicity in rats: a preliminary study. Ren. Fail. 25 (1), 15–20.

Ali, B.H., Al-Salam, S., Al Husseni, I., Kayed, R.R., Al-Masroori, N., Al-Harthi, T., et al., 2010. Effects of gum Arabic in rats with adenine-induced chronic renal failure. Exp. Biol. Med. (Maywood) 235 (3), 373–382.

Bollinger, H., 2001. Functional drinks with dietary fibre. Fruit Process 12, 252–254.

Calame, W., Weseler, A.R., Viebke, C., Flynn, C., Siemensma, A.D., 2008. Gum Arabic establishes prebiotic functionality in healthy human volunteers in a dose-dependent manner. Br. J. Nutr. 100 (6), 1269–1275.

Chau, C.F., Huang, Y.L., 2003. Comparison of the chemical composition and physicochemical properties of different fibers prepared from peel of the *Citrus sinensis* L. Cv. Liucheng. J. Agric. Food Chem. 51, 2615–2618.

Cherbut, C., Michel, C., Raison, V., Kravtchenko, T., Severine, M., 2003. Acacia gum is a bifidogenic dietary fiber with high digestive tolerance in healthy humans. Microb. Ecol. Health Dis. 15, 43–50.

Dhingra, D., Michael, M., Rajput, H., Patil, R.T., 2012. Dietary fiber in foods: a review. J. Food Sci. Technol. 49 (3), 255–266.

Gashua, I.B., Williams, P.A., Yadav, M.P., Baldwin, T.C., 2015. Characterisation 499 and molecular association of Nigerian and Sudanese Acacia gum exudates. Food 500 Hydrocoll. 51, 405–413.

Khalid, S.A., Musa, A.M., Saeed, A.M., Abugroune, E.A., Sid Ahmed, E.O., Ghalib, M.B., Elnima, E.I., Alkarib, S.Y., Abdelsalam, T.M., d Abdelgader, A., Phillips, G.O., Phillips, A.O., 2014. Manipulating dietary fibre: gum Arabic making friends of the colon and the kidney. Bio. Carbohydr. Diet. Fiber 3, 71–76.

Lopez-Torrez, L., Nigen, M., Williams, P., Doco, T., Sanchez, C., 2015. Acacia 502 senegal vs. Acacia seyal gums—Part 1: Composition and structure of 503 hyperbranched plant exudates. Food Hydrocoll. 51, 41–53.

Mahmoud, M.F., Diaai, A.A., Ahmed, F., 2012. Evaluation of the efficacy of ginger, Arabic gum, and Boswellia in acute and chronic renal failure. Ren. Fail. 34 (1), 73–82.

McLean-Ross, H.A., Eastwood, M.A., Brydon, W.G., 1983. A study of the effects of dietary gum Arabic in humans. Am. J. Clin. Nutr. 37 (3), 368–375.

Nassar, A.G., AbdEl-Hamied, A.A., El-Naggar, E.A., 2008. Effect of citrus by-products flour incorporation on chemical, rheological and organoleptic characteristics of biscuits. World J. Agric. Sci. 4 (5), 612–661.

Nelson, A.L., 2001. High-fiber Ingredients. Eagan Press Handbook Series. Eagan Press, St. Paul.

Sharif, M.K., Masood, S.B., Faqir, M.A., Nawaz, H., 2009. Preparation of fiber and mineral enriched defatted rice bran supplemented cookies. Pak. J. Nutr. 8 (5), 571–577.

Suliman, S.M., Hamdouk, M.I., Elfaki, M.B., 2000. Gum Arabic fiber as a supplement to low protein diet in chronic renal failure patients. Sudan Association of Physicians 17th Conference, Khartoum. .

Williams, P.A., Phillips, G.O., 2000. Gum arabic. In: Phillips, G.O., Williams, P.A. (Eds.), Handbook of Hydrocolloids. Woodhead, Cambridge, pp. 155–168.

Further Readings

Ali, B.H., Ziada, A., Blunden, G., 2009. Biological effects of gum Arabic: a review of some recent research. Food Chem. Toxicol. 47, 1–8.

Elleuch, M., Bedigian, D., Roiseux, O., Besbes, S., Blecker, C., Attia, H., 2011. Dietary fiber and fiber-rich by-products of food processing: characterization, technological, functionality and commercial applications: a review. Food Chem. 124 (2), 411–421.

Gashua, I.B., Williams, A., Yadav, M.P., Baldwin, T.C., 2009. Characterization and molecular association of Nigerian and Sudanese Acacia gum exudates. Food Hydrocoll. 51, 405–4013.

Patel, S., Goyal, A., 2012. The current trends and future perspectives of prebiotics research: a review. 3 Biotech 2 (2), 115–125.

Phillips, G.O., Ogasawara, T., Ushida, K., 2008. The regulatory and scientific approach to defining gum Arabic (*Acacia senegal* and *Acacia seyal*) as a dietary fiber. Food Hydrocoll. 22, 24–35.

Thevenet, F., 2012. Acacia gum. Carbohydrate-Based Polymer Building Blocks and Biopolymers. Elsevier B.V, Nexira, Rouen, France, pp. 205–213.

Gum Arabic and Kidney Failure: An Exceptional Panacea

21

Isam Mohamed Abu Zeid, Abdul Rahman FarajAllah***
*King Abdulaziz University, Jeddah, Saudi Arabia; **Batterjee Medical College for Science & Technology, Jeddah, Kingdom of Saudi Arabia

Introduction

The poor savannah belt is the expanse of virtual nothingness of a huge land that extends from the Atlantic Ocean through the middle west of Africa until it reaches Sudan. The belt is situated in the south of the African Sahara Desert; it has already been depicted as the most harsh, dreadful, and inhospitable domain, characterized by nonabiding, sparse vegetation. Also, due to the ensuing unpredictable climate change, there has been an unprecedented, unrecorded encroachment of desertification in this poor savannah belt, which is further extended through the southern extremity in an eastward direction of the desert fringes and has led to an imbalance of life processes. Therefore, as a result of this, all life processes of biota (plants and animals, including human beings) have paid a heavy price and a great toll that exceeded beyond the limits of their tolerance and, even worse, is continually shattering whole aspects of life. Pessimistically and tragically, the worst is yet to come.

Nevertheless, with all these constraints, man is able to survive and to overcome these unsurmountable hurdles, but these unacceptable conditions, combined within-hospitable situations, yet the fundamentals of life need to thrive, and successfully life progresses toward ticking once more.

Gum Arabic (GA), or acacia gum (AG), is obtained from the acacia trees that grow in this region from Senegal to Sudan, where these hardy bushes and trees are able to withstand and tolerate all the harsh vagaries of weather. They produce a precious commodity that oozes out from these heavy, thorny shrubs and trees. Their sticky material exudates/flows out very slowly and has become the mainstay and the bloodline for most of the inhabitants of this region.

The precious GA produced by these trees becomes the most highly valued commodity, and its demand worldwide is never ebbed and considered a desert prize, which brings lots of opportunities and happiness to those savannah inhabitants. With the importance of this commodity worldwide and with is inherent versatility, many initiatives have been tried to preserve and protect those trees, including the Great Green Wall Initiative (GGWI) during the year 2007, as a pan-African effort to battle desertification, heal and tackle poverty, and blast land degradation across the Sahel–Saharan region of the continent.

The monumental steps to achieve the GGWI in the dire quest for alleviating constraints have, to some degree, restored the degraded ecosystems and improved rural livelihoods while helping to shake the encroachment of the Sahara Desert by

Gum Arabic. http://dx.doi.org/10.1016/B978-0-12-812002-6.00021-X

addressing the ecological and socioeconomic vulnerability of the disadvantaged rural populations. It is of great urgency to put emphasis on the adoption of sustainable landscape management techniques so as to give a helping hand to Mother Nature. Also, GA provided a socioeconomic forum and a great potential because *Acacia* has high drought resistance and is deeply rooted in the history, culture, and heritage of the region since the onset of exporting GA to Europe, beginning back in the 17th century.

Gum Arabic as a global commodity

The global demand for GA as a precious international commodity is never matched with any other comparable commodity, while it keeps its own specific heritage, traditions, and rituals. Based on the valuable versatility of its ingredients, it is utilized to an unimaginable extent as foods, in pharmaceutical industries, as stabilizer in beverages, medicine, writing, learning, confectionery, printmaking, cosmetics, industrial applications, textiles, paintings, arts, ceramics, photography, pyrotechnics, smoking and as an emulsifier.

The procedure of obtaining the gum after mechanical injury of the trees by incision of the bark involves removing sections of the bark with a sharp tool like an axe with great care taken not to damage the trees. Again, GA is also obtained by removal of branches to cover systematic wounds on the bark of trees, which prevents attack by insects or microorganisms, and also after invasion by bacteria to shape hard modules that are picked by hand and put away, as indicated by their color and size. Based on information reported by, GA is defined by the Food and Agriculture Organization (FAO)/World Health Organization Joint Expert Committee on Food Additives as a dried exudation obtained from the stems of *Acacia senegal* (L.) or closely related species of *Acacia* (FAO, 1999). The prime physical characteristics of GA on which related the quantitative and qualitative parameters were based to include moisture content, total ash, volatile matter, and internal energy, which were the exact international specifications proposed by FAO.

Health benefits and welcome scenarios of gum Arabic

GA is a polysaccharide with branched chains of (21, 22 and 23) linked β-D-galactopyranosyl units. GA is rich in calcium, potassium, and magnesium. It is actually produced from the stems and branches of *A. senegal*. It is digested in the colon by microorganisms to short-chain fatty acids (SCFAs); therefore, it is reported as one of the safest dietary fibers and hence has been an important food additive (Ali et al., 2003). The supplementation of the eating routine with GA has been appeared to increase high nitrogen discharge and lower serum urea nitrogen concentration in patients with constant renal failure. The data from Polonsky (2012) demonstrated that dietary supplementation with GA lowers serum butyrate, which has good impact on renal profibrotic cytokine generation (Bliss et al., 1996).

Prior reviews have yielded confirmations for and against the antioxidant impact of GA and the defensive impacts in trials of hepatic, renal, and cardiovascular danger (Ali et al., 2009). GA has appeared to lessen blood pressure, plasma cholesterol focuses in rats, to foster dental remineralization, to show that it can make an environment hostile to microbial movement, and to animate intestinal assimilation, in this manner preventing diarrhea (Al Mosawi, 2002; Ibrahim et al., 2004; Nasir et al., 2008). Current trials have shown that the extra-renal impacts of GA application in mice reduced expression of intestinal sodium coupled with a glucose carrier with subsequent delay of intestinal glucose transport, glucose-induced hyperglycemia, hyperinsulinemia, and body weight gain (Nasir, 2013). Soluble fibers are effective in reducing cholesterol by reducing the absorption of fats and cholesterol from the small intestine and absorption of bile acids from the terminal ileum (Younes et al., 1995). Finally, renal function was also improved after GA supplementation with a clear reduction in blood urea nitrogen (BUN) and creatinine concentration in diabetics and diabetic nephropathy patients. All patients under the investigation have recorded an overall health improvement; therefore, this study has revealed that GA supplementation has effects on type 2 diabetes mellitus patients and has impaired the prognosis of the disease (Nasir et al., 2016).

Since a long time ago, AG has had medical values and has been used to treat a variety of illnesses, including bleeding, respiratory tract infections, and pain. Due to the availability of a rich, soluble fiber in AG, it improves the patient's digestive system, reduces the risk of constipation, and improves appetite, where for kidney disease patients, it helps them to get enough energy from their food. This eventually strengthens their bodies and enhances immunity, an added parameter that better aids in a fight against kidney problems. Moreover, unfortunately, kidney patients are subjected to some common complications, including headaches and leg and back pains; hence for remedy, physicians prescribe the use of painkillers, which adversely induce side effects on the diseased kidneys. A new cure is heralded by using AG as a natural painkiller that is used to ease pain in kidney disease patients.

AG is used to support a low-protein diet in children with end-stage renal disease (ESRD), where most die without renal replacement therapy (RRT). AlKhodor and Shatat investigated the possibility of using AG supplementation to improve the quality of life and give a sigh of relief for children with ERSD with a dialysis-free period. Their results showed that the supplementation with AG could be a contrasting option to renal replacement therapy (RRT) to enhance the personal satisfaction and decrease or wipe out the requirement for dialysis in kids with end organize renal sickness.

Other studies conducted on the impact of AG on oxidative stress and aggravation in adenine-prompted perpetual renal disappointment were considered by Badreldin et al., focusing on the inflammation and oxidative anxiety that was known to be required in the pathogenesis of interminable kidney infection in people as ceaseless renal failure in rats. They found that adenine essentially expanded the convergence of urea and creatinine in plasma, altogether diminished the creatinine clearance, and actuated critical increments in the concentration of the measured inflammatory mediators. In addition, it brought on oxidative stress and DNA damage, but the treatment with GA fundamentally improved these activities. Consequently, the system of detailed salutary impact of GA in adenine-incited CRF is related to the moderation of the adenine-initiated inflammation and the generation of free radicals. The bidirectional relationships in the alteration of the human microbiome affect the kidney disease progression, and how kidney disease may disrupt a balanced microfilm has been reported by AlKhodor and Shatat. This is due to the fact that the human body contains a combined group of microorganisms of both prokaryotic and eukaryotic microbes that occupy intuitive interfaces presented or associated with the outer condition (Hamady and Knight, 2009). Moreover, it was reported that these trillions of microbial cells called human microbiota and the genes encoding these microbes collectively from the human microbiome all represent an integral part of the human genetic landscape, based on information reported by Solt et al. The knowledge of the microbial communities colonizing various human niches was advanced by recent research work by Kuczynski et al. (2012); however, the ecological communities of microbial species exist in symbiotic mutualistic relationship with the human host, though it might be unique to each person (Cho and Blaser, 2012). The abundance and diversity of microbial communities are almost individual-specific (Morgan et al., 2013), albeit much of the interpersonal diversity was attributed to a variety of differences, including host genetics, geographical origin, location, age, lifestyle, eating habits, early microbial exposure as well as antibiotic and probiotic intake. Moreover, a major plethora of other factors has been reported to contribute to the basic microbial diversity, such as mode of birth delivery, gestational age, hospitalization, diet, and nature of feeding (Johnson and Versalovic, 2012).

Chronic kidney disease (CKD) has become a global health epidemic associated with the grave loss of kidney functions, cardiovascular diseases, infectious diseases, and awful premature death (Levey et al., 2005).The loss of kidney function in CKD has resulted in major alternations in the blood concentration of numerous molecules, especially the substances that are normally excreted or metabolized by the kidney and accumulate as renal function declines, resulting in increased blood concentrations (Meyer and Hostetter, 2007). These uremic retention molecules (URM) constitute an expanding list of substances that, upon accumulation, substantially contribute to the uremic syndrome; however, it is increasingly recognized that intestinal microbial metabolism also contributes to the generation of numerous URMs (Wikoff et al., 2009).

Kidney disease is associated with decreased consumption of dietary fibers, frequent use of antibiotics, slow colonic transit, metabolic acidosis, volume overload with intestinal wall congestion, intestinal wall edema, and oral iron intake (Fouque et al., 2014; Jakobsson et al., 2010; Kalantar-Zadeh et al., 2002; Nakao, 2012). Therefore, targeting the large intestine and understanding the composition of the gut microbial community might be a promising approach to tackle the high morbidity and mortality in patients with CKD.

The very unique properties of GA have been researched and reported recently by a huge number of scientists and research facilities, proving that it is a branched-chain complex polysaccharide, either neutral or slightly acidic, found as mixed calcium, magnesium, and potassium salt of a polysaccharidic acid. It has a variety of uses in industry as a stabilizer, thickening agent, and an emulsifier, mainly in the food industry, i.e., in soft drinks, syrup, gummy candies, and marshmallows, but also in textile, pottery, lithography, cosmetics, and pharmaceutical industries (Verbeken et al., 2003). In folk medicine, it has been reported to be used internally for the treatment of inflammation of the intestinal mucosa and externally to cover inflamed surfaces (Gamal el-din et al., 2003). More experiments have been conducted on GA to show its ability to be used as a vehicle for drugs, especially in these studies about physiological and pharmacological reactions with the assumption that it is an inert substance. Moreover, recent studies have claimed that GA possesses antioxidant, nephroprotective, and other valuable beneficial effects, especially those that go along with kidney problems (Ali et al., 2008; Rehman et al., 2001). With the widespread occurrence of kidney ailments, clinical studies have been conducted on patients with chronic renal failure and have claimed to reduce urea and creatinine plasma concentrations; however, it might reduce the need for dialysis from three to two times weekly (Suliman et al., 2000).

As it is well-documented that GA is primarily indigestible to both humans and animals, however, it is not degraded in the small intestine, but fermented in the large intestine by microorganisms' SCFAs, specifically propionic acid (Kishimoto et al., 2006; Phillips, 1998). Those degraded products are absorbed by the human colon and subsequently utilized energetically in the metabolism. More work conducted by Kishimoto et al. (2006) showed that a *Prevotella ruminicola*–like bacterium was the predominant organism that was most likely to be responsible for the fermentation of GA.

Panacea of gum Arabic: The medical dimensions of gum Arabic

The usual medical intervention by using dietary fiber in protecting against any disease incidence or curing an already attacked person was adopted and practiced in the past 100 years. This is because a high intake of dietary fiber has been associated with numerous health benefits. It has been shown that a high intake of fiber from cereals and high consumption of whole grain foods are significantly associated with a reduced risk of colorectal cancer (Aune et al., 2011).

Many varied observations were reported suggesting that high-fiber diets may offer protection from the risk of cancers beyond the colon (La Vecchia et al., 2003). Other suggestions gave supportive evidence of the effect of whole-grain foods on cardiovascular disease (Truswell, 2002) and myocardial infarction (Rimm et al., 1996). Also, the consumption of whole grains and the risk of type 2 diabetes with added observational studies led to an inverse association between fiber intake and occurrence of type 2 diabetes, and suggested a role of added fiber intake and occurrence of type 2 diabetes associated with lower incidence of obesity (Cho et al., 2013). The more welcome news was reported by Krishnamurthy et al. (2012), who suggested that high dietary fiber intake provided clinical benefits for chronic disease patients. Such promising news has given a sigh of relief for patients and for the health community, where chronic renal diseases have become an expensive asset that are growing by the day, where the current estimates of incidence reached 10% of the population with impaired renal dysfunction (Coresh et al., 2007).

Manipulated dietary fiber

The heated debate about the definition of the dietary fiber ran simultaneously into its effect and related to chemical or physiological category or function of the fiber itself. The internationally acceptable definition in 2009 was adopted and reported by Phillips and Cui (2011). Dietary fiber means carbohydrate polymers with 10 or more monomeric units that are not hydrolyzed by the endogenous enzymes in the small intestine of humans. They may belong to the following categories: either edible carbohydrate polymers naturally occurring in the consumed food, or carbohydrate polymers obtained from raw material by physiological or chemical means. These might reflect physiological effects or be beneficial to health demonstrated by accepted scientific evidence to competent authorities or synthetic carbohydrate polymers that did show physiological effect (Cummings and Stephen, 2007). The outcome of the debate relied on the conversion in the colon to SCFAs, which is a necessary requirement if the material is to be accepted as a dietary fibrosis. But to make a health claim needed critical characterization in the chemical, biological, physiological, and clinical domains. Therefore, while the general identification of dietary fiber can be made on a chemical basis, any specific health claim must be based upon its physiological and clinical performance (Phillips, 2013).

Dietary fiber and renal disease

Currently the dietary advice for chronic kidney disease (CKD) is the intake of salt, phosphate potassium, and protein, which showed that dietary fiber is well-defined, where fruits and vegetables often restricted in CKD to prevent or correct hyperkalemia are an important source of dietary fiber. A recent landmark study has energized the interest in using the Third National Health and Nutritional Examination Survey (NHANES III), a database that identified a positive association between fiber intake and mortality in patients with kidney disease (Krishnamurthy et al., 2012).

Recent investigations have suggested reducing of the load of colon-derived solutes, which was postulated to be a potential therapeutic option in CKD in order to delay the progression of renal disease and also to reduce the associated cardiovascular morbidity and fatalities. One of the promising approaches was to administer sorbents that bind to microbial metabolites. Evidences have been accumulated that carbon-based sorbent AST-120 may reduce circulating levels of inxodyl sulfate and ameliorate decline in renal function (Lida et al., 2006; Nakamura et al., 2011; Owada et al., 2010; Ueda et al., 2007, 2008). This reduces the intestinal absorption to tryptophan-derived indoles, which subsequently reduce the hepatic conversion of indoles to indoxyl sulfate. Dietary fiber is postulated to provide health benefits through the production of SCFAs and a product fiber of fermentation, since SCFAs are known to have numerous biologically important effects, including modulation of cell proliferation, apoptosis, and regulation of angiogenesis and inflammation (Smith et al., 1998).

The effect of SCFAs was linked with the improvement of both colonic and systemic health. Also, other studies of the pathogenesis of the progression of renal fibrosis suggest that SCFAs have antifibrotic activities in reducing the generation of biological activity of the key profibrotic cytokine transforming growth factor beta 1 (Matsumoto et al., 2006). The amount and type of SCFAs formed by colonic bacteria depend on the chemical structure, composition, and the amount of the available substrate. This has important implications, such as health benefit claims cannot be assumed to be generic for all types of fiber. One spectacular dietary fiber that has generated an interest in the context of CKD is GA.

Gum Arabic (*Acacia senegal*) as a source of useful dietary fiber

GA has been in the scientific community worldwide because of successful trials and cures for kidney disease patients, even at the end stage. In reality, GA is an edible tree gum exudate that has a multitude of uses as a stabilizer, thickening agent, and an emulsifier, mainly in the food industry; however, it has a lot of benefits in the food industry, in addition to textile, pottery, cosmetics, and pharmaceutical industries. It has been extensively tested for its properties as nondigestible polysaccharides. GA has been shown to conform with the definition of dietary fiber adopted by Codex as a nonstarch polysaccharide, since GA is not digested in the intestine, but is fermented in the colon to give SCFAs, leading to a wide range of potential health benefits. The overall definition of GA allows for the two tree species, *A. senegal* and *Acacia seyal*. Based on its growth in a wide geographical region all over the Sahelian belt of Africa, the commercial product can show extremely wide chemical and physical variability (Idris et al., 1998; Mocak et al., 1998).

Based on its physical properties, it retards glucose absorption, increases stool mass, traps bile acids, and has a potential to beneficially modify the physiological status of human subjects (Adiotomre et al., 1990; Annison et al., 1995). The GA reaches the large intestine undigested in the small intestine. Therefore, it is categorized as a nondigestible carbohydrate or dietary fiber. However, it is fermented by intestinal bacteria to SCFAs in the large intestine (Annison et al., 1995; May et al., 1994). It has also been shown to be bifidogenic dietary fiber (Cherbut et al., 2003; Michel et al., 1998). They also reported that it has been too easy and selectively raised the proportions of lactic acid bacteria and *Bifidobacteria* in healthy subjects. It is fermented slowly; hence, it produces SCFAs, and the fecal digestibility is almost 95%; however, it increases stool output by augmenting the water content of stools. It has a real intestinal tolerance with high dietary doses that can be consumed without any adverse intestinal events. All the mentioned properties would support a potential role for dietary supplementation with Arabic gum in renal diseases by changing the proteolytic to saccharolytic fermentation pattern of colonic bacteria.

Gum Arabic and renal function

Currently, many research works have been carried over in so many multitudes of rational parameters in equating the credibility and significance of GA with the objectives of its critical validity in the remedy and amelioration of kidney dysfunction and end-stage kidney failure. Those promising aspects have been based on its chemistry, chemical composition, degradation in the small intestine (Al-Assaf et al., 2005; Kishimoto et al., 2006), and its pharmacological properties as an antioxidant (Ali and Al Moundhri, 2006), and more recently, Al-Assaf et al. (2005) reported that GA was ineffective in ameliorating hepatocellular damage in cholestasis induced by fenofibrate in rats. The effect of GA on renal function has been tackled by many scientists on patients with end-stage renal failure (ESRF) disease that requires renal replacement therapy (RRT) in the form of either dialysis or renal transplantation for survival. The ESRF patients requiring RRT have increased dramatically throughout the world during the last decade (Locatelli et al., 2006); hence the cost of treating uremia represents a growing demand on the health-care systems of both rich and poor people. Most dietary attempts to treat chronic renal failure (CRF) and to decrease uremia use a protein restriction regimen (Fouque et al., 2006), although Chaturvedi and Jones (2007) recently have proposed alternative dietetic approaches and some suggestions based on fermentable carbohydrate supplementation of the diet (Winchester and Salsburgh, 2004), which has led them to claim that these results in a similar urea-lowering effect by increasing urea nitrogen (N) excretion in stools with a concomitant decrease in the total N excreted in the urine of adults. Ali et al. (2008) have also reported no explanation of many results recovered, but the inclusion of iron and folic acid supplements have not been justified, although it was concluded that GA could alleviate "adverse effects of CRF."

The effect of GA on blood glucose concentration is when the infusion of meals containing starch showed that a decrease in the digestion rate of starch in the upper small intestine accounted for part of the viscosity of glycemic response, whereas the main effect of gum appeared to slow gastric emptying (Leclère et al., 1994). The other major predominant investigations on the overall effects of GA on the gastrointestinal tract were conducted by Codipilly et al. (2006), Karrasch and Jobin (2008), and Wapnir et al. (2008), and the degradation of GA in the intestine was done by Kishimoto et al. (2006).

In addition to the effect of GA on lipid metabolism (Pasquier et al., 1996) and by mimicking pH conditions in the stomach, and evaluating the effect of several soluble fibers, including GA, on dietary fat emulsification and lipolysis in vitro, where it was found that GA strongly inhibited lipolysis at pH 4.5, but without having any effect on the emulsion droplet size. However, the inhibition was attributed to the possible strong interaction of the protein moiety of GA with the available droplet (lipid/water) interface (where gastric lipase acts), blocking it or preventing the release of the non-esterified fatty acid generated at the droplet surface.

Moreover, the total effect of GA on tooth mineralization, where dental caries occur when tooth enamel is lost due to an imbalance of the demineralization and remineralization phases, if prevention can be achieved by remineralization phase is enhanced (Aoba, 2004). Several agents have been used, including fluoride. Using histopathological methods showed that GA has the ability to enhance remineralization (Onishi et al., 2008) or by supporting other remineralization activities may be due to rich content of calcium, magnesium, and potassium salts of polysaccharides in GA to the effect of the gum on the metabolism of calcium and possibly phosphate. It was also known that GA contains cyanogenetic glycosides and enzymes such as oxidases, peroxidases, and pectinases that inhibit antimicrobial properties against organisms such as *Prophyromonas gingivalis* and *Prevotella intermedia*.

The effect of GA on hepatic macrophages came from earlier studies that GA is capable of completely blocking the macrophage function (Fujiwara et al., 1995; Mochida et al., 1990). Therefore, such effects of GA merit consideration in the therapy of chronic liver disease as deranged from the function of Kupffer cells and hepatic macrophages occur in this disease and is involved in its complications, such as endotoxemia. Some extra studies were conducted on what might be adverse effects and toxicity of GA allergy, contact dermatitis, and asthma; however, in all reports, the offending allergen has not yet been identified. Moreover, in 1998, carbohydrate-specific immunoglobulin E (IgE) was identified in a chocolate confectionery, along with cough and dyspnea (Fötisch et al., 1998). More recently, Sander et al. (2006) found that sensitization to GA carbohydrate structures occurs casually in atopic patients with pollen sensitization without obvious exposure to GA, and the allergy to GA is mediated preferentially by IgE antibodies directed to polypeptide chains of GA. The addition of 10% GA in the drinking water for either 3 or 14 days was found to decrease urinary excretion of inorganic phosphate and increased that of calcium and magnesium; it also decrease plasma concentration of 1,25-dihydroxyvitamin D in healthy mice (Nasir et al., 2008).

Role of gum Arabic in renal disease

In many parts of Africa and Middle Eastern countries, it is believed that GA has many attributes to be used in folk medicine and the cure of kidney disease. Lowering of serum urea nitrogen was reported almost 20 years ago in a cohort of patients with CRD. Encouraging positive results on treating renal disease was attributable work performed over a decade in Khartoum (Sudan), suggesting beneficial effects on blood biochemistry following dietary supplementation with 50 grams/day in patients with CKD, predominantly under RRT (Suliman et al., 2000). More welcome news came from central Sudan from a cohort of patients undergoing regular hemodialysis led to an improvement of their biochemical profile (Ali et al., 2008). This news is welcomed by needy people, as it offers a potential supplementary therapy, which, when combined with dialysis, reduces the cost burden and improves patients symptomatically and their well-being. GA may also have a direct effect in modifying renal disease prior to the need of RRT. Trials showed that with GA supplementation of 7.5 g/100 mL in drinking water in a rat model of acute nephrotoxicity (gentamicin), renal decline led to the reduction in renal injury independently of its action of fecal bacterial ammonia metabolism (Ali et al., 2003). This direct renoprotective effect of GA is supported by observation in models of cisplatin-induced renal injury (Al-Majed et al., 2003) and a model of ischemic acute renal injury (Mahmoud et al., 2012). The experimental evidence in models of chronic renal injury is less clean, with the studies providing both positive and negative findings. With two different rodent models of progressive renal injury, Ali et al. (2010) and Ali et al. (2004) have shown their major effect on these rodents. Previous studies on GA did not specify its species nor the molecular parameters of the materials used. Some wide variation could be related to its geographical origin, age of the tree, and nature of the soil (Al-Assaf et al., 2005).

Usually, a random commercial product may not be pure and could be made for more than one species of *Acacia*. However, other researchers have embarked in clinical programs to evaluate the potential health benefits using characterized standardized preparations of GA, specifically *A. senegal* var. *senegal* using the same material already described by Shao-Ping et al. (2013) This work has demonstrated that the dietary supplementation with GA has a significant effect on blood pressure in a cohort of patients with renal disease (Glover et al., 2009); how this may translate into alteration in renal functions in the longer term remains to be explored in depth, which has generated fervor and energized a multitude of scientists to delve and deeply address focusing on the ongoing clinical trials.

Herbal therapy medicine and hemodialysis

Awfully, the end-stage disease has been described as the worst kidney function; hence, an urgent priority is the improvement of its prognosis for better thriving human life. Tragically speaking, the ESRD is the worst condition because most kidney functions are lost, and this is coupled with a bad prognosis for patients who are tethered to dialysis machines and characterized with low urine output. For humans, the two kidneys represent part of the bloodline for healthy life or part of the elixir of life because both kidneys are vital motors that kick away all sorts of wastes and toxic substances. Therefore, the loss of the kidney function or part of them represents an unmatched predicament to a fundamentally precious human life. Most of the sufferings that emanate from anyone with a kidney disease comes from the presence of toxic substances and excessive water buildup in the blood. Hence, the deposition of toxins contributes to further damage to kidney cells and tissues and negates the kidney-treating medications. Currently, many of the renal experts all over the world, including Shijiazhuang Kidney Disease Hospital in China, work in conjugation with the concerted efforts to develop natural treatment under the banner of toxins-removing treatment. The adoption of herbal medicine therapy for kidney disease treatment becomes rational assets with the end result of elimination of various waste products and excessive toxic substances due to the presence of active ingredients that can penetrate inside the affected kidney lesions directly. It is reported that GA can help slow down the progression of kidney damage by inhibiting inflammation and oxidative stress. However, it cannot repair damaged kidney tissues, therefore GA cannot help kidney failure patients to avoid dialysis. Hence, a sigh of relief might be achieved by recovering all kidney functions due to the improved prognosis naturally and gradually by herbal medications (Anon., 2015b).

The vitality of the human body and the kidneys have multifactorial benefits, including the maintenance of water balance, electrolytes, hormones, regulating blood pressure, strengthening bone, and, in certain cases, generating enough red blood cells in hypoxic kidneys. The prime function of the kidneys is the production of urine coupled with the discharge of wasteful products of metabolism out of the body, although the kidneys are very fragile, easily injured, or damaged; hence the renal parenchyma, renal vasculature, damaged part, or even all of the kidney functions will be lost. Miraculously, the kidneys are endowed with great compensatory abilities because often, kidney disease patients show no obvious symptoms until, ironically, more than 50% of the kidney function is already lost. In case the kidneys are damaged, the patients will endure and mention a few symptoms, including swelling, electrolyte abnormalities, high blood pressure, urine changes, bone problems, and anemia.

Normally, the ESRF patients are not unfamiliar with kidney transplants because many are put on the waiting list for a donor kidney while they are on maintenance dialysis; hence, they can easily get rid of dialysis and rebuild the kidney functions, as successful transplant of one kidney can improve 5% of kidney functions.

Intrinsically, kidney transplant is very effective, but usually fails within the years because not every kidney patient is able to do kidney transplant. Hence, Micro-Chinese Osmotherapy (Anon., 2016) is a reliable choice; although it takes longer time than the kidney transplant to have curative effects, it is more natural and safer. GA cannot eliminate the need for dialysis where a better resort is Micro-Chinese medicine Osmotherapy, which helps to repair kidney damage and helps to improve renal function. This is the radical way to treat kidney failure and it is possible to eliminate dialysis. The active ingredients in Chinese medicine can help eliminate blood qi stasis, promote blood circulation so as to provide adequate blood oxygen to renal tissues. Besides, Chinese medicine can also provide many nutrients such as vitamins, essential amino acids, trace elements for restoring injured kidney tissues. The active ingredients in the medicines have the functions of vasodilation, and they increase blood flow anticoagulation, antiinflammation, and antidegradation of extracellular matrix (Anon., 2015a). This automatically promotes blood circulation and increases blood flow to the kidneys and other vital organs. It also contains many nutrients to nourish the renal tissues and can refresh damaged renal tissue and generate new renal functioning tissues so as to gradually restore renal structure and improve kidney functions (Anon., 2013). It is well-documented that AG can reverse part of kidney damage, but not enough for curing kidney failure and evidently no way for elimination of dialysis. Recent advancement in the treatment of kidney failure and the elimination of dialysis has been offered by Micro-Chinese Osmotherapy that helps repair kidney damage and improves renal function. The active ingredients in Chinese medicine can help eliminate blood stasis and promote blood circulation to provide adequate blood and oxygen to needy, hungry renal tissues. In addition to that, Chinese medicine can provide essential nutrients such as vitamins, essential amino acids, and trace elements for restoring injured kidney tissues. Moreover, it can promote the replication of DNA of renal intrinsic cells so as to produce new renal functioning cells and rebuild renal functions.

Kidney failure means that the kidneys cannot work well, and people have to live with dialysis and kidney transplants, however, dialysis is not always necessary. Still, some patients might still have a chance to avoid dialysis. It was reported that the unique properties in GA can help slow down the progression of kidney damage by inhibiting inflammation and oxidative stress. Moreover, GA is known to reduce urea through utilizing the bowel as a "substitute kidney" and increasing BUN excretion in stools, but GA cannot repair damaged kidney tissues; even though it is helpful, it cannot stop the progression. The best treatment for kidney failure patients is in Shijizhuang Kidney Disease Hospital in China, which, indeed, helps patients avoid dialysis with the secret of traditional Chinese medicine (Anon., 2015b). In the Micro-Chinese Osmotherapy, which is a popular herbal therapy where the effective ingredients can permeate into kidney lesions directly, which have the function of purifying the blood and increasing blood circulation, bringing enough blood volume and oxygen to function adequately since the kidneys were normally, patients won't need dialysis (Anon., 2015b).

Therefore, in order to summarize all activities, it is fair to say that the potential of GA in kidney dysfunction is very promising, but more work is needed to decipher in active performance of this natural product with its impeccable active ingredients. This emanates from the realization that GA is a nondigestible food ingredient that has found many applications in the food and pharmaceutical industries. The gums are claimed to have therapeutic usefulness in hepatic and renal failures, and this awaits further verification in animal models and, automatically, the concern of humans. No significant adverse or toxic actions have been associated with the use of GA, which is a certificate and witness that has been said and documented by many human patients who have been tied to hemodialysis from three to four times weekly.

References

Adiotomre, J., Eastwood, M.A., Edwards, C.A., Brydon, W.G., 1990. Dietary fibre: in vitro methods that anticipate nutrition and metabolic activity in humans. Am. J. Clin. Nutr. 52 (1), 128–134.

Al-Assaf, S., Phillips, G.O., Williams, P.A., 2005. Studies on acacia exudate gums. Part 1: the molecular weight of *Acacia senegal* gum exudates. Food Hydrocoll. 19, 647–660.

Al-Majed, A.A., Abd-Allah, A.R., Al-Rikabi, A.C., Al-Shabanah, O.A., Mostafa, A.M., 2003. Effect of oral administration of Arabic gum on cisplatin-induced nephrotoxicity in rats. J. Biochem. Mol. Toxicol. 17 (3), 146–153.

Al Mosawi, A.J., 2002. The challenge of chronic renal failure in the developing world: possible use of acacia gum. Pediatr. Nephrol. 17, 390–391.

Ali, B.H., Ziada, A., Blunden, G., 2009. Biological effects of gum Arabic: a review of some recent research. Food Chem. Toxicol. 47, 1–8.

Ali, B.H., Al Moundhri, M.S., 2006. Agents ameliorating or augmenting the nephrotoxicity of cisplatin and other platinum compounds: a review of some recent research. Food Chem. Toxicol. 44, 1173–1183.

Ali, A.A., Ali, K.E., Fadlalla, A.E., Khalid, K.E., 2008. The effects of gum Arabic oral treatment on the metabolic profile of chronic renal failure patients under regular haemodialysis in Central Sudan. Nat. Prod. Res. 22 (1), 12–21.

Ali, B.H., Al-Qarawi, A.A., Haroun, E.M., Mousa, H.M., 2003. The effect of treatment with gum Arabic on gentamicin nephrotoxicity in rats: a preliminary study. Ren. Fail. 25 (1), 15–20.

Ali, B.H., Al-Salam, S., Al Husseni, I., Kayed, R.R., Al-Masroori, N., Al-Harthi, T., 2010. Effects of gum Arabic in rats with adenine-induced chronic renal failure. Exp. Biol. Med. (Maywood) 235 (3), 373–382.

Ali, B.H., Alqarawi, A.A., Ahmed, I.H., 2004. Does treatment with gum Arabic affect experimental chronic renal failure in rats? Fundam. Clin. Pharmacol. 18 (3), 327–329.

Annison, G., Trimble, R.P., Topping, D.L., 1995. Feeding Australian Acacia gums and gum Arabic leads to non-starch polysaccharide accumulation in the cecum of rats. J. Nutr. 125, 283–292.

Anonymous, 2016. What is the Prognosis of End Stage Renal Disease Patients with Dialysis. http://www.kidneyfailureweb.com/esrd/4769.html.

Anonymous, 2013. How To Improve Kidney Failure Functions. http://www.kidneyfailureweb.com/treatment/289.html.

Anonymous, 2015a. Can Acacia Gum Cure Kidney Failure and Eliminate the Need for Dialysis. http://www.kidney-cares.org/kidney-failure-treatment/5142.html.

Anonymous, 2015b. Kidney Failure Treatment. http://www.kidney-symptom.com/kidney-failure-treatment/1494.html.

Aoba, T., 2004. Solubility properties of human tooth mineral and pathogenesis of dental caries. Oral Dis. 10, 249–257.

Aune, D., Chan, D.S., Lau, R., Vieira, R., Greenwood, D.C., Kampman, E., 2011. Dietary fibre, whole grains, and risk of colorectal cancer: systematic review and dose-response meta-analysis of prospective studies. BMJ 343, d6617.

Bliss, D.Z., Stein, T.P., Schleifer, C.R., Settle, R.G., 1996. Supplementation with G.A fiber increases fecal nitrogen excretion and lowers serum urea nitrogen concentration in chronic renal failure patients consuming a low-protein diet. Am. J. Clin. Nutr. 63, 392–398.

Chaturvedi, S., Jones, C., 2007. Protein restriction for children with chronic renal failure. Cochrane Database Syst. Rev. 17, CD006863.

Cherbut, C., Michel, C., Raison, V., Kravtchenko, T., Meance, S., 2003. Acacia gum is a bifidogenic dietary fibre with high digestive tolerance in healthy humans. Microb. Ecol. Health Dis. 15, 43–50.

Cho, S.S., Qi, L., Fahey, Jr., G.C., Klurfeld, D.M., 2013. Consumption of cereal fibre, mixtures of whole grains and bran, and whole grains and risk reduction in type 2 diabetes, obesity, and cardiovascular disease. Am. J. Clin. Nutr. 98 (2), 594–619.

Cho, I., Blaser, M.J., 2012. The human microbiome: at the interface of health and disease. Nat. Rev. Genet. 13, 260–270.

Codipilly, C.N., Teichberg, S., Wapnir, R.A., 2006. Enhancement of absorption by gum Arabic in a model of gastrointestinal dysfunction. J. Am. Coll. Nutr. 25, 307–312.

Coresh, J., Selvin, E., Stevens, L.A., Manzi, J., Kusek, J.W., Eggers, P., 2007. Prevalence of chronic kidney disease in the United States. JAMA 298 (17), 2038–2047.

Cummings, J.H., Stephen, A.M., 2007. Carbohydrate terminology and classification. Eur. J. Clin. Nutr. 61 (Suppl. 1), S5–S18.

FAO, 1999. Gum Arabic. Food and Nutrition Paper, No. 52, addendum 7. .

Fötisch, K., Fäh, J., Wüthrich, B., Altmann, F., Haustein, D., Vieths, S., 1998. IgE antibodies specific for carbohydrates in a patient allergic to G.A. (*Acacia senegal*). Allergy 53, 1043–1051.

Fouque, D., Laville, M., Boissel, J.P., 2006. Low protein diets for chronic kidney disease in non diabetic adults. Cochrane Database Syst. Rev. 1999, CD001892.

Fouque, D., Cruz, C.M., Lindley, E., Rogers, S., Pancirova, J., Kernc, J., Copley, J.B., 2014. Dietary trends and management of hyperphosphatemia among patients with chronic kidney disease: an international survey of renal care professionals. J. Ren. Nutr. 24, 110–115.

Fujiwara, K., Mochida, S., Nagoshi, S., Iijima, O., Matsuzaki, Y., Takeda, S., Aburada, M., 1995. Regulation of hepatic macrophage function by oral administration of xiao-chai-hu-tang (sho-saiko-to,TJ-9) in rats. J. Ethnopharmacol. 46, 107–114.

Gamal el-din, A.M., Mostafa, A.M., Al-Shabanah, O.A., Al-Bekairi, A.M., Nagi, M.N., 2003. Protective effect of Arabic gum against acetaminophen-induced hepatotoxicity in mice. Pharmacol. Res. 48, 631–635.

Glover, D.A., Ushida, K., Phillips, A.O., Riley, S.G., 2009. Acacia (sen) SUPERGUM: an evaluation of potential health benefits in human subjects. Food Hydrocoll. 23 (8), 2410–2415.

Hamady, M., Knight, R., 2009. Microbial community profiling for human microbiome projects: tools, techniques, and challenges. Genome Res. 19, 1141–1152.

Ibrahim, M.A., Kohn, N., Wapnir, R.A., 2004. Proabsorptive effect of gum Arabic in isotonic solutions orally administered to rats: effect on zinc and other solutes. J. Nutr. Biochem. 15, 185–189.

Idris, O.H.M., Williams, P.A., Phillips, G.O., 1998. Characterisation of the gums from *Acacia senegal* trees of different age and location using multidetection gel permeation chromatography. Food Hydrocoll. 12, 379–389.

Jakobsson, H.E., Jernberg, C., Andersson, A.F., Sjolund-Karlsson, M., Jansson, J.K., Engstrand, L., 2010. Short-term antibiotic treatment has differing long-term impacts on the human throat and gut microbiome. PLoS One 5, e9836.

Johnson, C.L., Versalovic, J., 2012. The human microbiome and its potential importance to pediatrics. Pediatrics 129, 950–960.

Kalantar-Zadeh, K., Kopple, J.D., Deepak, S., Block, D., Block, G., 2002. Food intake characteristics of hemodialysis patients as obtained by food frequency questionnaire. J. Ren. Nutr. 12, 17–31.

Karrasch, T., Jobin, C., 2008. NF-kappaB and the intestine: friend or foe? Inflamm. Bowel Dis. 14, 114–124.

Polonsky, K.S., 2012. The past 200 years in diabetes. N. Engl. J. Med. 367, 1332–2134.

Kishimoto, A., Ushida, K., Phillips, G.O., Ogasawara, T., Sasaki, Y., 2006. Identification of intestinal bacteria responsible for fermentation of gum Arabic in pig model. Curr. Microbiol. 53, 173–177.

Krishnamurthy, V.M., Wei, G., Baird, B.C., Murtaugh, M., Chonchol, M.B., Raphael, K.L., 2012. High dietary fibre intake is associated with decreased inflammation and allcause mortality in patients with chronic kidney disease. Kidney Int. 81 (3), 300–306.

Kuczynski, J., Lauber, C.L., Walters, W.A., Parfrey, L.W., Clemente, J.C., Gevers, D., Knigh,t, R, 2012. Experimental and analytical tools for studying the human microbiome. Nat. Rev. Genet. 13, 47–58.

La Vecchia, C., Chatenoud, L., Negri, E., Franceschi, S., 2003. Session: whole cereal grains, fiber and human cancer wholegrain cereals and cancer in Italy. Proc. Nutr. Soc. 62 (1), 45–49.

Leclère, C.J., Champ, M., Boillot, J., Guille, G., Lecannu, G., Molis, C., Bornet, F., Krempf, M., Delort-Laval, J., Galmiche, J.P., 1994. Role of viscous guar gums in lowering the glycemic response after a solid meal. Am. J. Clin. Nutr. 59, 914–921.

Levey, A.S., Eckardt, K.U., Tsukamoto, Y., Levin, A., Coresh, J., Rossert, J., De Zeeuw, D., Hostetter, T.H., Lameire, N., Eknoyan, G., 2005. Definition and classification of chronic kidney disease: a position statement from Kidney Disease: Improving Global Outcomes (KDIGO). Kidney Int. 67, 2089–2100.

Lida, S., Kohno, K., Yoshimura, J., Ueda, S., Usui, M., Miyazaki, H., 2006. Carbonic-adsorbent AST-120 reduces overload of indoxyl sulphate and the plasma level of TGF-beta1 in patients with chronic renal failure. Clin. Exp. Nephrol. 10 (4), 262–267.

Locatelli, F., Del Vecchio, L., Pozzoni, P., Manzoni, C., 2006. Nephrology: main advances in the last 40 years. J. Nephrol. 19, 6–11.

Mahmoud, M.F., Diaai, A.A., Ahmed, F., 2012. Evaluation of the efficacy of ginger, Arabic gum, and Boswellia in acute and chronic renal failure. Ren. Fail. 34 (1), 73–82.

Matsumoto, N., Riley, S., Fraser, D., Al-Assaf, S., Ishimura, E., Wolever, T., Phillips, G.O., Phillips, A.O., 2006. Butyrate modulates TGF-beta1 generation and function: potential renal benefit for Acacia (sen) SUPERGUM (G.A.)? Kidney Int. 69, 257–265.

May, T., Mackie, R.I., Fahey, G.C., Cremin, J.C., Garleb, K.A., 1994. Effect of fiber source on short-chain fatty acid production and on the growth and toxin production by Clostirudium difficile. Scand. J. Gastroenterol. 29, 916–922.

Meyer, T.W., Hostetter, T.H., 2007. Uremia. New Engl. J. Med. 357, 1316–1325.

Michel, C., Kravtchenko, T.P., David, A., Gueneau, S., Kozlowski, F., Cherbut, C., 1998. In vitroprebiotic effects of Acacia gums onto the human intestinal microbiota depends on both botanical origin and environmental pH. Anaerobe 4 (6), 257–266.

Mocak, J., Jurasek, P., Phillips, G.O., Vaarga, S., Casedei, E., Chikemai, B.N., 1998. Chemometric characterization of exudate gums that conform to the revised specification of gum Arabic for food use, and the identification of adulterants. Food Hydrocoll. 12, 141–150.

Mochida, S., Ogata, I., Hirata, K., Ohta, Y., Yamada, S., Fujiwara, K., 1990. Provocation of massive hepatic necrosis by endotoxin after partial hepatectomy in rats. Gastroenterology 99, 771–777.

Morgan, X.C., Segata, N., Huttenhower, C., 2013. Biodiversity and functional genomics in the human microbiome. Trends Genet. 29, 51–58.

Nakamura, T., Sato, E., Fujiwara, N., Kawagoe, Y., Suzuki, T., Ueda, Y., 2011. Oral adsorbent AST-120 ameliorates tubular injury in chronic renal failure patients by reducing proteinuria and oxidative stress generation. Metabolism 60 (2), 260–264.

Nakao, T., 2012. Chronic kidney disease (CKD) – recent progress. Topics: VII. Management of chronic kidney disease (CKD) and treatment; 1. Dietary treatment and lifestyle management in CKD patients. Nihon Naika Gakkai Zasshi 101, 1340–1346.

Nasir, O., Artunc, F., Saeed, A., Kambal, M.A., Kalbacher, H., Sandulache, D., Boini, K.M., Jahovic, N., Lang, F., 2008. Effects of gum Arabic (*Acacia senegal*) on water and electrolyte balance in healthy mice. J. Ren. Nutr. 18, 230–238.

Nasir, O., 2013. Renal and extra renal effects of gum Arabic (*Acacia senegal*). what can be learned from animal experiments? Kidney Blood Press Res. 37, 269–279. doi: 10.1159/000350152.

Nasir, O., Babiker, S., Salim, A.M., 2016. Protective effect of gum Arabic supplementation in for type 2 diabetes mellitus and its complications. Int. J. Multidisciplinary Curr. Res. 4, 288–294.

Onishi, T., Umemura, S., Yanagawa, M., Matsumura, M., Sasaki, Y., Ogasawara, T., Ooshima, T., 2008. Remineralization effects of gum Arabic on caries-like enamel lesions. Arch. Oral. Biol. 53, 257–260.

Owada, S., Maeba, T., Sugano, Y., Hirayama, A., Ueda, A., Nagase, S., 2010. Spherical carbon adsorbent (AST-120) protects deterioration of renal function in chronic kidney disease rats through inhibition of reactive oxygen species production from mitochondria and reduction of serum lipid peroxidation. Nephron Exp. Nephrol. 115 (4), e101–e111.

Pasquier, B., Armand, M., Castelain, C., Guillon, F., Borel, P., Lafont, H., Lairon, D., 1996. Emulsification and lipolysis of triacylglycerols are altered by viscous soluble dietary fibres in acidic gastric medium in vitro. Biochem. J. 314, 269–275.

Phillips, G.O., 1998. Acacia gum (G.A.): a nutritional fibre; metabolism and caloric value. Food Addit. Contam. 15, 251–264.

Phillips, G.O., 2013. Dietary fibre: a chemical category or a health ingredient? Bioact. Carbohydr. Dietary Fibre, 3–9.

Phillips, G.O., Cui, S.W., 2011. An introduction: evolution and finalisation of the regulatory definition of dietary fibre. Food Hydrocoll. 25, 139–143.

Rehman, K., Wingertzahn, M.A., Harper, R.G., Wapnir, R.A., 2001. Proabsorptive action of G.A.: regulation of nitric oxide metabolism in the basolateral potassium channel of the small intestine. J. Pediatr. Gastroenterol. Nutr. 32, 529–533.

Rimm, E.B., Ascherio, A., Giovannucci, E., Spiegelman, D., Stampfer, M.J., Willett, W.C., 1996. Vegetable, fruit, and cereal fibre intake and risk of coronary heart disease among men. JAMA 275 (6), 447–451.

Sander, I., Raulf-Heimsoth, M., Wiemer, K., Kespohl, S., Brüning, T., Merget, R., 2006. Sensitization due to G.A. (*Acacia senegal*): the cause of occupational allergic asthma or cross-reaction to carbohydrates? Int. Arch. Allergy Immunol. 141, 51–56.
Shao-Ping, N., Wang, C., Cui, S.W., Wang, Q., Xie, M.Y., Phillips, G.O., 2013. A further amendment to the classical core structure of gum arabic (*Acacia senegal*). Food Hydrocoll. 31, 42–48.
Smith, J.G., Yokoyama, W.H., German, J.B., 1998. Butyric acid from the diet: actions at the level of gene expression. Crit. Rev. Food Sci. Nutr. 38 (4), 259–297.
Suliman, S.M., Hamdouk, M.I., Elfaki, M.B., 2000. Gum Arabic fibre as a supplement to low protein diet in chronic renal failure patients. Sudan Association of Physicians 17th Conference. Khartoum.
Truswell, A.S., 2002. Cereal grains and coronary heart disease. Eur. J. Clin. Nutr. 56 (1), 1–14.
Ueda, H., Shibahara, N., Takagi, S., Inoue, T., Katsuoka, Y., 2007. AST-120, an oral adsorbent, delays the initiation of dialysis in patients with chronic kidney diseases. Ther. Apher. Dial. 11 (3), 189–195.
Ueda, H., Shibahara, N., Takagi, S., Inoue, T., Katsuoka, Y., 2008. AST-120 treatment in pre-dialysis period affects the prognosis in patients on haemodialysis. Ren. Fail. 30 (9), 856–860.
Verbeken, D., Dierckx, S., Dewettinck, K., 2003. Exudate gums: occurrence, production, and applications. Appl. Microbiol. Biotechnol. 63, 10–21.
Wapnir, R.A., Sherry, B., Codipilly, C.N., Goodwin, L.O., Vancurova, I., 2008. Modulation of rat intestinal nuclear factor NF-kappaB by gum arabic. Rat small intestine by gum arabic. Dig. Dis. Sci. 53, 80–87.
Wikoff, W.R., Anfora, A.T., Liu, J., Schultz, P.G., Lesley, S.A., Peters, E.C., Siuzdak, G., 2009. Metabolomics analysis reveals large effects of gut microflora on mammalian blood metabolites. Proc. Natl. Acad. Sci. U.S.A. 106, 3698–3703.
Winchester, J.F., Salsburgh, J.A., 2004. Sorbents in the treatment of renal failure. Minerva Urol. Nefrol. 56, 215–221.
Younes, H., Garleb, K., Behr, S., Remesy, C., Demigne, C., 1995. Fermentable fibers or oligosaccharides reduce urinary nitrogen excretion by increasing urea disposal in the rat caecum. J. Nutr. 125, 1010–1016.

Further Readings

Al Mosawi, A.J., 2007. The use of acacia gum in end stage renal failure. J. Trop. Pediatr. 53, 362–365.
Cindoruk, M., Kerem, M., Karakan, T., Salman, B., Akin, O., Alper, M., Erdem, O., Unal, S., 2007. Peroxisome proliferators-activated alpha agonist treatment ameliorates hepatic damage in rats with obstructive jaundice: an experimental study. BMC Gastroenterol. 28, 44.
Hinson, J.A., Reid, A.B., McCullough, S.S., James, L.P., 2004. Acetaminophen-induced hepatotoxicity: role of metabolic activation, reactive oxygen/nitrogen species, and mitochondrial permeability transition. Drug Metab. Rev. 36, 805–822.
Rehman, K.U., Wingertzahn, M.A., Teichberg, S., Harper, R.G., Wapnir, R.A., 2003. Gum arabic. (GA) modifies paracellular water and electrolyte transport in the small intestine. Dig. Dis. Sci. 48, 755–760.

Dietary Gum Arabic as Animal Feed Additive

22

Abdelaziz A. Fadlelmoula
University of Jeddah, Alkamel, Kingdom of Saudi Arabia; University of Khartoum, Shambat, Sudan

Introduction

Gum Arabic (GA), which is also known as acacia gum, is an exudate from acacia trees in the form of nodules or tears. Chemically, GA is defined as a high-molecular-weight complex polysaccharide, and it includes neutral monosaccharide sugars such as galactose, rhamnose, and arabinose. It contains also organic acid, mainly glucuronic acid, and some metallic minerals such as calcium, potassium, magnesium, phosphorus, and sodium.

GA is known to be widely utilized in the food industry as stabilizer, emulsifier, and thickening agent. GA was reported in folk medicine as an internal inflammatory agent for treatment of intestinal mucosa inflammation as well as an external inflammatory agent that assists in healing of inflamed surfaces. The antioxidant effect of GA was attained in some effect; some reported the effect of GA in renal diseases as nephron protector.

GA was tried in many poultry feeding trials; it was found that some components of GA, mainly the soluble dietary fibers oligofructose and inulin, were subjected to fermentation in the caecum. In rabbits, it was tried to enhance growth performance and to study its effects on serum constituents. Adverse effects of feeding GA in diets were claimed in some reports; Ilchyshyn and Smith (1985) stated that diets consisting of GA may cause contact dermatitis and asthma, and the causing agents were not identified. Foetish et al. (1998) were able to recognize a carbohydrate-specific Immunoglobin E immune agent in a chocolate that caused clear symptoms of cough and dyspnea. Another investigation concluded that GA has no toxic effect when fed at dietary levels to male and female rat in an experiment, with no clear clinical signs between the two treated sexes and the control (Doi et al., 2006). Anderson (1986) stated that GA utilization as a feed additive at any level has no hazardous effects and meets international food safety requirements. In poultry, it has been claimed that different levels of GA supplemented to layers expressed no effect on serum and yolk albumin values, and consequently, no upper limits were stated (Cehen et al., 2005). The Food and Agriculture Organization/World Health Organization joint committee recognized GA as one of the safest feed additives with no recommended upper limitation (FAO/WHO, 1969).

Gum Arabic. http://dx.doi.org/10.1016/B978-0-12-812002-6.00022-1

Chemical composition and biochemical properties of gum Arabic

GA was found to be composed of mainly two polysaccharide components in accordance with the molecular weight: high-molecular-weight glycoprotein, with 90% carbohydrate, and low-molecular-weight heterogeneous polysaccharide sugars (Lamport and Fong, 1991). The carbohydrate fraction of GA is galactose, rhamnose, glucuronic acid, and Arabinose. However, the protein part of GA was suggested to be pounded to carbohydrate with a covalent bond (Akiyama et al., 1984). It has also been reported that GA is rich with essential and trace elements like aluminum, phosphorus, magnesium, copper, zinc, and iron (Buffo et al., 2001). In some other investigations, GA was chemically classified as a variable complex mixture of arabinogalactan, monosaccharide, and glycoprotein (William and Philips, 2000). The authors also stated that depending on the source of GA, the glycan portion of GA included more L-arabinose, which is relative to D-galactose. The chemical as well as nutritional components of GA can be clearly shown in Table 22.1 (Elkhalifa, 1998; Siham et al., 2015). GA includes both essential and nonessential amino acids in variable quantities, as presented in Table 22.2.

Regarding the properties of GA, it is a sugar crystal that can act as an emulsifier, stabilizer, and preventive of chemical breakdown; the copper content of GA acts as a firming agent in fermentation reactions to promote the precipitation and coagulation of protein and tannin (Gabb, 1997).

Gum Arabic as dietary fibers

Dietary fibers were earlier defined by Trowell et al. (1976) as "plant products such as lignin and polysaccharides that are not subject to digestion by the digestive tract secretions." Since this definition is mainly used in humans, it can commonly be

Table 22.1 Chemical and biochemical components and energy contents in gum Arabic (Elkhalifa, 1998; Siham et al., 2015)

Gum Arabic component and energy content	Percentage
DM	87
Crude protein	2.5
Arabinose	25
Lactose	14
Rhamnose	14
Methoxy	2.5
Methylglucuronic acid	1.5
Ash	3.6
Ph	4.4
ME	3000 (kcal/kg feed)

DM, Dry matter.

Table 22.2 Amino acid composition of gum Arabic (mg/kg)

Amino acid	Elkhalifa (1998)	Ibrahim et al. (2013)	Gashua (2016)
Alanine	27	31	30
Arginine	13	7.0	21
Cysteine	2.0	1.0	1.5
Glutamate	42	36	73
Glycine	50	49	50
Histidine	44	51	124
Hydroxyproline	304	274	453
Isoleucine	12	14	24
Leucine	66	75	139
Lysine	33	26	51
Proline	63	72	114
Serine	129	123	217

used for nonruminant animals as well. The dietary fibers effect of GA was noticed during treatment of digestive tract conditions. Turvil et al. (2000) stated that soluble fiber of GA reduced the cholera toxins that has been induced in rats. Some studies showed that the effect of GA dietary fibers was noticed in stopping diarrhea faster when supplemented to oral rehydration salts, enhancing the net water and sodium absorption, and the conclusion reported was that GA fibers can effectively enhance sodium absorption when added to oral rehydration salts (Wapnir et al., 1996; Wingertzahn et al., 1999). It was found that nondigestible fiber containing GA, such as arabinogalactan, is able to resist the hydrolytic effect of the intestinal enzymes and enter the bowel intact to be fermented by the microflora, which resulted in increased total fecal anaerobes, *Lactobacillus* bacteria, decreased fecal ammonia level, and no notable change in plasma lipids and insulin (Robinson et al., 2001). In laying hens, significant increase of caecal fermentation of dietary GA was observed and hence a positive effect on absorption of minerals in the colon (Gultermirian et al., 2014).

Effect of gum Arabic on serum cholesterol level

GA included in animal feed was claimed to lower the serum cholesterol level, and the mechanism was suggested to be the interference with dietary cholesterol absorption (Kerri and Tsai, 1993). However, an earlier study of Kelly et al. (1978) stated that there is an association between GA intake and an increase in cholesterol biosynthesis; however, the result was different in broiler chicks, where GA was found to reduce the cholesterol level (Siham et al., 2015). It has been found that increasing GA 5%–15% in laying hen rations lowered the serum cholesterol and consequently low yolk cholesterol in eggs (El-Kheir et al., 2009).

In human, serum cholesterol was found to be reduced to the level of 6%–10% upon receiving GA in diets for a period of 21–30 days (Ross et al., 1983; Sharma, 1985). Another study declared that dietary GA could result in increasing total cholesterol biosynthesis (Kelly et al., 1978), but this suggestion was not in line with the claim that the diet and the genetic background on cholesterol synthesis are not intractable (Kerri and Tsai, 1993). In rabbits, feeding ration containing 5% GA resulted in an increased cholesterol level in serum compared to control. However, in the same study, when GA increased in the ration to 7%, the serum cholesterol level was tremendously increased, in addition to the total protein level and weight gain (Samia, 1993). In another study, the total serum was found to be significantly reduced in broiler chicks fed broiler ration supplemented with 0.6% GA (Tabidi and Ekram, 2015). GA supplemented in rats' diets in different ratios was also found to exert a significant positive effect in lowering serum cholesterol and triglyceride levels in rats (Omer, 2010). In an experiment conducted in rats, where diets containing 7.5% GA with 1% cholesterol was fed to one group and cholesterol-free diets were fed to a second group for 5 weeks, the result was a tremendously reduced plasma cholesterol in the group fed diets with cholesterol and GA compared to the other group (Al-Othman et al., 1998).

Effect of gum Arabic on growth performance

Studies declared no limitations to the effect to the amount of GA in animal feed, and it meets the safety demands recommended by the international food safety (Anderson, 1986). In poultry, GA was included in a dietary ration to broiler chicks to improve weight gain and to reduce low-density lipoprotein (Abdalla et al., 2015). Hence, Ali et al. (2013) and Kur et al. (2013) obtained a higher body weight gain for broilers fed GA in mixed ration. Feed intake was also increased in birds fed a high level of GA (El-Kheir et al., 2009) and justified that due to enhancement of palatability of the offered diet. Broiler poultry groups fed graded levels of GA powder (0, 2.5, 5, and 7.5%) for a period of 42 days ended up with increased feed intake, feed conversion ratio, and body weight gain rate (Siham et al., 2015). Final body weight and body weight gain were found to be increased in broiler chicks as well as laying hens when fed on rations supplemented with GA at levels ranged from 0.6% to 15% with differences being significant among GA levels; authors concluded that GA as prebiotics added to poultry feed could, in one way, improve the performance of the digestive tract, and consequently, increased feed intake and feed conversion ration resulted in high body weight gain (Abd-Razig et al., 2010; Piray et al., 2007; Tabidi and Ekram, 2015), improving the diets and hence modulating digestive tract health by interacting with the tract epithelium, mucus, and tract microflora and consequently protecting the digestive tract against enteric infections for the benefit of the individual (Montagne et al., 2003). However, some investigations revealed either no significant differences in growth performance due to the addition of GA in the diet, or it even exerted no effect on the trait at all (Midilli and Tuncer, 2001; Sims and Soften, 1999; Waldroup et al., 2003).

Conclusion

The review concluded that GA is among the natural polymer and feed additives products that play a great role in the health of animals when incorporated into the feed. Its interaction with digestive tract epithelium, mucosa, and microflora lead to control opportunistic bacteria and keep the tract healthy and maintain its functions. GA, with its fiber contents, can interfere with absorption of lipid from the intestine; this will lead to reduce the cholesterol and triglyceride levels in the body and consequently less fat deposits. GA could be a powerful growth promoter without adverse effects on the body and body organs. Currently, much more about the roles of the GA and interactions of its components with the body cells needs to be investigated.

References

Abdalla, S.A., Abdel Atti, K.A., Malik, H.E., Dousa, B.M., ElAmin, K.M., 2015. Effect of dietary inclusion of Gum Arabic (*Acasia Senegal*) on performance and blood chemistry of broiler chicks. Global J. Anim. Sci. Res. 3 (2), 305–310.

Abd-Razig, N.M., Sabahelkheir, M.K., Idris, O.F., 2010. Effect of gum Arabic on lipid and performance of laying hens. J. Appl. Biosci. 32, 2002–2007.

Akiyama, Y., Eda, S., Kato, K., 1984. Gum Arabic is a kind of arabinogalactan protein. Agric. Biol. Chem. 48 (1), 235–247.

Ali, F.M., Abdel Atti, K.A., Dousa, B.M., El Amin, K.M., 2013. Performance and blood chemistry of broiler birds fed different levels of dietary Rashad (*Lepidium sativum*). Wayamba J. Anim. Sci. Number 1368709198, 782–786.

Al-Othman, A.A., Al-Shagrawi, R.A., Hewedy, F.M., Hamdi, M.M., 1998. Plasma total, lipoprotein cholesterol, organs cholesterol and growth performance in rats fed dietary gum Arabic. Food Chem. 62 (1), 69–72.

Anderson, D.M., 1986. Evidence of the safety of gum Arabic (*Acasia senegal* (L. Wild) as a food additive – a brief review. Food Addit. Contam. 3, 225–230.

Buffo, R.A., Reineccius, G.A., Oehlert, G.W., 2001. Factors affecting the emulsifying and rheological properties of gum Arabic emulsions. Food Hydrocoll. 15, 53–66.

Cehen, Y.C., Nakthong, C., Chen, T., 2005. Improvement of laying hen performance by dietary prebiotic oligofructose and inulin. Int. J. Poult. Sci. 4, 103–108.

Doi, Y., Ichirhara, T., Hagiwara, A., Imai, N., Tamano, S., Orikashi, H., Ogasawara, K., Nakamura, M., Shirai, T., 2006. A ninety-days oral toxicity of a new type of processed gum Arabic from acacia tree (*Acacia senegal*) exudates in F344 rats. Food Chem. Toxicol. 44 (4), 560–566.

Elkhalifa, K.F., 1998. Gum Arabic. Agrotica, 30–31.

El-Kheir, M.K., Ishage, K.E., Yagoub, A.A., Abu Baker, A.A., 2009. Supplementation laying hen diet with gum Arabic (*Acacia senegal*): effect on egg production, shell thickness and yolk content of cholesterol, calcium and phosphorus. Asian J. Poul. Sci. 3 (1), 9–14.

FAO/WHO, 1969. FAO Nutrition Meeting Report Series, 46A, WHO/Food Add/70.36. Expert Committee on Food Additives, FAO, Rome, Italy.

Foetish, A., Faeh, J., Wuethrich, B., Altmann, F., Haustein, D., Vieths, S., 1998. IgE antibodies specific carbohydrate in a patient allergic to GA (*Acacia Senegal*). Allergy 53, 1043–1051.

Gabb, S., 1997. Gum Arabic Production in Sudan. Sudan Foundation, 212 London WC IV 9, UK.
Gashua, I.B., 2016. An Investigation of the Molecular Structure, Composition and Biophysical Properties of Gum Arabic (Ph.D. thesis). University of Wolverhampton, Wolverhampton, UK.
Gultermirian, M.L., Corti, H.R., Perez Chaia, A., Apella, M.C., 2014. Fermentation of in vitro of a mixture of dietary fibers and cane molasses by the cecal microbiota: application on mineral absorption through the laying hen's colonic epithelium. Anim. Feed Sci. Technol. 191, 76–82.
Ibrahim, O.B., Osman, M.E., Hassan, E.A., 2013. Characterization and simple fractionation of *Acacia senegal*. J. Chem. Acta 2, 11–17.
Ilchyshyn, A., Smith, A.G., 1985. GA sensitivity associated with epidemic hysteria dermatologica. Contact Dermatitis 13, 282–283.
Kelly, B., Cross, H.R., Pond, W.G., Mersmann, H.J., 1978. Effects of dietary fat cholesterol concentration of growing pig selected for high or low serum cholesterol. J. Anim. Sci. 71, 307–810.
Kerri, J., Tsai, A., 1993. Effects of pectin, gum Arabic and agar on cholesterol absorption, synthesis and turnover in rats. J. Nutr. 108, 630–639.
Kur, A.T., Abdel Atti, K.A., Dousa, B.M., Elagib, H.A., Malik, H.E., Elamin, K.M., 2013. Effect of treated cowpea seeds on broiler chicken. Global J. Anim. Sci. Res. 1 (1), 53–60.
Lamport, D.T., Fong, C., 1991. Gum Arabic glycoprotein is a twisted hairy rope. Plant Physiol. 96, 848–855.
Midilli, M., Tuncer, S.D., 2001. The effect of enzyme and probiotic supplementation to diets on broiler performance. Turk. J. Anim. Sci. 25, 895–903.
Montagne, L., Pluske, J.R., Hampson, D.J., 2003. A review of interactions between dietary fiber and the intestinal mucosa, and their consequences on digestive health in young non-ruminant animals. Anim. Feed Sci. Technol. 108, 95–117.
Omer, N.A., 2010. Effect of feeding gum Arabic on serum total and lipoprotein cholesterol in hypercholestrolemic rats (M.Sc. thesis). University of Khartoum, Sudan.
Piray, A.H., Reman Shahi, A., Tahmasbi, A.M., Bahram Pour, J., 2007. Effect of cecal cultures and as pergillus meal prebiotic (Fermacto) on growth performance and organ weight of broiler chickens. Int. J. Poult. Sci. 6, 340–344.
Robinson, R.R., Feirtag, J., Salvin, J.L., 2001. Effect of dietary arabinogalactan on gastrointestinal and blood parameter in healthy human subjects. J. Am. Coll. Nutr. 20 (4), 4377–4382.
Ross, A.H., Eastwood, M.A., Brydon, W.G., Anderson, J.R., Anderson, D.M., 1983. A study of the effects of dietary gum Arabic in humans. Am. J. Clin. Nutr. 37, 368–375.
Samia, I.T., 1993. Response of rabbits to gum Arabic (M.Sc. thesis). University of Khartoum, Sudan.
Sharma, R.D., 1985. Hypocholesterolaemic effect of gum acacia in men. J. Nutr. 5 (12), 1321–1326.
Siham, A.A., Khadiga, A.A., Huwaida, E.E., Bakheit, M.D., Khalid, M.E., 2015. Effect of dietary inclusion of gum Arabic (*Acacia senegal*) on performance and blood chemistry of broiler chicks. Global J. Anim. Sci. Res. 3 (2), 305–310.
Sims, M.D., Soften, A.E., 1999. Comparative effects of amannanoligo saccharide and on antibiotic growth promoter on performance of commercial tom turkeys. 48th Western Poultry Disease Conf., Vancouver, Canadapp. 78–82.
Tabidi, M.H., Ekram, K.A., 2015. Effect of feeding gum Arabic with or without commercial xylem enzyme 500 on the performance of broiler chicks. World J. Pharm. Pharm. Sci. 4 (9), 1863–1872.

Trowell, H., Southgate, D.A., Wolever, T.M., Leeds, A.R., Gassull, M.A., Jenkins, D.J., 1976. Letter: dietary fiber refined. Lancet 1, 967.

Turvil, J.L., Waprir, R.A., Wingertzahn, M.A., Teichberg, S., Farthing, M., 2000. Cholera toxin-induce secretion in rats is reduced by a soluble fiber, gum Arabic. Dig. Dis. Sci. 45 (5), 946–951.

Waldroup, P.W., Fritts, C.A., Fengland, Y., 2003. Utilization of Bb-Mas mannanoligo saccharide and Bioplex copper in broiler diets. Int. J. Poult. 2, 44–52.

Wapnir, R.A., Teichberg, S., Go, J.T., Wingertzahn, M.A., Harper, R.G., 1996. Oral rehydration solution: enhanced sodium absorption with gum Arabic. J. Am. Coll. Nutr. 15 (4), 377–382.

William, P.A., Philips, G.O., 2000. Handbook of Hydrocolids. CRC Press, Cambridge, England, pp.155–168.

Wingertzahn, M.A., Teichberg, S., Mark, A., Moyse, J., Wapnir, R.A., 1999. Effect of gum Arabic in an oral rehydration solution in recovery from diarrhea in Rats. J. Pediatr. Gastroenterol. Nutr. 29 (4), 411–417.

Applications of Gum Arabic in Medical and Health Benefits

23

Noura K.M. Salih[*,**]
*University of Bahri, Khartoum Bahri, Sudan; **Infrastructure University Kuala Lumpur, Kajang, Malaysia

Introduction

Since the ancient era, gum Arabic (GA) (*Acacia senegal*, *Acacia seyal*, and *Acacia nilotica*) is one of the indigenous trees that has nutrients and therapeutic ingredients that have been used in traditional medicine as sedative, relaxing, softening, and healing agents. It is involved in some internal and external traditional medicine interventions such as cough, diarrhea, dysentery, hemorrhage, and inflammation of the intestinal mucosa. It is also applied for treatment of skin diseases and inflammations in veterinary medicine (Gamal el-din et al., 2003; Mariana et al., 2012).

GA contains a high percentage of nonviscous soluble fiber (Wiliams and Phillips, 2000). The Food and Agriculture Organization/World Health Organization Joint Expert Committee for Food Additives defined GA as "a dry exudation attained from the stems of *A. senegal* or closely related species of *Acacia*" (FAO, 1999).

Verbeken et al. (2003) describes GA as a means of cosmetics and pharmaceutical industries. The Ebers described it as contraception when utilized with dates. GA also acts as an antioxidant and has a nephron-protectant effect (Ali et al., 2008; Gamal el-din et al., 2003; Rehman et al., 2001). Clinically, it ensures reduction of urea and creatinine plasma concentrations in patients with chronic renal failure; it was claimed that it helps and reduces the need for dialysis from three to two times per week (Suliman et al., 2000). In kidneys, GA has a positive effect on nephrotoxicity induced by *Aristolochia bracteata* aqueous extract (Omer et al., 2013).

The antimicrobial activity of *Acacia arabica* bark extracts, prebiotic aqueous extract, and nanoemulsion powder against selected antibiotic-resistant bacterial and fungal pathogens was examined, and variable gradation of antimicrobial activity was recorded (Chaubal and Tambe, 2006; Johny et al., 2010; Lawrence et al., 2015; Luo et al., 2014; Sharma et al., 2014).

GA, or acacia gum species, are mixtures of polysaccharides and glycoproteins [polysaccharides with split chains of (1–3) connected to β-D-galactopyranosyl units that contain β-D-glucuronopyranosyl, α-L-rhamnopyranosyl, 4-O-methyl-β-D-glucuronopyranosyl units, and α-L-arabinofuranosyl], which are the main sources of ribose and arabinose (Deckwer et al., 2006).

In this chapter, the role of GA soluble fiber and its medical and health benefits are reviewed.

Gum Arabic. http://dx.doi.org/10.1016/B978-0-12-812002-6.00023-3

Antioxidants

GA (*A. nilotica*) is one source of natural antioxidants that can be used as supplemental supporting therapy for many free radical diseases, such as diabetes, cancer, inflammation, and some others (Amos et al., 1999). In *A. nilotica,* the hydroxyl groups existing in the phenolic compounds can hunt the free radicals (Kalaivani and Mathew, 2010; Vadivel and Biesalski, 2012). Ahmed et al. (2015) stated that the GA, *A. seyal* and *A. senegal* are branched-chain polysaccharides that contain strong antioxidant properties and have been used for the reduction of the experimental toxicity. The results of their test experiment reduced the lipid peroxidation and enhanced the activities of antioxidant enzymes and their mRNA expression in the diabetic rats' liver. Thus, they concluded that GA extracts are useful to protect against hepatic oxidative stress in diabetic rats.

Different parts of the tree can be used, as traditionally, people used the bark, leaves, pods, and flowers against cancer, cold, congestion, cough, dysentery, diarrhea, gallbladder disorder, hemorrhoid, bleeding piles, fever, ophthalmia, sclerosis, tuberculosis, small pox, leprosy, leukoderma, and menstrual problems (Ambasta, 1994; Bhargava et al., 1998). In West Africa, the bark and gum of these trees are used against cancer, tumors, and recuperation of the liver and spleen; the wood for smallpox; the root for tuberculosis; and the leaves for ulcer (Kalaivani and Mathew, 2010). The Egyptians consume powdered pods of *A. nilotica* (Rahman, 2010) to treat diabetics.

Hypolipidemic and antidiabetic

The effect of GA pods extracted with aqueous methanol is significantly decreasing body fat and blood sugar; this has been reported by Ahmad et al. (2008). They tested the aqueous methanol extracted from *A. nilotica* pods on various parameters, such as high-density lipoproteins (HDLs), triglycerides, total cholesterol, blood glucose levels, serum glutamate oxaloacetate, pyruvate transaminase, and serum creatinine removal in alloxan-induced diabetic rabbits, which exposed to effective dose of 400 mg/kg body mass outstandingly. All the researches done in GA's pods in animal and human revealed that polyphenols are hypoglycemic and hypolipidemic (Ahmad et al., 2008; Hirosh et al., 2004; Sabu et al., 2002; Zaki and Al-Mustafa, 2000).

Priyanka and Maya (2013) reported that the addition of Indian GA pod powder in the food of people having cardiovascular disease (CVD) and type 2 diabetes will decrease the risk factors associated with those diseases. Nasir et al. (2008) found that the effect of GA on intestinal glucose transport may be beneficial in the prophylaxis and treatment of type 2 diabetes.

When adding the pod powder as a dietary fiber, mainly a 4 g/day dose, it reduces the fasting blood glucose by 10%–19%, lipid profile by 6%–18%, low-density lipoprotein (LDL) by 7%–10%, total cholesterol by 5%–11%, very low-density lipoprotein by 7%–15%, HDL cholesterol by 5%–10%, and blood pressure by 8%–13%, and it is indicated as beneficial for diabetes and coronary artery diseases. Pods and tender leaves are considered very helpful in folk medicine for treatment of diabetes mellitus (Gilani et al., 1999; Priyanka and Choudhry, 2013).

Blood pressure and cholesterol

Nowadays, coronary heart disease (CHD) is adversely affecting the health of women over 50 years old and men over 45 years old in Europe. The cholesterol level is high, above the desirable level (normal level is 5.2 mmol/L) in 79% of men and 65% of women aged between 35 and 50 years (Pyorala et al., 1994). Research findings from Betteridge et al. (1993) showed that soluble fiber effectively lowers the cholesterol by significant amounts. Many research studies have shown that the risk of mortality from CHD increases with rising average levels of plasma cholesterol (Simons, 1986). Soluble fiber can be effective in dropping the plasma cholesterol level by significant amounts (Coats, 1998; Xuan et al., 2010). Supplementation of *A. nilotica* proanthocyanidins was tested to be a powerful nutritional style for reducing the CVD risk in individuals with both raised LDL and metabolic syndrome. GA decreases the plasma total cholesterol and triglyceride in rabbits, and also, it significantly increases the plasma HDL levels (Lerman et al., 2010).

The effects of GA in lipid lowering were investigated by Zaki and Al-Mustafa (2000), Anderson et al. (2004), and Tomura et al. (2007); they found that induced toxicity of *A. nilotica* feeding in rats with the phenolic compounds from GA may lessen the risk of CVD by antioxidative properties.

Research has shown that adding 14 g of soluble fiber per 1000 kcals of the consumed energy to the diet can lower cholesterol levels significantly and protect against CVD (Slavin, 2008).

Many epidemiologic researches suggest that regular intake of fiber lowers the risk of CVD and CHD through a reduction in LDL levels. The results suggested that fiber may play a helpful role in decreasing C-reactive protein levels, blood pressure, and apolipoprotein levels; all are indicators for heart disease. The water-soluble fibers were most effective for lowering serum LDL cholesterol concentrations without affecting HDL concentrations.

Antibacterial activity

The antibacterial activity of acacia gum was assessed, as the infectious diseases represent a significant cause of morbidity and mortality among the general population. The ability of microorganisms to acquire and spread resistance against antibiotics causes nosocomial- and community-acquired infections (Mattana et al., 2012). The development of antibiotic resistance in many microorganisms has led many scholars to search for new antimicrobial agents (Negi and Dave, 2010). Eighty percent of the population in developing countries use medicinal plants for the treatment of infectious diseases (Kim et al., 1987).

The antimicrobial activity of *A. arabica* bark extracts, prebiotic aqueous extract, and the nanoemulsion powder against selected multidrug-resistant Gram-positive and Gram-negative bacterial and fungal pathogens was investigated, and varying degrees of antimicrobial activity were recorded (Chaubal and Tambe, 2006; Johny et al., 2010; Luo et al., 2014; Okoro et al., 2014; Rubina et al., 2015; Sharma et al., 2014). In a study conducted by Nagumanthri et al. (2012), no antimicrobial activity was observed in the methanolic bark extract of *A. nilotica* against the test bacteria with the exemption of *Bacillus circulans*.

The activity was tested against nine bacterial isolates, including *Klebsiella pneumonia, Escherichia coli, Pseudomonas aeruginosa, Proteus vulgaris, Salmonella typhi, Shigella dysenteriae, Staphylococcus aureus, Streptococcus pneumoniae*, and *Streptococcus pyogenes*, using different extracts with different concentrations and different solvents (Xuan et al., 2010). The antimicrobial activity of the nanoemulsion powder of GA was determined against *Listeria monocytogenes* and *Salmonella enteritidis* (Johny et al., 2010; Luo et al., 2014), and in another research, Hu et al. (2016) stated that the nanoemulsions of GA with eugenol have the potential to be applied in the food industry as a sanitizer against *L. monocytogenes* and *S. enteritidis*, as they adversely affect their antimicrobial activity.

The fine aqueous suspension of whole gum was used at concentrations of 0.5%–1.0% w/v in Columbia Agar and tested against antibacterial activity using fresh isolates and reference strains of *Actinobacillus actinomycetemcomitans, Capnocytophaga* spp., *Porphyromonasgingivalis, Prevotella intermedia*, and *Treponema denticola*; cultures were inhibited and the protease activities also reduced. The action of acacia gum against suspected periodontal pathogens and their enzymes suggests that it may be of clinical value (Clark et al., 1993).

Antiparasitic (antimalarial)

According to the in vivo experiments of gum Arabic in mice, the administration has been approved that it is slightly but significantly reduces the parasitaemia and decreases the clinical progress of murine malaria and its effects on angiogenin expression and dendritic cells. GA may be valuable in curing inflammatory diseases and malaria (Ballal et al., 2011; Nasir et al., 2008).

Renal and extra-renal effects

Study of GA (*A. senegal*) effect on kidneys was done in healthy mice by addition of 10% GA to the drinking water, which was shown to enhance creatinine clearance and renal ADH excretion in addition to its effect on both intestinal and renal excretion of Mg^{2+} and Ca^{2+}. GA treatment decreases the urinary Na^+ and Cl^- excretion and increases the plasma Na^+ and Cl^- with an obvious reduction in plasma phosphate concentration, plasma urea concentration, urinary flow rate, natriuresis, phosphaturia, glucosuria, proteinuria, blood pressure as well as urinary excretion of Na^+ and phosphate (Nasir et al., 2008). GA treatment enhances urinary and fecal Ca^{2+} excretion, decreases plasma concentrations of 1, 25-dihydroxyvitamin D [1, $25(OH)^2D3$], and tends to decrease plasma parathyroid hormone concentration.

In a recent study by Omer et al. (2013), it was reported that the GA significantly decreases the nephrotoxicity induced by *A. bracteata* aqueous extract on albino rats, as it elevates urea, creatinine, AST, and ALT levels caused by *A. bracteata*, and it also reduces the damage of the kidney tubules. GA is used in Middle Eastern countries for the treatment of chronic kidney disease and end-stage renal disease (Al Majed et al., 2002).

The studies of GA treatment of extra-renal effects in mice decrease the expression of intestinal Na^+ coupled glucose carrier SGLT1 with subsequent delay of electrogenic intestinal glucose transport, hyperinsulinemia, glucose-induced hyperglycemia, and body weight gain. GA treatment decreases colonic transcription of the angiogenetic factors angiogenin 1, angiogenin 3, and angiogenin4, of CD38 antigen, interleukin18, aquaporin4, vav-3-oncogene, y^+-amino acid-transporter, ubiquitin D, sulfatase 1, and chemokine ligand 5. Moreover, GA treatment decreases angiogenin and β-catenin protein expression. It also modifies the maturation and cytokine release. Accordingly, the effects of the GA treatment on blood pressure, plasma phosphate concentration, and proteinuria may show helpful effects in chronic renal failure and diabetic nephropathy.

Intestinal transport effect

GA pods and tender leaves stimulate intestinal sodium ions and water absorption favoring dehydration, which counteracts chronic-osmotic diarrhea (Gilani et al., 1999; Teichberg et al., 1999; Xuan et al., 2010). GA treatment has been shown to decrease intestinal sodium ions coupled with glucose transport, which determines the rate of intestinal glucose absorption and thus affects glucose-induced insulin discharge and the development of obesity (Nasir et al., 2010; Wright et al., 2003).

Dietary fibers were known to decrease body mass (Koh-Banerjee et al., 2004), inhibit the metabolic syndrome (McKeown et al., 2004), and progress glycemic control as well as hyperinsulinemia in type 2 diabetes (Chandalia et al., 2000). The effect of dietary fibers has been attributed to interaction with food intake and body weight through satiety, glycemia, insulinemia, blood pressure, and blood lipids (Delzenne and Cani, 2005). In the colon, GA components are degraded by fermentation microorganisms to short-chain fatty acids, which may prevent the inflammation (Phillips, 1998). According to the US Food and Drug Administration (FDA)'s comment on GA, it is one of the safest dietary fibers (Anderson, 1986).

Antiinflammatory responses

GA has been known by its excellent inflammatory responses in a wide range of tissues (Abdel-Rahman et al., 2006; Suzumori et al., 2004), including inflammatory bowel disease (Koutroubakis et al., 2004). It may impact the inflammation by interacting with the intestinal dendritic cells, which have an important role in regulating the immune response (Adler and Steinbrink, 2007; Zhao et al., 2012). Treatment by GA stimulates formation of many types of interleukins, which are important regulators of the immune response (Blanco et al., 2008; Li and Flavell, 2008).

GA treatment also decreases the phagocytic activity of dendritic cells and increases the percentage of T cells and B lymphocytes in the spleen, which reflect an influential effect of GA on the development and maturation of cytokine that releases from dendritic cells (Xuan et al., 2010).

Anticarcinogenic effect

GA has a direct effect on some colonic tissue genes, which are responsible for proliferation and/or tumor growth, such as CD38 antigen (Deaglio et al., 2008), interleukin 18 (Dinarello, 2006), solute carrier family 7, member 9 (Kanai and Endou, 2001), sulfatase 1 (Aka et al., 2009; Lai et al., 2008), ubiquitin D (Orlowski and Kuhn, 2008), and chemokine (C–C motif) ligand 5 (Sancho-Martinez and Martin-Villalba, 2009). Moreover, the treatment by GA reduces the protein of β-catenin, an influential oncogene in colonic tumors (Jin et al., 2008; Kim and Milner, 2007). Regular intake of dietary fibers (GA) has previously been claimed to confer protection against colonic, prostate, and rectal cancer (Cummings et al., 1992).

The observation by Phillips (1998) show that GA strongly influences the expression of angiogenins, which provides clarification for the protective effect of it on tumor growth. The fermentation of GA under the influence of microorganisms in the colon to its short-chain fatty acids may suppress and neutralize tumor growth (Finley et al., 2007; Usami et al., 2008). Additionally, short-chain fatty acids have been discovered influencing oncogene expression (Basson et al., 1998; Chen et al., 2003).

Dental effect

Early research suggests that acacia controls gum disease, as it has antibacterial properties that help control harmful bacteria in the mouth that cause gum disease. One study found that chewing gum made with acacia for 7 days was more helpful in reducing plaque when compared with regular gum. Also, it is a substitute for dental remineralization (Xuan et al., 2010).

Effect on sore throat

Acacia is a sedative that relieves irritation in the mucus membranes of the mouth by creating a protective film. Historically, this is why it was used to heal wounds, mouth sores, and cold symptoms. Today, it's an ingredient in many cold medications, such as throat lozenges and cough medicines.

Boosting the immunity

Recent researches revealed a powerful immunomodulary effect of GA on dendritic cells (Nasir et al., 2008; Xuan et al., 2010). Antigen-presenting cells initiate both innate and adaptive immunity and thus play an essential role in the regulation of the immune response of the body (Alder and Steinbrink, 2007).

Prebiotic

Roberfroid et al. (2011) updated the definition of prebiotics. It is a nondigestible food such as garlic, leeks, chicory, Jerusalem asparagus, onions, wheat, oats, soybeans, and artichokes (Van Loo et al., 1995). The caloric value of nondigestible oligosaccharides has been estimated between 1 and 2 kcal/g (Livesey, 1992). Some identified prebiotics (inulin) are low digestible carbohydrates and are associated with impaired gastrointestinal tolerance, especially when consumed in large amounts (Grabitske and Slavin, 2009; Lied et al., 2011), while other prebiotics, e.g., wheat dextrin and polydextrose, exhibit high gastrointestinal acceptability, about 30–45 g per day (Pasman et al., 2006).

GA also acts as a prebiotic: nondigestible food ingredients that favorably and positively affect the host by stimulating the growth and/or activity of probiotic in the colon, therefore improving host healthiness (Gibson and Roberfroid, 1995), or it is fermented with some ingredient that makes certain changes in the composition and/or activity in the gastrointestinal microflora that confers benefits, e.g., *Lactobacilli* and *Bifidobacteria*, which are usual target genera for prebiotics (Gibson et al., 2004).

GA efficacy as a prebiotic was determined by Calame et al. (2008); they stated that consumption by man for up to 4 weeks boosts the beneficial bacteria; they established the dose–effect relationship. When comparing with the negative control, the numbers of *Bifidobacteria and Lactobacilli* 4 weeks after consumption were significantly higher for GA: the ideal dose is around 10 g. Moreover, at this dose, the numbers of *Lactobacilli*, *Bifidobacteria*, and *Bacteroides* were significantly larger for GA than for inulin.

Effects on body mass index

The GA polysaccharide has been considered as a safe dietary fiber by the US FDA since the 1970s. Many studies were conducted to determine effects of it on body mass index (BMI) and body fat percentage among healthy adult females. GA ingestion causes significant reduction in BMI and body fat percentage among healthy adult females; this may refer to the benefit of taking a diet rich in fiber, which aids in staying full for long time. The effect of GA on intestinal glucose transport may be beneficial in the prophylaxis and decreases of obesity (Nasir et al., 2008). The effect could be useful in the treatment of obesity following fixed intake of 30 g per day of GA for 6 weeks.

In humans, GA treatment indeed adjusts body weight; taking GA as a treatment decreases the BMI and body fat percentage among healthy adult females, which could be used in the treatment of obesity (Babiker et al., 2012).

The GA is giving the feeling of satiety, which directly affects the amount of food intake and may possibly influence obesity in humans. Also, Calame et al. (2011) reported that the GA treatment decreases the caloric intake and increases the sensation of being satisfied.

Side effects

Side effects that are caused by ingestion of GA were practiced in the first week only. They involved unfavorable viscous feelings in the mouth, hypersensitivity, morning nausea, slight diarrhea, and also bloating in the abdomen in a few cases (Babiker et al., 2012).

In conclusion, GA has been found in many applications in medicine and health care because it is composed of polysaccharides and glycoproteins. GA is claimed to have therapeutic usefulness in the treatment of many diseases and is conventionally used in treatment of skin wounds such as leprosy, burns, and cuts, or digestive issues like gonorrhea, dysentery, and coughs. GA is still applied in some of these diseases, especially the coughs.

Recent research on GA recommends its application for its effective treatment of chronic hepatic and renal failure, reducing the proteinuria, as well as reduction of plasma phosphate concentration, then lowering the blood pressure, decreases the plasma total cholesterol, triglyceride, and LDLs, then reduces coronary artery diseases and fall in arterial blood pressure. Moreover, GA has several extra-renal effects, such as reducing the intestinal glucose transport, and it also reduces the lipid profile, which is useful in the prophylaxis and treatment of obesity and diabetes as well as decrease of angiogenin and β-catenin expression, which may affect the development of colon cancer and inflammatory conditions. GA also contains phenolic compounds, which have antioxidative properties. It is also acts as a prebiotic as well. GA has no significant adverse or toxic effects that have been allied with its use.

References

Abdel-Rahman, A.M., El Sahrigy, S.A., Bakr, S.I., 2006. A comparative study of two angiogenic factors: vascular endothelial growth factor and angiogenin in induced sputum from asthmatic children in acute attack. Chest 129, 266–271.

Adler, H.S., Steinbrink, K., 2007. Tolerogenic dendritic cells in health and disease: friend and foe! Eur. J. Dermatol. 17, 476–491.

Ahmad, M., Zaman, F., Sharif, T., Zabta, C.M., 2008. Antidiabetic and hypolipidemic effects of aqueous methanolic extract of *Acacia nilotica* pods in alloxan-induced diabetic rabbits. Scand. J. Lab. Anim. Sci. 35 (1), 29–34.

Ahmed, A.A., Fedail, J.S., Musa, H.H., Kamboh, A.A., Sifaldina, A.Z., Musa, T.H., 2015. Gum Arabic supplementation improved antioxidant status and alters expression of oxidative stress gene in ovary of mice fed high fat diet. Middle East Fertil. Soc. J. Pathophysiol. 22, 189–194.

Aka, J.A., Mazumdar, M., Lin, S.X., 2009. Reductive 17β-hydroxysteroid dehydrogenases in the sulfatase pathway: critical in the cell proliferation of breast cancer. Mol. Cell Endocrinol. 301, 183–190.

Al Majed, A.A., Mostafa, A.M., Al Rikabi, A.C., Al Shabanah, O.A., 2002. Protective effects of oral Arabic gum administration on gentamicin-induced nephrotoxicity in rats. Pharmacol. Res. 46, 445–451.

Ali, A.A., Ali, K.E., Fadlalla, A., Khalid, K.E., 2008. The effects of gum Arabic oral treatment on the metabolic profile of chronic renal failure patients under regular haemodialysis in Central Sudan. Nat. Prod. Res. 22, 12–21.

Ambasta, S.P., 1994. The Useful Plants of India. Publication and Information Directorate, CSIR, New Delhi, pp. 4.

Amos, S., Akah, P.A., Odupw, C.J., Gamaniel, K.S., Wambede, C., 1999. The pharmacological effects of an aqueous extract from *Acacia nilotica* seeds. Phytother. Res. 13, 683–685.

Anderson, D.M., 1986. Evidence for the safety of gum Arabic (*Acacia senegal* (L.) Willd) as a food additive: a brief review. Food Addit. Contam. 3, 225–230.

Anderson, R.A., Broadhurst, C.L., Polonsky, M.M., Schmidt, W.F., Khan, A., Flanagan, V.P., Schoene, N.W., Graves, D.J., 2004. Isolation and characterization of polyphenol type-A polymers from cinnamon with insulin like biological activity. J. Agric. Food Chem. 52 (1), 65–70.

Babiker, R., Merghani, T.H., Elmusharaf, K., Badi, R.M., Lang, F., Saeed, A.M., 2012. Effects of Gum Arabic ingestion on body mass index and body fat percentage in healthy adult females: two-arm randomized, placebo controlled, double blind trial. Nutr. J. 11, 111.

Ballal, A., Bobbala, D., Qadri, S.M., Föller, M., Kempe, D., Nasir, O., Saeed, A., Lang, F., 2011. Anti-malarial effect of gum Arabic. Malar. J. 10 (139), 1–7.

Basson, M.D., Emenaker, N.J., Hong, F., 1998. Differential modulation of human (Caco-2) colon cancer cell line phenotype by short chain fatty acids. Proc. Soc. Exp. Biol. Med. 217, 476–483.

Betteridge, D.J., Dodson, P.M., Durrington, P.N., 1993. Management of hyperlipidaemia: guidelines of the British Hyperlipidaemia Association. Postgrad. Med. J. 69, 359–369.

Bhargava, A., Srivastava, A., Kumbhare, V.C., 1998. Antifungal activity of polyphenolic complex of *Acacia nilotica* bark. Indian For. 124, 292–298.

Blanco, P., Palucka, A.K., Pascual, V., Banchereau, J., 2008. Dendritic cells and cytokines in human inflammatory and autoimmune diseases. Cytokine Growth Factor Rev. 19, 41–52.

Calame, W., Weseler, A.R., Viebke, C., Flynn, C., Siemensma, A.D., 2008. Gum Arabic establishes prebiotic functionality in healthy human volunteers in a dose-dependent manner. Br. J. Nutr. 100 (6), 1269–1275.

Calame, W., Thomassen, F., Hull, S., Viebke, C., Siemensma, A.D., 2011. Evaluation of satiety enhancement, including compensation, by blends of gum Arabic. A methodological approach. Appetite 57, 358–364.

Chandalia, M., Garg, A., Lutjohann, D., von Bergmann, K., Grundy, S.M., Brinkley, L.J., 2000. Beneficial effects of high dietary fibre intake in patients with type 2 diabetes mellitus. N. Engl J. Med. 342, 1392–1398.

Chaubal, R., Tambe, A., 2006. Isolation of new straight chain compound from *Acacia nilotica.* Indian J. Chem. 45 (B), 1231–1233.

Chen, J.S., Faller, D.V., Spanjaard, R.A., 2003. Short-chain fatty acid inhibitors of histone deacetylases: promising anticancer therapeutics. Curr. Cancer Drug Targets 3, 219–236.

Clark, D.T., Gazi, M.I., Cox, S.W., Eley, B.M., Tinsley, G.F., 1993. The effects of Acacia arabica gum on the in vitro growth and protease activities of periodontopathic bacteria. J. Clin. Periodontol. 20 (4), 238–243.

Coats, A.J., 1998. The potential role of soluble fibre in the treatment of hyper cholesterolaemia. Postgrad. Med. J. 74 (873), 391–394.

Cummings, J.H., Bingham, S.A., Heaton, K.W., Eastwood, M.A., 1992. Fecal weight, colon cancer risk, and dietary intake of non-starch polysaccharides (dietary fibre). Gastroenterology 103, 1783–1789.

Deaglio, S., Aydin, S., Vaisitti, T., Bergui, L., Malavasi, F., 2008. CD38 at the junction between prognostic marker and therapeutic target. Trends Mol. Med. 14, 210–218.

Deckwer WD, Dill B, Eisenbrand E, Bornscheuer U, Pühler A, Heiker FR, Kirschning A, Schreier P, Fugmann B, Pohnert G. Römpp Online. Georg-Thieme-Verlag; 2006.

Delzenne, N.M., Cani, P.D., 2005. A place for dietary fibre in the management of the metabolic syndrome. Curr. Opin. Clin. Nutr. Metab. Care 8, 636–640.

Dinarello, C.A., 2006. The paradox of pro-inflammatory cytokines in cancer. Cancer Metastasis Rev. 25, 307–313.

FAO, 1999. Gum Arabic. Food and Nutrition Paper, No. 52, Rome, Italy. Fincher, G.B., Stone, B.A., Clarke, A.E, 1983. Arabinogalactan proteins: structure, biosynthesis and function. Annu. Rev. Plant Physiol. 34, 47–70.

Finley, J.W., Burrell, J.B., Reeves, P.G., 2007. Pinto bean consumption changes SCFA profiles in fecal fermentations, bacterial populations of the lower bowel, and lipid profiles in blood of humans. J. Nutr. 137, 2391–2398.

Gamal el-din, A.M., Mostafa, A.M., Al-Shabanah, O.A., Al-Bekairi, A.M., Nagi, M.N., 2003. Protective effect of Arabic gum against acetaminophen-induced hepatotoxicity in mice. Pharmacol. Res. 48, 631–635.

Gibson, G.R., Roberfroid, M.B., 1995. Dietary modulation of the human colonic microbiota: introducing the concept of prebiotics. J. Nutr. 125, 1401–1412.

Gibson, G.R., Probert, H.M., van Loo, J., Rastall, R.A., Roberfroid, M.B., 2004. Dietary modulation of the human colonic microbiota: updating the concept of prebiotics. Nutr. Res. Rev. 17, 259–275.

Gilani, A.H., Shaheen, F., Zamanm, M., 1999. Studies on antihypertensive antispasmodic activities of methanol extract of *Acacia nilotica* pods. Phytother. Res. 13, 665–669.

Grabitske, H.A., Slavin, J.L., 2009. Gastrointestinal effects of low-digestible carbohydrates. Crit. Rev. Food Sci. Nutr. 49, 327–360.

Hirosh, T., Mitsuy, I., Mi, T., Ew, J., Tosiyasu, S., Ikuko, K., 2004. Effect of green tea on blood glucose levels and serum proteomics patterns in diabetic mice and on glucose metabolism in healthy humans. BMC Pharmacol. 4, 18–30.

Hu, Q., Gerhard, H., Upadhyaya, I., Venkitanarayanan, K., Luo, Y., 2016. Antimicrobial eugenol nano emulsion prepared by gum Arabic and lecithin and evaluation of drying technologies. Int. J. Biol. Macromol. 87, 130–140.

Jin, T., George, F.I., Sun, J., 2008. Multiple mechanisms control the transcriptional property of betacatenin. Cell Signal 20, 1697–1704.

Johny, A.K., Hoagland, T., Venkitanarayanan, K., 2010. Effect of subinhibitory concentrations of plant-derived molecules in increasing the sensitivity of multidrug-resistant *Salmonella enterica* Serovar Typhimurium DT104 to antibiotics. Foodborne Pathog. Dis. 7, 1165–1170.

Kalaivani, T., Mathew, L., 2010. Free radical scavenging activity from leaves of *Acacia nilotica* (L.) Wild. ExDelile, an Indian medicinal tree. Food Chem. Toxicol. 48, 298–305.

Kanai, Y., Endou, H., 2001. Heterodimeric amino acid transporters: molecular biology and pathological and pharmacological relevance. Curr. Drug Metab. 2, 339–354.

Kim, Y.S., Milner, J.A., 2007. Dietary modulation of colon cancer risk. J. Nutr. 137, 2576S–2579S.

Kim, H.S., Han, S.S., Oh, K.W., Jeong, T.S., Nam, K.Y., 1987. Effects of ginseng saponin on antimicrobial activities of some antibiotics. Korean J. Mycol. 15, 87–91.

Koh-Banerjee, P., Franz, M., Sampson, L., Liu, S., Jacobs, Jr., D.R., Spiegelman, D., Willett, W., Rimm, E., 2004. Changes in whole-grain, bran, and cereal fiber consumption in relation to 8-y weight gain among men. Am. J. Clin. Nutr. 80, 1237–1245.

Koutroubakis, I.E., Xidakis, C., Karmiris, K., Sfiridaki, A., Kandidaki, E., Kouroumalis, E.A., 2004. Serum angiogenin in inflammatory bowel disease. Dig. Dis. Sci. 49, 1758–1762.

Lai, J.P., Thompson, J.R., Sandhu, D.S., Roberts, L.R., 2008. Heparin-degrading sulfatases in hepatocellular carcinoma: roles in pathogenesis and therapy targets. Future Oncol. 4, 803–814.

Lawrence, R., Jeyakumar, E., Gupta, A., 2015. Antibacterial activity of Acacia Arabica (Bark) extract against selected multi drug resistant pathogenic bacteria. Int. J. Curr. Microbiol. Appl. Sci. 1, 213–222.

Lerman, R.H., Minich, D.M., Darland, G., Lamb, J.J., Chang, J., His, A., 2010. Subjects with elevated LDL cholesterol and metabolic syndrome benefit from supplementation with soy protein phytosterols, hops *rho* iso-alpha acids, and *Acacia nilotica* proanthocyanidins. J. Clin. Lipidol. 4, 59–68.

Li, M.O., Flavell, R.A., 2008. Contextual regulation of inflammation: a duet by transforming growth factor-beta and interleukin-10. Immunity 28, 468–476.

Lied, G.A., Lillestol, K., Lind, R., Valeur, J., Morken, M.H., Vaali, K., Gregersen, K., Florvaag, E., Tangen, T., Berstad, A., 2011. Perceived food hypersensitivity: a review of 10 years of interdisciplinary research at a reference centre. Scand. J. Gastroenterol. 40, 1169–1178.

Livesey, G., 1992. The energy values of dietary fibre and sugar alcohols for man. Nutr. Res. Rev. 5, 61–84.

Luo, Y., Zhang, Y., Pan, K., Critzer, F., Davidson, P.M., Zhong, Q., 2014. Self-emulsification of alkaline-dissolved clove bud oil by whey protein, gum Arabic, lecithin, and their combinations. J. Agric. Food. Chem. 62 (19), 4417–4424.

Mariana, A.M., María, L.B., Lorena, V., Claudio, D.B., 2012. Gum Arabic: More than an Edible Emulsifier. Available from: http://www.intechopen.com/books/products-and-applications-of-biopolymers/gum-arabic-more-than-an-edible-emulsifier.

Mattana, C.M., Satorres, S.E., Escobar, F., Sabini, C., Sabini, L., Fusco, M., Alcaraz, E., 2012. Antibacterial and cytotoxic activities of *Acacia aroma* extracts. Emir. J. Food Agric. 24 (4), 308, 313.

McKeown, N.M., Meigs, J.B., Liu, S., Saltzman, E., Wilson, P.W., Jacques, P.F., 2004. Carbohydrate nutrition, insulin resistance, and the prevalence of the metabolic syndrome in the Framingham Offspring Cohort. Diabetes Care 27, 538–546.

Nagumanthri, V., Rahiman, S., Tantry, B.A., Nissankararao, P., Kumar, M.P., 2012. In vitro antimicrobial activity of *Acacia nilotica*, *Ziziphus mauritiana*, *Bauhinia variegate* and *Lantana camara* against some clinical isolated strains. Iran. J. Sci. Technol. A2, 213–217.

Nasir, O., Artunc, F., Saeed, A., Kambal, M.A., Kalbacher, H., Sandulache, D., Boini, K.M., Jahovic, N., Lang, F., 2008. Effects of gum Arabic (*Acacia senegal*) on water and electrolyte balance in healthy mice. J. Ren. Nutr. 18, 230–238.

Nasir, O., Artunc, F., Wang, K., Rexhepaj, R., Foller, M., Ebrahim, A., Kempe, D.S., Biswas, R., Bhandaru, M., Walter, M., Mohebbi, N., Wagner, C.A., Saeed, A.M., Lang, F., 2010. Down regulation of mouse intestinal Na^{+}-coupled glucose transporter SGLT1 by gum Arabic (*Acacia senegal*). Cell Physiol. Biochem. 25, 203–210.

Negi, B.S., Dave, B.P., 2010. In vitro antimicrobial activity of *Acacia catechu* and its phytochemical analysis. Indian J. Microbiol. 50 (4), 369–374.

Okoro, S.O., Kawo, A.H., Arzai, A.H., 2014. Phytochemical screening, antibacterial and toxicological activities of *Acacia nilotica* extracts. Bayero J. Pure Appl. Sci. 7 (1), 105–115.

Omer, A.E., Ayed, I.A.M., El Badwi, S.M.A., 2013. Effect of Gum Arabic on Nephrotoxicity Induced by *Aristolochia bracteata* in Rats. Scholars Acad. J. Biosci. 1 (7), 377–380.

Orlowski, R.Z., Kuhn, D.J., 2008. Proteasome inhibitors in cancer therapy: lessons from the first decade. Clin. Cancer Res. 14, 1649–1657.

Pasman, W., Wils, D., Saniez, M.H., Kardinaal, A.F., 2006. Long-term gastrointestinal tolerance of NUTRIOSE FB in healthy men. Eur. J. Clin. Nutr. 60, 1024–1034.

Phillips, G.O., 1998. Acacia gum (Gum Arabic): a nutritional fibre; metabolism and calorific value. Food Addit. Contam. 15, 251–264.

Priyanka, P., Choudhry, M., 2013. Management of type 2 diabetics by indian gum arabic (*Acacia nilotica*) pods powder. Int. J. Food Nutr. Sci. 2 (2), e-ISSN: 2320 -7876 www.ijfans.com.

Pyorala, K., De Backer, G., Graham, I., Wilson, P., Wood, D., 1994. Prevention of coronary heart disease in clinical practice: recommendations of the Task Force of the European Society of Cardiology, European Atherosclerosis Society and European Society of Hypertension. Atherosclerosis 110, 121–161.

Rahman, O., 2010. A review of uses of *Acacia nilotica* (Booni) In Alternative Medicine. Sofowara EA 1986. The State of Medicinal Plant Research in Nigeria. Nigeria Society of Pharmacognosy. University Press, Ibadan, pp. 13 – 65.

Rehman, K., Wingertzahn, M.A., Harper, R.G., Wapnir, R.A., 2001. Proabsorptive action of Gum Arabic: regulation of nitric oxide metabolism in the base lateral potassium channel of the small intestine. J. Pediatr. Gastroenterol. Nutr. 32, 529–533.

Roberfroid, M., Gobson, G.R., Hoyles, L., McCartney, A.L., Rastall, R., Rowland, I., Wolvers, D., Watzl, B., Szajewska, H., Stahl, B., 2011. Prebiotic effects: metabolic and health benefits. Br. J. Nutr. 104, 1–63.

Rubina, L., Ebenezer, J., Akshi, G., 2015. Antibacterial Activity of *Acacia arabica* (Bark) Extract against selected Multi Drug Resistant Pathogenic Bacteria. Int. J. Curr. Microbiol. Appl. Sci. (1), 213–222, ISSN: 2319-7706.

Sabu, M., Smitha, K., Kuttan, R., 2002. Antidiabetic activity of green tree phenols and their role in reducing oxidative stress in experimental diabetes. J. Ethnopharmacol. 83, 109–116.

Sancho-Martinez, I., Martin-Villalba, A., 2009. Tyrosine phosphorylation and CD95: a fascinating switch. Cell Cycle 8, 838–842.

Sharma, C., Anejam, K.R., Surain, P., Dhiman, R., 2014. In vitro, evaluation of anti-microbial spectrum of *Acacia nilotica* leaves and bark extracts against pathogens causing otitis infection. J. Innov. Biol. 1 (1), 51–56.

Simons, L.A., 1986. Interrelations of lipids and lipoproteins with coronary artery disease mortality in 19 countries. Am. J. Cardiol. 57, 5–10.

Slavin, J.L., 2008. Position of the American Dietetic Association: health implications of dietary fibre. J. Am. Diet. Assoc. 108, 1716–1731.

Suliman, S.M., Hamdouk, M.I., Elfaki, M.B., 2000. Gum Arabic fibre as a supplement to low protein diet in chronic renal failure patients. Sudan Association of Physicians, 17th Conference, Friendship Hall, Khartoum, Sudan, March 21–23.

Suzumori, N., Zhao, X.X., Suzumori, K., 2004. Elevated angiogenin levels in the peritoneal fluid of women with endometriosis correlate with the extent of the disorder. Fertil. Steril. 82, 93–96.

Teichberg, S., Wingertzahn, M.A., Moyse, J., Wapnir, R.A., 1999. Effect of gum Arabic in an oral rehydration solution on recovery from diarrhea in rats. J. Pediatr. Gastroenterol. Nutr. 29, 411–417.

Tomura, M., Takano, N., Osakabe, H., Yasuda, A., Inove, K.I., Yangisawa, R., Ohwatari, T., Uematsu, H., 2007. Dietary supplementation with cacao liquor proanthocyanaldins prevents elevation of blood glucose levels in diabetic obese mice. Nutrition (Elsevier) 10, 10–16.

Usami, M., Kishimoto, K., Ohata, A., Miyoshi, M., Aoyama, M., Fueda, Y., Kotani, J., 2008. Butyrate and trichostatin A attenuate nuclear factor kappa B activation and tumor necrosis factor alpha secretion and increase prostaglandin E2 secretion in human peripheral blood mononuclear cells. Nutr. Res. 28, 321–328.

Vadivel, V., Biesalski, H.K., 2012. Total phenolic content, in vitro antioxidant activity and type II diabetes relevant enzyme inhibition properties of methanolic extract of traditionally processed underutilized food legume, *Acacia nilotica* (L.) Willd ex Delile. Int. Food Res. J. 19 (2), 593–601.

Van Loo, J., Coussement, P., de Leenheer, L., Hoebregs, H., Smits, G., 1995. On the presence of inulin and oligofructose as natural ingredients in the Western diet. Crit. Rev. Food. Sci. Nutr. 35, 525–552.

Verbeken, D., Dierckx, S., Dewettinck, K., 2003. Exudate gums: occurrence, production, and applications. Appl. Microbiol. Biotechnol. 63, 10–21.

Wiliams, P.A., Phillips, G.O., 2000. In: Williams, P.A., Phillips, G.O. (Eds.), Handbook of Hydrocolloids. CRC Press, Cambridge, UK, pp. 155–168.

Wright, E.M., Martin, M.G., Turk, E., 2003. Intestinal absorption in health and disease—sugars. Best Pract. Res. Clin. Gastroenterol. 17, 943–956.

Xuan, N.T., Shumilina, E., Nasir, O., Bobbala, D., Gotz, F., Lang, F., 2010. Stimulation of mouse dendritic cells by gum Arabic. Cell Physiol. Biochem. 25, 641–648.

Zaki, H., Al-Mustafa, 2000. A study on the toxicology of *Acacia nilotica*. Am. J. Chin. Med. 28 (01), 123–129.

Zhao, H., Li, M., Wang, L., Su, Y., Fang, H., Lin, J., Mohabeer, N., Li, D., 2012. Angiotensin II induces TSLP via an AT1 receptor/NF-KappaB pathway, promoting Th17 differentiation. Cell. Physiol. Biochem. 30, 1383–1397.

Further Readings

Ali, B.H., Ziada, A., Blunden, G., 2009. Biological effects of gum Arabic: a review of some recent research. Food Chem. Toxicol. 47, 1–8.

Bnuyan, I.A., Hindi, N.K.K., Jebur, M.H., Mahdi, M.A., 2015. In vitro antimicrobial activity of gum Arabic (Al Manna and Tayebat) prebiotics against infectious pathogens. Human J. Int. J. Pharmacogn. Phytochem. Res. (IJPPR) 3 (3), 77–85.

Pareek, P., Choudhry, M., 2013. Management of type 2 diabetics by Indian gum Arabic (*Acacia nilotica*) pods powder. Int. J. Food Nutr. Sci. 2 (2).

Slavin, J.L., Savarino, V., Parades-Diaz, A., Fotopoulos, G., 2009. A review of the role of soluble fibre in health with specific reference to wheat dextrin. J. Int. Med. Res. 37, 1–17.

Functional Properties of Gum Arabic

24

Abdalbasit A. Mariod[*,**,†]
[*]University of Jeddah, Alkamil, Kingdom of Saudi Arabia; [**]Sudan University of Science & Technology, Shambat, Sudan; [†]Ghibaish College of Science & Technology, Ghibaish, Sudan

Introduction

Gum Arabic (GA) is a palatable biopolymer obtained as exudates of mature trees of *Acacia senegal* and *Acacia seyal*, which are found in the African district of the Sahel in Sudan. The exudate is a nonthick liquid, rich in fiber, and it spreads from the stems and branches and commonly occurs under anxiety conditions, for instance, dry season, poor soil fertility, and injury (Williams and Phillips, 2000).

Gum Arabic chemical structure

GA's useful properties are firmly identified with its structure, which decides, for instance, solubility, viscosity, the level of collaboration with water and oil in an emulsion, and microencapsulation capacity, among others (Montenegro et al., 2012). The chemical structure of *A. senegal* and *A. seyal* gums is contrasting. Actually, both gums have similar sugar residues, yet *A. seyal* gum has a lower substance of rhamnose and glucuronic acid and a higher substance of arabinose and 4-O-methyl glucuronic acid than *A. senegal* gum. Rather, *A. seyal* gum contains a lower extent of nitrogen, and specific rotations are additionally totally different. The evaluation of the last parameters may clearly detect the contrast between the two species (Osman et al., 1993). GA is a group of polysaccharides acquired as a blended calcium, magnesium, and potassium salt. Its structure comprises of 1,3-linked β-D-galactopyranosyl units. The side chains are made out of two to five 1,3-linked β-D-galactopyranosyl units, joined to the primary chain by 1,6-linkages. Both the principle and the side chains contain units of β-L-arabinofuranosyl, β-L-rhamnopyranosyl, β-D-glucuronopyranosyl, and 4-Omethyl-β-D-glucuronopyranosyl, the last two, for the most part, as end units. The attributes of GA may change fundamentally, contingent upon the geological origin and age of the trees, climatic conditions, soil condition, and even the place of exudation on the tree (Masuelli, 2013; Fig. 24.1).

The distinctive GAs demonstrate some useful properties, for example, emulsification, stabilization, and microencapsulation. GA has the ability to settle oil-in-water (O/W) emulsions in low-consistency, concentrated solutions. GA is broadly utilized

Gum Arabic. http://dx.doi.org/10.1016/B978-0-12-812002-6.00024-5

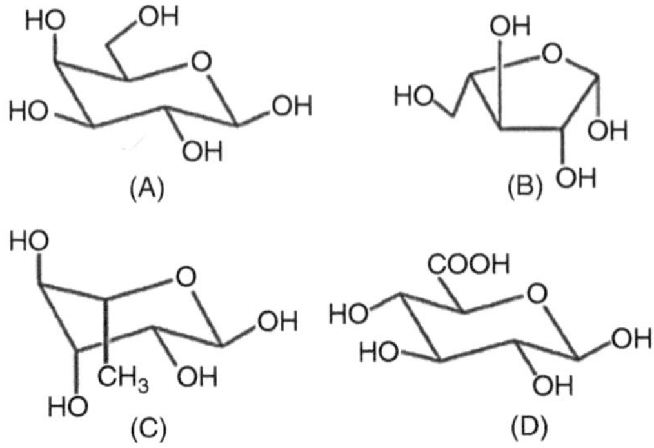

Figure 24.1 The main chemical building blocks of the plant hydrocolloid gum arabic. (A) D-Galactopyranose, (B) L-arabinofuranose, (C) L-rhamnopyranose, and (D) D-glucopyranosuronic acid.
Source: Gils, P.S., Ray, D., Sahoo, P.K., 2010. Designing of silver nanoparticles in gum Arabic based semi-IPN hydrogel. Int. J. Biol. Macromol. 46 (2) 237–244.

as a part of the industry as a stabilizer, a thickener, and an emulsifier and in additive encapsulation, not only in food processing, but also in textiles, ceramics, lithography, cosmetic, and pharmaceutical industries. With respect to industry, it is utilized as a stabilizer and a thickener as well as an emulsifying specialist (e.g., soft drink syrup, sticky confections, and creams) (Verbeken et al., 2003).

GA is widely utilized in pharmaceutical industries as a bearer of medications since it is viewed as a physiologically harmless substance. Also, late reviews have highlighted GA's antioxidant properties, its part in the digestion of lipids, and its positive outcomes while being utilized as a part of treatments for several degenerative illnesses, for example, kidney failure and cardiovascular and gastrointestinal disease (Ali et al., 2008; Glover et al., 2009; Montenegro et al., 2012). Other than all the sensory and texturizing properties, GA has fascinating antioxidant properties, for example, an effective limit with regard to deactivation of excited electronic states and directed radical scavenging capacity. There is expanding exploratory proof that the antioxidant function works with its protein portion, fundamentally by amino acid residues, for example, histidine, tyrosine, and lysine, which are, for the most part, considered as antioxidant molecules (Park et al., 2005).

Functional properties

Emulsifying stability

GA is an extremely powerful emulsifying factor as a result of its defensive colloid uses and has discovered a wide use in the preparation of O/W food emulsions, with most oils over a wide pH range and within the sight of electrolytes without the requirement for an optional stabilizing agent (Phillips et al., 1988). Emulsifiers have been classified as a gathering of surface-active agents that can balance out a dispersion of two liquids, for example, water and oil, which are fundamental for emulsion formation and stabilization to happen. Sabah El-Kheir et al. (2008) investigated the strength of acacia gum emulsion with respect to the kind of refined oil (sesame, groundnut, cottonseed, sunflower, and corn), temperature, stirring time, concentration, and gum grade. Their outcomes uncovered that emulsion stability is essentially influenced by the kind of oil utilized. Cotton seed oil gave the most stable emulsion, while groundnut brought about the least stable emulsion. These researchers likewise revealed that the expansion in the length of the stirring time has significantly increased the stability of the emulsion. Additionally, the emulsion stability was influenced by gum grades. Different components of concentration and temperature did not altogether impact emulsion stability.

Omer et al. (2015) studied the emulsifying stability of GA (*A. senegal*) using sesame oil. The effects of different factors have been studied (concentration, temperature, stirring time, and blending). It has been observed that gum Arabic showed good results of emulsifying stability, and when concentration, the time of stirring increased the emulsifying stability also increased. When the temperature rises, the emulsifying stability of the gum is declined. A blend of *A. senegal* and *Acacia polyacantha* gums (50:50) was found to possess better emulsifying properties compared to parent, *A. senegal* gum.

Kheir (2005) reported a highly significant difference of character of emulsifying stability between various GA formulations (GAF). The maximum emulsifying stability was found in kibbled 107 and spray-dried GAF, and the minimum was in handpicked. Yao et al. (2013) utilized O/W emulsions as a delivery system to secure conjugated linoleic acid (CLA), a polyunsaturated fatty acid, from oxidation. These researchers utilized a customary GA and two developed GA specimens (EM2 and EM10) as emulsifiers to prepare CLA-in-water emulsions. Their outcomes showed an ideal physical and chemical stability of the emulsion at gum concentrations of 5% for every one of the three gums. Emulsions with higher gum concentrations are more vulnerable to lipid oxidation. This is ascribed to decreased physical stability at higher gum concentrations due to the mixture- and consumption-actuated flocculation of the emulsion droplets. The prooxidants iron and copper inherently contained in the gums could likewise add to this instability. Among the three gums, EM10 gives the best protection for CLA, both physically and chemically, as a result of its better interfacial properties over GA and EM2. The impacts of GA addition (0%–4%, w/w) on the steadiness of O/W emulsion balanced out by flaxseed protein concentration (FPC) and soybean protein concentrate (SPC) were reported by Wang et al. (2011). The outcome demonstrated that emulsions stabilized by both proteins within the presence of the 2% GA (w/w) have preferred stability over its absence by expanding the emulsion viscosity of the FP-stabilized emulsion and bringing on competitive adsorption between the GA and SPC layers to give a steric repulsion for the SPC-stabilized emulsion, respectively. At that point, the impacts of ionic strength (0–200 mM NaCl) and temperature (25–95°C for 20 min) on these emotions within the presence of GA were determined. The GA adsorbed at SPC-stabilized oil–water interface gave stability against NaCl concentration. Within the presence of GA, the SPC–balanced-out emulsions likewise indicated better stability at higher temperatures contrasted with the FPC stabilized emulsions because of the denaturation of SPC and competitive adsorption among GA and SPC at higher temperatures (Wang et al., 2011).

Water-holding capacity

Water-holding capacity (WHC) is the ability of the material to hold water against gravity. The range value of WHC for *A. senegal* gum was 65.40–65.80 (Omer, 2004). Due to the high water-holding properties of the gum, the gum imparts a smooth texture to the frozen product by inhibiting the formation of ice crystals (Omer, 2004). Kheir (2005) studied some functional properties (WHC and the emulsifying stability) of eight GA formulations, namely, handpicked selected, cleaned, sifting, kibbled 105, kibbled 107, kibbled 119, kibbled 121, and spray dried in an attempt to set standard specifications for each. She noticed that the formulations were varied between raw and processed GA of *A. senegal* as the influence of processing on properties of GA. The author found that WHC expressed as percentages was highly significantly different between eight GAF. The maximum WHC was recorded in the kibbled 107 (69.7%) and spray-dried (69.5%) formulations. The least WHC was recorded in kibbled 105 forms. The water solubility showed some variations among the different gum formulations, with the highest for the spray dried (98.80%) and the lowest for the sifting formulation (97.40%). However, the solubility in organic solvents (ethanol, acetone, and chloroform) had been generally very low yet with variations among the different formulations. Kheir (2005) reported that the WHC expressed as percentages was highly significantly different between eight GAF. Soibe et al. (2015) researched the impact of partial replacement of wheat with 10%–40% plantain and 1, 2, and 3% GA on composite bread quality attributes. They reported that the difference between the control and the composites was statistically significant ($p < 0.05$) at all levels of plantain incorporation of WHC, foaming capacity, emulsion capacity, and bulk density. GA, as used in this study, was found to improve the bread's textural qualities, as a significantly softer and more springy bread was produced.

Encapsulating agent

In food processing, microencapsulation is a vital procedure to enhance the chemical stability of delicate mixes, to give the controlled release of microencapsulated compounds, and to give a free-flowing powder with enhanced handling properties. The encapsulating material must preserve and secure the encapsulated mixes during manufacture, storage, and taking care to discharge them into the last product during production or utilization. Solubility and low-viscosity emulsion properties have encouraged the utilization of GA as an encapsulation factor for retention and assurance of chemically reactive and volatile commercial food flavoring (Montenegro et al., 2012). GA is fundamentally utilized for fat microencapsulation, since it produces stable emulsions on account of most oils in a wide pH range, and it has the ability to form films. Barbosa et al. (2005) investigated the photostability of the microencapsulated carotenoid bixin in the various edible polysaccharides. They discovered that microencapsulated bixin in GA was three to four times more stable than the one microencapsulated with maltodextrin, and around ten times more than in homogeneous solvents. The major use for encapsulation in the food industry is for liquid flavors. Encapsulation has been able to mask bitter tastes of compounds, reduce volatility and flammability of liquids, control release of materials, provide protection to compounds, reduce toxicity, separate reactive materials, and to make liquids behave like solids. Microencapsulation by spray drying is the most economical and flexible way for the food ingredient to retain the needed properties in the final food products. This technique also has been used in pharmaceutical industries, e.g., vitamins and minerals. Yet it is not widely used in the food industry. Chranioti and Tzia (2014) investigated the encapsulation of fennel oleoresin by freeze-drying technique utilizing binary and ternary blends of GA joined with modified starch, maltodextrin, and chitosan. The encapsulation capacity of the last blends was assessed in light of their microencapsulating productivity and capacity stability toward the primary compounds of fennel oleoresin. The incomplete replacement of GA with modified starch gave the best encapsulating mixture [GA–modified starch (MS)] as far as storage stability, microencapsulating productivity (74.88%), and redispersibility ($D4,3 = -2.74$ μm). The properties of the underlying shaped emulsions impacted the stop-dried last items chiefly in regard dampness substance and redispersibility. GA in blends turned out to be effective to encapsulate and protect fennel oleoresin compounds (fenchone, estragole, trans-anethole, and D-limonene) amid storage. Cano-Higuita et al. (2015) examined the spray-drying procedure of turmeric oleoresin microencapsulated with double and ternary blends of various wall materials: GA, maltodextrin, and modified corn starch. The encapsulation efficiency, curcumin retention, prepare yield, water content, solubility, and molecule morphology were tested. Wall matrices containing higher extents of modified starch and GA brought about higher encapsulation effectiveness and curcumin retention, while the procedure yield and water content expanded with higher proportions of maltodextrin and GA, respectively.

Adsorption isotherms

Isotherms have been checked for the adsorption of GA onto polystyrene latex; the sum adsorbed was found to increase with expanding ionic quality. Gel permeation chromatography (GPC) tests demonstrated that fractionation happened on adsorption with higher molecular mass particles adsorbing preferentially. It was discovered additionally that this molecular mass portion contained the majority of the proteinaceous material. Electron spin resonance spectroscopy proposed that the polymer adsorbs with roughly 50% of its segments in loops and tails extending far from the surface into solution in 0.5 mol/drrr′ NaCl (Snowden et al., 1987).

In water, the GA was found to enlarge the electrophoretic mobility of the particles, and the latex scatterings were appeared to be steady against flocculation on account of electrostatic repulsions. In 0.5 mol/drrr′ NaCl, where charge aversions are ineffective, the scatterings flocculate at low-polymer surface coverage; however, they end up plainly stable once the particle surface is completely coated. Stabilization is accomplished in view of the repulsive interactions between adsorbed polymer layers. The low molecular mass, nonproteinaceous portion of the gum can't settle the scatterings under these conditions (Snowden et al., 1987). Rodríguez-Bernal et al. (2015) investigated the sorption mechanisms of borojó mash (BPP), three borojó phases [liquid phase (LP), medium phase (MP), and solid phase (SP)], GA, and a blend of these (BP) delivered by freeze-drying were applied through adsorption isotherms. The adsorption models satisfactorily summarize the trial information over the whole measured range of water activity (aw); R^2 was near 1 for the investigated conditions. For the thermodynamic properties, the differential and integral enthalpy of BP demonstrated a synergic impact of the consolidated components. This demonstrates that utilizing a mix of polymers expands the interaction energy of water molecules with the surface of the material. The monolayers from the Brunauer–Emmett–Teller (BET) and Guggenheim, Ander- son, and de Boer (GAB) models indicated comparative moisture content values that related to the base fundamental entropy. At last, the performance of the enthalpy–entropy settlement of water adsorption for the SP, MP, and GA powders at low moisture constituents exhibited that the content is controlled by an entropic mechanism; for BP, BPP, and LP powders, the water adsorption might be considered as principally enthalpy driven (Rodríguez-Bernal et al., 2015).

Viscosity

GA can upgrade not just the viscosity of the watery medium like a large portion of other polysaccharide gums, but it likewise forms a thick macromolecular layer around the emulsion droplets, bringing about great stabilization (Ibanoglu, 2002). Gómez-Díaz et al. (2008) studied the inner viscosity and stream conduct of watery solutions of various concentrations of GA. They calculated the molecular weight of this polymer, utilizing the estimation of the inherent viscosity, utilizing both Huggins and Kraemer and Mark–Houwink equations. They determined the estimation of the evident viscosity and the impact of the shear rate on the rheological conduct of various polymer concentrations in a watery solution. They likewise investigated the impact of temperature over the flow behavior. They utilized the Ostwald model to fit the trial data and to acquire the model parameters.

Watery solutions of Arabic gum utilized in the Gómez-Díaz et al. (2008) study exhibit a nonNewtonian stream behavior because of the impact of the shear rate upon the clear viscosity. An expansion in the shear rate creates a diminishing in the evident viscosity value, with a higher reduction at low values of shear rate applied to the liquid stage. This behavior shows that the nonNewtonian behavior is a pseudoplastic sort. These researchers presumed that Arabic gum watery solutions demonstrated a nonNewtonian and pseudoplastic behavior. The power law model has been utilized to summarize this behavior under shear in the rotational concentric chamber framework. An increase in temperature creates an expansion in Newtonian behavior of the GA watery solutions, with estimations of behavior recorded close to the Newtonian behavior ($n = 1$). A direct trend between behavior index and temperature was found, and a liquid phase polymer concentration has no impact in the slope of the plot (Gómez-Díaz et al., 2008).

Rheological behavior

GA promptly dissolves in water to create solutions of low viscosity and henceforth discovers little application as a thickening factor. In the perspective of this, it is not amazing, maybe, that there are few detailed studies and reviews announced concerning its rheological behavior (Islam et al., 1997). GA solution is generally regarded as a Newtonian liquid with an anomalous low viscosity. However, recent research on the rheological characteristics of GA in water has shown that it is not simply Newtonian. In this work, rheological characteristics of GA solutions were investigated, with emphasis on its nonNewtonian behavior of shear thinning and thixotropy, and interpreted based on molecular association in the solution. It was found that the arabinogalactan protein complex (AGP) in GA might be the main source of molecular association, though it only represents about 10% of the total gum. The occurrence of AGP micelles endowed the GA solution with nonNewtonian behavior of shear thinning and thixotropy. This fact was evidenced by using single-walled carbon nanotubes to trap AGP. The nonNewtonian behavior of the GA solution vanished after AGP was removed from the solution (Li et al., 2011).

Tanaka et al. (2006) investigated the rheological behavior of watery solutions of GA. Their results indicated that the GA solutions at consolidations of 10 and 30 weight % gave shear thinning flow behavior. The viscosity of the noncentrifuged solution was higher than that of the centrifuged solutions at low shear rate region, proposing that the air rises in the solution add to expanding of viscosity. The Newtonian administration can be seen at high shear rate region in the both noncentrifuged and centrifuged solutions, and the viscosity of the noncentrifuged solution was the same as that of centrifuged solutions. In dynamic rheological practices, the capacity modulus of the noncentrifuged solution was higher than that of the centrifuged solution at low-recurrence region. The capacity modulus of both noncentrifuged and centrifuged solutions increments with time when the temperature is changed from 10°C to 32°C, as in the gelling procedure (Tanaka et al., 2006). Singh et al. (2017) analyzed the impacts of adding GA (0.1%–1.0%) on solubility, pasting, and rheological characteristics of 5% native tapioca starch (TS). They determined the dynamic viscoelasticity, and their outcomes showed that the elastic modulus (G′) and the viscous modulus (G″) increased in the presence of GA, suggesting a strong interaction between GA–TS in the composite system. All concentrations of GA resulted in higher elasticity ($G' > G''$). Pasting properties revealed that the peak viscosity (PV) and setback (SB) of the TS–GA dispersion decreased with increasing concentration of GA, suggesting the importance of GA in controlling overall viscosity. The flow behavior index (n) decreased from 0.99 to 0.23 as the concentration of GA in the dispersion increased, indicating pseudoplastic behavior ($n < 1$). However, the consistency coefficient increased by five times, from 0.006 to 0.283 Pa.s. with increasing GA concentration (0.1–1.0), showing the synergistic effect of GA. These results show that low concentrations of GA can beneficially improve the rheological behavior of tapioca starch. In addition, GA has advantages of being low cost, clean label, and fewer calories, thus broadening the application spectra of native starches.

Castel et al. (2017) studied the emulsifying characteristics of brea gum (BG), the exudate from *Cercidium praecox*, in contrast with GA. Droplet size disseminations, rheological features, and permanence of corn oil emulsion balanced out with BG and GA solutions were tested. The outcomes demonstrated that an increase in BG addition prompted a decline in Z-normal diameter and an increase in emulsion-evident viscosity and stability. All emulsion flow curves presented shear-thinning behavior at low shear rates and Newtonian plateau at high shear rates. The mechanical spectra showed droplets tending to arrange as a network in the emulsions, which was related to the high stability. BG emulsion presented higher viscosity and stability than GA emulsion at the same concentration, suggesting that BG could replace GA in some industrial utilization (Castel et al., 2017).

Chen et al. (2015) studied the impact of GA on swelling power (SP), solubility index (SI), freeze–thaw stability, and sticking and rheological attributes of native, anionic, and cationic tapioca starch (TS, ATS, and CTS). They found that SP and SI of TS, ATS, and CTS were reduced by GA. They said that this might be because of alluring communications between oppositely charged particles happening when anionic GA experienced positively charged CTS. The addition of GA reduced SB estimations of ATS yet had no impact on CTS ($p < 0.05$). All of the three starch–GA blends showed lower PVs contrasted with their controls. The degree of the reduction in PV brought about by GA addition took after the order: ATS > TS > CTS. For ATS–GA blends, the decrease of PV can be interpreted by the repulsing power among ATS and GA. Also, these researchers demonstrated that GA played an accelerative impact on syneresis of CTS, a hindering impact on syneresis of ATS, and no impact on syneresis of TS. Rheological reviews demonstrated that all starch and starch–GA gels were named ordinary weak gels. The capacity modulus (G') of TS and CTS could be upgraded by GA, while G' of ATS–GA glue was debilitated by GA. The prior information showed that the impact of GA on physical properties of starch relied on upon its charge.

Stabilization

Yao et al. (2013) utilized conventional GA and two developed GA specimens (EM2 and EM10) as O/W emulsions as a delivery system to preserve CLA from oxidation. The emulsions have ideal physical and chemical dependability at gum additions of 5% for every one of the three gums. Emulsions with higher gum percentages are more vulnerable to lipid oxidation. This is ascribed to decreased physical stability at higher gum percentages as a result of the combination and depletion-induced flocculation of the emulsion droplets. The prooxidants iron and copper characteristically contained in the gums could likewise add to this instability. Among the three gums, EM10 gives the best preservation to CLA both physically and chemically as a result of its better interfacial properties over GA and EM2 (Yao et al., 2013).

Djordjevic et al. (2008) studied the preservation of limonene in O/W emulsion droplets utilizing whey protein isolate (WPI) and GA. They found that limonene degradation and development of the limonene oxidation items, limonene oxide and carvone, were less in the WPI- than GA-stabilized emulsions at both pHs 3.0 and 7.0. This information recommended that WPI could hinder the oxidative weakening of limonene in O/W emulsions. The capability of WPI to reduce oxidative responses could be because of the formation of a cationic emulsion droplet interface at pH 3.0, which can repulse prooxidative metals, as well as the capacity of amino acids in WPI to scavenge free radical and chelate prooxidative metals. Proteins have been found to deliver both oxidatively and physically stable emulsions containing triacylglycerols.

Kattumuri et al. (2007) displayed a basic, productive, and organically based technique to get little GA-stabilized gold nanoparticles (GA-AuNPs) with noteworthy stability in physiological pHs. The in vitro stability tests in intestinal (pH 6.8) and gastric (pH 1.2) recreated pHs revealed that GA-AuNPs display a surprisingly high preservation, even close to the zero zeta potential. At the point when subjected to GA-AuNPs, changes in the viability, expansion, and morphology were specifically initiated in the B16-F10 melanoma cell line, though no modifications in the macrophage cell line, RAW 264.7, or in the fibroblast cell line, BALB/3T3, were seen at similar concentrations. Therefore, considering the astounding stability and particular impact on cell lines, we demonstrate that GA-AuNPs display characteristics that could give an alternative to melanoma treatment.

Achouri et al. (2012) investigated the capacity of emulsifiers [GA, starch, and soy protein isolate (SPI)] and texture modifiers (gum tragacanth and carrageenan) alone and as a blend with SPI to balance out O/W emulsions at high oil concentrations. Parameters studied included droplet estimate distribution, color ΔE consistency, and creaming stability. At 10% (w/v) emulsifier-to-oil proportion, emulsions settled with GA, starch, and SPI had a normal oil droplet sizes lower than 1 μm. Expanding the oil content by twofold (emulsifier/oil proportion of 10:20), separately, while expanded the normal droplet size of the GA and SPI emulsions to 1.5 and 4.2 m that of the starch emulsion stayed unchanged. At 5% oil concentration, gum tragacanth– and carrageenan-stabilized emulsions yielded two normal droplet sizes of 0.5 μm (52.7%) and 1.8 μm (46.5%), and 0.5 μm (48.8%) and 1.7 μm (46.6%), respectively. Expanding the oil percentage to 10% expanded the molecule estimate ΔE than starch and SPI (except at with fairly similar proportions). GA stabilized emulsions had less ΔE of carrageenan and tragacanth emulsion and diminished essentially when utilized as a blend (20% oil content) with SPI. Droplet size emphatically influenced color. Every single concentrated emulsion showed shear-diminishing behavior regardless of oil, emulsifier sort, and concentration. SPI, followed by starch and GA emulsions, indicated great dependable qualities over the 15-day storage period. Gum tragacanth and carrageenan emulsion stability qualities were just enhanced when joined with SPI, especially at 5% oil focus (Achouri et al., 2012).

Antioxidant properties

Selenium nanoparticles (SeNPs) were set up by utilizing GA as the stabilizer in a simple manufactured approach. The size, morphology, solidness, and antioxidant action in vitro of the GA–selenium nanocomposites (GA-SeNPs) were described by transmission electron microscopy, dynamic light dissipating, Fourier-transform infrared spectroscopy (FTIR), nuclear drive microscopy, and ultraviolet–visible spectrophotometry. SeNPs (molecule size of ~34.9 nm) can be settled in GA aqueous solutions for around 30 days. FTIR outcomes demonstrated that SeNPs were joined to the hydroxyl groups of GA. In the present work, the soluble base hydrolyzed GA (AHGA) was also arranged, and its proficiency in settling SeNPs was compared with GA. It was concluded that the branched structure of GA was a critical variable for the functionality. The hydroxyl radical scavenging capacity and 2,2-diphenyl-1-picrylhydrazyl (DPPH) scavenging capacity of GA-SeNPs were higher than those of AHGA-SeNPs and could achieve 85.3 ± 2.6% and 85.3 ± 1.9% at a concentration of 4 mg/ml, respectively (Kong et al., 2014).

Obesity is a worldwide well-being concern related to high death and mortality. Therapeutic methodologies incorporate surgery and synthetic drugs, and may bring about high expenses and genuine confusions. GA (*A. senegal*) is considered as a dietary fiber that lessens body fat accumulation. However, the impact of the GA on reproductive capacities in high-fat diet regimen stays unclear. In this study, we fed mice either a normal diet (control), low-fat diet (low), high-fat diet (high), or a high-fat diet supplemented with 10% w/w GA (high + gum) for 12 weeks. Body weight, visceral adipose tissue, plasma lipid profile, and blood glucose were determined. The ovarian antioxidant limit was assessed by the estimation of antioxidant enzymes, malondialdehyde (MDA), and antioxidant enzyme activity. Moreover, ovarian histopathological changes and oxidative stress–related genes' mRNA were measured. GA treatment significantly ($p < 0.05$) increased activities of superoxide dismutase (SOD), catalase, and glutathione peroxidase (GPx) compared to low, HFD, and control groups. The treatment of GA significantly ($p < 0.05$) decreased ovary MDA, plasma total cholesterol, LDL-c, and triglyceride concentrations, whereas it increased HDL-c concentrations compared to low, high fat diet (HFD), and control groups. SOD and GPx mRNA expression were significantly increased in GA group compared to low, HFD, and control groups. Ovaries of all HFD mice showed marked degeneration, whereas slight degeneration was observed in GA-treated mice compared to low, HFD, and control groups. Our findings suggest that GA may protect ovaries by improvement of antioxidant capacity; thus, it may be useful to ameliorate the fertility complications in obese patients. GA supplementation enhanced cancer prevention agent status and modifies the expression of oxidative stress gene in the ovary of mice sustained with a high-fat food (Ahmed et al., 2016). Catechin was oligomerized utilizing free laccase and laccase–GA conjugate. The procedure of oligomerization was determined as for solvent, the proportion of solvent to buffer (0.2:10 to 1:10), pH of buffer (3–10), enzyme (575–18,400 U/mg) and substrate concentration (1–7 mM). The highest extreme output of oligomer was seen in methanol at proportion 0.6:10 of methanol: buffer of pH 5 utilizing 2300 U/mg of laccase and 5 mM of catechin. The laccase-GA conjugate demonstrated a lower response, however, higher permanence than free laccase in methanol. Free laccase delivered cross-connected water-insoluble oligomer, though conjugated laccase created straight water-soluble oligomer. The direct water-solvent oligomer indicated higher antioxidant activity, as controlled by the DPPH test, and lessening power when contrasted with monomer, making it appropriate for biological applications. The molecular weight of the linear oligomer was observed to be 13.14 kDa, which recommended it to be made out of 45 monomer units. Promote properties of linear and cross-linked oligomer were done utilizing FTIR and differential scanning calorimetry (Jadhav and Singhal, 2014).

References

Achouri, A., Zamani, Y., Boye, J.I., 2012. Stability and physical properties of emulsions prepared with and without soy proteins. J. Food Res. 1 (1), 254–267.

Ahmed, A.A., Fedail, J.S., Musa, H.H., Musa, T.H., Sifaldi, A.Z., 2016. Gum Arabic supplementation improved antioxidant status and alters expression of oxidative stress gene in ovary of mice fed high fat diet. Middle East Fertil. Soc. J. 21 (2), 101–108.

Ali, A.A., Ali, K.E., Fadlalla, A., Khalid, K.E., 2008. The effects of GA oral treatment on the metabolic profile of chronic renal failure patients under regular haemodialysis in Central Sudan. Nat. Prod. Res. 22 (1), 12–20.

Barbosa, M.I., Borsarelli, C.D., Mercadante, A.Z., 2005. Light stability of spray-dried bixin encapsulated with different edible polysaccharide preparations. Food Res. Int. 38 (8–9), 989–994.

Cano-Higuita, D.M., Vélez, H.A.V., Teli, V.R.N., 2015. Microencapsulation of turmeric oleoresin in binary and ternary blends of gum Arabic, maltodextrin and modified starch. Ciênc. Agrotec. Lavras 39 (2), 173–182.

Castel, V., Rubiolo, A.C., Carrara, C.R., 2017. Droplet size distribution, rheological behavior and stability of corn oil emulsions stabilized by a novel hydrocolloid (Brea gum) compared with gum Arabic. Food Hydrocoll. 63, 170–177.

Chen, H., Fu, X., Luo, Z., 2015. Effect of gum Arabic on freeze-thaw stability, pasting and rheological properties of tapioca starch and its derivatives. Food Hydrocoll. 51, 355–360.

Chranioti, C., Tzia, C., 2014. Arabic gum mixtures as encapsulating agents of freeze-dried fennel oleoresin products. Food Bioproc. Tech. 7 (4), 1057–1065.

Djordjevic, D., Cercaci, L., Alamed, J., McClements, D.J., Decker, E.A., 2008. Chemical and physical stability of protein and gum Arabic-stabilized oil-in-water emulsions containing limonene. J. Food Sci. 73 (3), C167–C172.

Gils, P.S., Ray, D., Sahoo, P.K., 2010. Designing of silver nanoparticles in gum Arabic based semi-IPN hydrogel. Int. J. Biol. Macromol. 46 (2), 237–244.

Glover, D.A., Ushida, K., Phillips, A.O., Riley, S.G., 2009. *Acacia* (sen) SUPERGUM (gum Arabic): an evaluation of potential health benefits in human subjects. Food Hydrocoll. 23 (8), 2410–2415.

Gómez-Díaz, D., Navaza, J.M., Quintáns-Riveiro, L.C., 2008. Intrinsic viscosity and flow behaviour of Arabic gum aqueous solutions. Int. J. Food Prop. 11, 773–780.

Ibanoglu, E., 2002. Rheological behavior of whey protein stabilized emulsions in the presence of gum Arabic. J. Food Eng. 52, 273–277.

Islam, A.M., Phillips, G.O., Sljivo, A., Snowden, M.J., Williams, P.A., 1997. A review of recent developments on the regulatory, structural and functional aspects of gum Arabic. Food Hydrocoll. 11 (4), 493–505.

Jadhav, S.B., Singhal, R.S., 2014. Laccase–gum Arabic conjugate for preparation of water-soluble oligomer of catechin with enhanced antioxidant activity. Food Chemistry 150, 9–16.

Kattumuri, V., Katti, K., Bhaskaran, S., Boote, E.J., Casteel, S.W., Fent, G.M., Robertson, D.J., Chandrasekhar, M., Kannan, R., Katti, K.V., 2007. Gum Arabic as a phytochemical construct for the stabilization of gold nanoparticles: in vivo pharmacokinetics and x-ray-contrast-imaging studies. Small 3, 333–341.

Kheir, M.E.M., 2005. Physicochemical, functional and microbial properties of crude and processed gum Arabic (Acacia Senegal). Gum Arabic. Faculty of Agriculture, University of Khartoum, Khartoum, Sudan, MSc. Thesis.

Kheir, M.K., Yagoub, A.A., Abu Baker, A.A., 2008. Emulsion-stabilizing effect of gum from *Acacia senegal* (L) Willd. The Role of quality and grade of gum, oil type, temperature, stirring time and concentration. Pak. J. Pharm. Sci. 7 (3), 395–399.

Kong, H., Yang, J., Zhang, Y., Fang, Y., Nishinari, K., Phillips, G.O., 2014. Synthesis and antioxidant properties of gum Arabic-stabilized selenium nanoparticles. Int. J. Biol. Macromol. 65, 155–162.

Li, X., Zhang, H., Fang, Y., Al-Assaf, S., Phillips, G.O., Nishinari, K., 2011. Rheological properties of gum Arabic solution: the effect of arabinogalactan protein complex (AGP). In: Kennedy, J.F., Philips, G.O., Williams, P.A. (Eds.), Gum Arabic. RSC, UK, pp. 1–4.

Masuelli, M.A., 2013. Hydrodynamic properties of whole Arabic gum. Am. J. Food Sci. Technol. 1 (3), 60–66.

Montenegro, M.A., Boiero, M.L., Valle, L., Borsarelli, C.D., 2012. Gum Arabic: more than an edible emulsifier. In: Verbeek, J. (Ed.), Products and Applications of Biopolymers. InTech, London, United Kingdom, ISBN: 978-95351-0226-7.

Omer, E.A. 2004. Characterization and Analytical Studies of *Acacia Polyacantha* gum (Ph.D. Thesis). Sudan University of Science &Technology, Khartoum, Sudan.

Omer, E.A., AL-Omari, A.A., Elgamidy, A.H., Elgamidy, A.A., Elgamidy, A.M., 2015. The emulsifying stability of gum Arabic using the local sesame oil obtained from AL-BAHA area. Int. J. Eng. Res. Technol. 4 (2), 1172–1175.

Osman, M.E., Williams, P.A., Menzies, A.R., Phillips, G.O., 1993. Characterization of commercial samples of gum Arabic. J. Agric. Food Chem. 41 (1), 71–77.

Park, E.Y., Murakami, H., Matsumura, Y., 2005. Effects of the addition of amino acids and peptides on lipid oxidation in a powdery model system. J. Agric. Food Chem. 53 (21), 8334–8341.

Phillips, G.O., Wedlock, D.J., Williams, P.A. (Eds.), 1988. Gums and stabilisers for the food industry. Royal Society of Chemistry, Cambridge, UK, pp. 507–522, vol. 4.

Rodríguez-Bernal, J.M., Flores-Andrade, E., Lizarazo-Morales, C., Bonillab, E., Pascual-Pineda, L.A., Gutierréz-López, G., Quintanilla-Carvajal, M.X., 2015. Moisture adsorption isotherms of the borojó fruit (*Borojoa patinoi* Cuatrecasas) and gum Arabic powders. Food Bioprod. Process. 9 (4), 187–198.

Singh, A., Geveke, D.J., Yadav, M.P., 2017. Improvement of rheological, thermal and functional properties of tapioca starch by using gum Arabic. LWT—Food Sci. Technol. 80, 155–162.

Snowden, M. l., Phillips, G.O., Williams, P.A., 1987. Functional characteristics of gum Arabic. Food Hydrocoll. 1 (4.), 291–300.

Soibe, L.G., Shitand, A., Mahungu, S., Ben, C., Agasa, L., 2015. Effect of the gum Arabic on the physico-chemical and baking properties of the wheat-plantain composite flour. J. Food Res. Technol. 3 (3), 98–105.

Tanaka, S., Fang, Y., Nishinari, K., 2006. Anomalous rheological behaviour of gum Arabic solutions. Foods Food Ingredients J. Jpn. 211 (3), 216.

Verbeken, D., Dierckx, S, Dewettinck, K., 2003. Exudate gums: occurrence, production, and applications. Appl. Microbiol. Biotechnol. 63 (1), 10–21, ISSN: 0175-7598.

Wang, B., Wang, L., Li, D., Adhikari, B., Shi, J., 2011. Effect of gum Arabic on stability of oil-in-water emulsion stabilized by flaxseed and soybean protein. Carbohydr. Polym. 86, 343–351.

Williams, P.A., Phillips, G.O. (Eds.), 2000. Handbook of Hydrocolloids. CRC Press, Cambridge, pp. 155–168.
Yao, X., Xu, X., Tian, D., Wang, N., Fang, Y., Deng, Z., Phillips, G.O., Lu, J., 2013. Physical and chemical stability of gum Arabic-stabilized conjugated linoleic acid oil-in-water emulsions. J. Agric. Food Chem. 61 (19), 4639–4645.

Further Reading

De Barros, H.R., Cardoso, M.B., de Oliveira, C.C., Franco, C.R.C., Belan, D.L., Vidotti, M., Riegel-Vidotti, C.I., 2016. Stability of gum Arabic-gold nanoparticles in physiological simulated pHs and their selective effect on cell lines. RSC Adv. 6, 9411–9420.

Effect of Gum Arabic on the Fecal Bacterial Mass in Healthy Human Volunteers

25

Rasheeda Hamid Abdalla Ahmed, Abdalbasit Adam Mariod**,†, Sulieman Mohamed El Sanousi‡*
*Prince Sattam Bin Abdulaziz University, Al-Kharj, Kingdom of Saudi Arabia;
**University of Jeddah, Alkamil, Kingdom of Saudi Arabia; †Sudan University of Science & Technology, Shambat, Sudan; ‡University of Khartoum, Khartoum, Sudan

Introduction

Gum Arabic is an ancient natural product used by man since the days of the pharaohs, who used it for making coloring matter. *Acacia senegal* trees, the main source of GA, are spread through what is known as the "African Gum Belt." This belt is located north of the equator and south of the Sahara Desert, extending from east to west continuously from Somalia through Ethiopia, Sudan, and many other countries to Senegal. *A. senegal* is also found in parts of Africa south of the equator. In Asia, *A. senegal* has been found in Saudi Arabia and India. In the Sudan, GA is produced in two main regions, namely eastern and western Sudan. The west comprises North Kordofan, Darfur, and part of the White Nile region, while the east includes the Blue Nile and Kassala regions. The economic significance of GA for the Sudan is that it is one of the main export commodities, commanding about 10%–20% of the total national income. GA was earlier used by ancient Egyptians for making paint colors. In modern times, GA has been extensively used in the food, pharmaceuticals, cosmetics, and other industries, such as textile, painting, and ink. GA has also been used in medicine, such as in the treatment of nephritic edema and in plastic surgery.

Medicinal uses

GA has been used for the treatment of problems in blood pressure flow caused by hemorrhagic or surgical shock. Intravenous saline injections alone were not successful because the saline escapes too rapidly from the blood vessels. Alkarib et al. (2016) determined hypertension in kidney disease patients; they evaluated the changes that occurred in the blood pressure in relation to GA. They used 24 volunteers in a randomized control trial study, performed using different blood pressure readings and different kidney functions with 6 months past history. They reported a significant correlation between GA component and blood pressure as monitored with consideration

Gum Arabic. http://dx.doi.org/10.1016/B978-0-12-812002-6.00025-7

of avoidance of hypo-tension; it seems to depend on the gum component ratios. The study showed significant lowering association with GA administration and adjusting the level of serum sodium to normal values ($p = 0.003$) and serum potassium increase ($p = 0.000$); these two factors seem to be of great value in controlling the blood pressure. In plastic surgery, a 50% GA adhesive has been used successfully in grafting destroyed peripheral nerves.

Metabolic products of gum Arabic

GA is a complex polysaccharide, primarily indigestible to both humans and animals; it is not degraded in the intestine, but fermented in the colon under the influence of microorganisms. Despite a range of animal studies, there are no usable data for humans that can quantify the utilizable energy of GA. Estimates in the literature from animal experiments vary from 0 to 4 kcal/g. After certain allowances are made for the energy losses from volatile and gaseous fermentation products, an upper level of 2 kcal/g for rats has been set. The situation in man is demonstrably different, with greatly reduced amounts of such products and the need to adapt for varying periods before GA is attacked by colonic bacteria. In the absence of an agreed scientific assignment, the FDA in the USA insists upon 4 kcal/g in nutritional labeling, whereas in Europe, no value has been assigned to soluble dietary fiber such as GA (Phillips, 1998).

Several epidemiological studies suggest that a high intake of dietary fiber, including GA, is associated with beneficial effects on fat metabolism. Dietary fiber promotes satiation and satiety, alters glycemic index, affects gastric emptying and gut hormone secretion, and thus helps to manage weight. Leptin promotes weight loss by two different mechanisms. It reduces appetite and thus food intake, and, at the same time, increases energy expenditure. Also, dietary fiber was inversely associated with leptin levels in young Japanese adults. In addition to that, a study has shown that GA inhibits intestinal glucose absorption via interaction with membrane abundance of SGLT1 in mice. GA significantly blunted the increase in body weight, fasting plasma glucose, and fasting insulin concentrations during high-fat diets (Babiker et al., 2012; Kuroda et al., 2010; Murakami et al., 2007). Babiker et al. (2012) determined the effects of regular GA ingestion on body mass index and body fat percentage among healthy adult females. Pre- and postanalysis among the study group showed significant reduction in body mass index (BMI) by 0.32 and body fat percentage by 2.18% following regular intake of 30 gm/day GA for 6 weeks. Side effects caused by GA ingestion were experienced only in the first week. They included that GA ingestion causes significant reduction in BMI and body fat percentage among healthy adult females.

Dietary fiber, which has marked effect on the function of the large intestine, increasing transit time, includes GA. GA degradation occurs in the cecum and is associated with increased methane excretion, increased volatile fatty acids concentration, and changes in the proportion of the volatile fatty acids in the feces. Sudanese individuals in western Sudan had been utilizing GA for a long time without impediments. It is indigestible to both people and animals, not debased in the digestive system, but rather aged in the colon to give short-chain unsaturated fats, prompting an expansive scope of conceivable medical advantages (Phillips and Phillips, 2011). One of these advantages is its prebiotic impact (Calame et al, 2008; Phillips et al., 2008). It has been guaranteed that 4-week supplementation with GA (10 g/day) prompted critical increases in *Bifidobacteria*, *Lactobacteria*, and *Bacteroides*, demonstrating a prebiotic impact (Calame et al., 2008). Different impacts include decrease in plasma cholesterol level in animals and people, anticarcinogenic impact (Nasir et al., 2010), and hostile to oxidant impact (Ali et al., 2003; Al-Majed et al., 2002), with a defensive part against hepatic and heart toxicities, notwithstanding that it has been asserted that GA mitigates impacts of interminable renal failure in humans; in any case, additional studies are required for affirmation (Babiker et al., 2012).

Gum Arabic and intestinal tract bacteria

Changes in human fecal flora have been detected following changes in the diet. Total short-chain fatty acids (SCFA) were significantly higher in the feces of animals fed the starch diet compared with those fed the sucrose diet. A high-starch diet can markedly modify the composition of fecal flora and alter considerably the fecal concentration of SCFAs, compounds that might have a health-promoting effect (Cresci et al., 1999). The proportion of the fecal flora able to degrade GA polymer rose from an initial level of 6.5 to more than 50% during GA ingestion.

Gum Arabic and stool weight

Many investigations were made to show the effect of dietary fiber, such is GA, in the stool weight. Also, stool weight is the most used indicator of dietary fiber metabolism in the human colon. Cherbut et al. (2003) suggested that acacia gum is a prebiotic fiber, as it was able to selectively raise the proportions of lactic acid bacteria and *Bifidobacteria* in healthy subjects. They reported that it was fermented slowly and produced SCFA, and its fecal digestibility was around 95%. As in the case of other soluble dietary fibers, it increased stool output by augmenting the water content of stools. In addition, its intestinal tolerance appeared to be excellent, and high daily doses could be consumed without inducing any adverse intestinal events (Cherbut et al., 2003). GA caused a significant increase in the stool weight in chronic renal failure patients when compared to the baseline period. The mechanism by which dietary fiber increases stool output is not fully understood, but the major mechanism is thought to be an increase in the water-holding capacity of the stool, which is either due to the presence of the fiber itself or to bacterial cells, which increase in number if the fiber is fermented. Fiber may influence fecal output by another mechanism. Colonic microbial growth may be stimulated by the ingestion of such fermentable fiber sources as apple, guar gum, or pectin. However, there is not always an increase in stool weight as the result of eating these fibers. There may also be an added osmotic effect of bacterial fermentation products on the stool mass, though this is not yet well-defined (Ahmed, 2002).

GA has many applications in the food industry, pharmaceuticals, medicine, and other miscellaneous applications. In medicine, GA has been used for the treatment of low blood pressure and nephritic edema. One of its recent and important applications in medicine is its use in the treatment of chronic renal failure. Studies showed that there was marked improvement in the management of patients who were given GA treatment for 4 or 8 months. Patients received GA became off of or had reduced hemodialysis frequency as well as improvement in their well-being. GA (10% in drinking water) significantly increased urinary excretion of Ca (2+) and significantly decreased plasma phosphate and urea concentrations, urinary flow rate, urinary Na (+), phosphate and glucose excretion, blood pressure, and proteinuria. GA treatments decrease blood pressure and proteinuria in diabetic mice and may thus prove beneficial in diabetic nephropathy (Nasir et al., 2012). As a dietary fiber, GA metabolism is through the action of the intestinal microflora, the net result of which is the lowering of serum urea nitrogen.

Ahmed (2002) investigated the effect of GA on the normal intestinal bacterial flora. The fecal bacterial mass (FBM) was measured as an index of the microbial population of the intestine. Two groups, male and female healthy human volunteers, were assigned to receive GA doses for 6 consecutive days, and stool samples were taken at the base, two times during the dosing period, and after cessation of the doses. In the female group, 50% of the volunteers showed an increase in their FBM at 2 days of treatment GA. Also, another 50% showed increase at 5 days of treatment when compared with their values at 2 days of treatment. These authors reported that 80% of the female group and 75% of the male group showed increases in their FBM, either at 2 or 5 days of the treatment (Table 25.1).

As shown in Tables 25.1 and 25.2, volunteers 1, 2, 3, and 7 showed decrease in their fecal bacterial mass 2 days after commencement of the treatment, with percentages of 36.3, 30.4, 24.9, and 41.9, respectively, when compared with the basal value.

Table 25.1 Fecal bacterial mass in ten apparently health female volunteers given 50 g gum Arabic for 6 consecutive days

Volunteers	Base values (before treatment) (mg/g)	Values 2 days after treatment (mg/g)	Values 5 days after treatment (mg/g)	Values 2 days after cessation of treatment (mg/g)
1	117.8	75	77.3	17.8
2	102.2	71.1	71	82.6
3	168.9	126.7	135.6	100
4	91.1	91.1	63	51.1
5	91.1	100	66.7	54
6	37.8	41.3	45.3	68.9
7	137.8	80	82.7	93.4
8	62.2	108.9	112.6	88.9
9	97.8	122.2	108.9	93.3
10	77.8	117.8	97.8	33.3

LSD (Least standard deviation) at $p = 0.05$ level =20.3963. No two groups are significantly different at $p = 0.05$.
Source: Ahmed, R.H. 2002. Effect of Gum Arabic on the Fecal Bacterial Mass in Healthy Human Volunteers. (MSc. Thesis), University of Khartoum, Khartoum, Sudan.

Table 25.2 Percentages of changes in the fecal bacterial mass at different stages of treatment with 50 g gum Arabic in 10 female volunteers

Volunteers	Percentage of change from base value at 2 days of treatment	Percentage of change from base value from 2-day value to 5-day treatment	Percentage of change from 5-day value to 2-day after cessation of treatment	Percentage of change from base value to 2 days after cessation of treatment
1	−36.3	3	−76.9	−84.8
2	−30.4	−4.5	16.3	−90.1
3	−24.9	7	−26.2	−5.2
4	0	−30.8	−18.8	−43.9
5	9.7	−33.3	−19	−40.7
6	9.2	9.6	52	52
7	−41.9	3.3	12.9	−32.2
8	75	3.3	−21	42.9
9	24.9	−10.8	−14.3	−4.6
10	51.4	−16.9	−65.9	−57.1

The minus sign (−) designates decrease in the percentages. No sign means increase in the percentage.
Source: Ahmed, R.H. 2002. Effect of Gum Arabic on the Fecal Bacterial Mass in Healthy Human Volunteers. (MSc. Thesis), University of Khartoum, Khartoum, Sudan.

Volunteers 5, 6, 8, 9, and 10 showed increase with the following percentages: 9.7, 9.2. 75, 24.9, and 51.4.Volunteer 4 showed no change (Ahmed, 2002).

Volunteers 1, 3, 4, 5, 8, 9, and 10 showed decrease in FBM with the following percentages: 76.9, 26.2, 18.8, 19, 21, 14.3, and 65.9, respectively. Volunteers 2, 6, and 7 showed increase in FBM with the following percentages: 16.3, 52, and 12.9, respectively, when compared with the sample examined after 5 days during the treatment. When compared with the basal values (prior to treatment), volunteers 1, 2, 3, 4, 5, 7, 9, and 10 showed decrease in FBM with percentages of 84.8, 90.1, 5.2. 43.9, 40.7, 32.2, 4.6, and 57.1, respectively. Volunteers 6 and 8 showed increase in FBM with 52% and 21%, respectively (Table 25.2).

Cherbut et al. (2003) suggested that consumption of gum raised the proportions of lactic acid bacteria and *Bifidobacteria* in healthy subjects. It was fermented slowly and produced SCFA, and its fecal digestibility was around 95%. These authors reported that it increased stool output by augmenting the water content of stools.

The proportion of the fecal flora able to degrade GA polymer rose from an initial level of 6.5 to more than 50% during GA ingestion period. Moreover, findings of Ahmed (2002) also agreed with those obtained by Bliss et al. (1996) in their experiment conducted on 16 chronic renal failure patients. The patients received GA doses for a period of 4 weeks. Bliss and his colleagues showed that FBM and fecal nitrogen contents were significantly increased during supplementation with pectin.

Increase in the FBM immediately (2 days after dosing) in some volunteers suggests that intestinal flora respond quickly to GA and showed increase accordingly. In those volunteers who showed decrease firstly (2 days after dosing) and then they showed increase (at 5 days of treatment), intestinal flora might have needed an adaptation period for the new diet. After cessation of GA treatment, most of the volunteers showed decrease in their FBM compared with their 5 days' value; this decrease range between 70% in the female group and 62.5% in the male group (Ahmed, 2002).

Ahmed (2002) found that the fecal bacterial count (FBC) results showed that there was an increase in 60% of female group at 2 days of treatment with GA (Table 25.3), while 40% showed an increase at 5 days of treatment when compared with their values at 2 days. In the male group, 25% of the volunteers showed increase in their FBC at two days of treatment, and 37.5% showed an increase in their FBC at 5 days of the treatment when compared with their values at 2 days of treatment. Two days after cessation GA treatment, 70% of the female group and 87.5% of the male group showed decreases in their FBC when compared with their values at 5 days (Table 25.4). Mohamed et al. (2014) found that in both healthy and patient volunteers, there was a significant increase in aerobic and anaerobic FBC during the GA feeding, a decrease in the count after cessation of GA treatment, and the increase was obvious after the fifth day of dosing.

Ahmed (2002) suggested that there is no variation between the males and females in response to GA dosing. They said that their findings of inconsistent results of FBC compared with the fecal bacterial mass might be attributed to what was found by other authors, who found that the principal GA fermenters are species of *Bacteriodes* and *Bifidobacterium*, which are anaerobic bacteria. The results of Ahmed (2002) generally agreed with the suggestions and findings of previous investigators. Shortness of the dosing period, individual variations between volunteers in their gut microflora, and absence of preparatory period all contribute to inconsistent results of some volunteers. The dosing period was short because it was difficult to persuade more than one person to take doses of the gum for a longer period, as they were in no need for this regimen, and they were not expecting benefit from it, as in the case of chronic renal patients. For the same reason, there was no preparatory period during which all volunteers should

Table 25.3 Fecal bacterial mass in eight apparently healthy male volunteers given 50 g gum Arabic for 6 consecutive days

Volunteers	Base values (before treatment) (mg/g)	Values 2 days after treatment (mg/g)	Values 5 days after treatment (mg/g)	Values 2 days after cessation of treatment (mg/g)
1	57.8	55.6	78.7	68.9
2	37.6	272	77.8	91.1
3	82.22	44.4	36.6	88.9
4	84.3	57.8	26.7	42.2
5	91.1	97.8	100	75.6
6	70	75.6	104.4	90.7
7	102.2	95.6	124	88.9
8	137.2	128.2	160	95

*LSD (Least standard deviation)at $p = 0.05$ level = 40.7816 **No two groups are significantly different at $p = 0.05$.
Source: Ahmed, R.H. 2002. Effect of Gum Arabic on the Fecal Bacterial Mass in Healthy Human Volunteers. (MSc. Thesis), University of Khartoum, Khartoum, Sudan.

Table 25.4 Percentages of changes in the fecal bacterial mass at different stages of treatment with 50 g gum Arabic in 8 male volunteers

Volunteers	Percentage of change from base value to 2-day treatment	Percentage of change from base value from 2-day value to 5-day treatment	Percentage of change from 5-day value to 2-day value after cessation of treatment	Percentage of change from base value to 2 days after cessation of treatment
1	−3.8	5.5	−12.5	19.2
2	269.5	−71.6	17.1	23.7
3	−45.9	−17.5	145.2	2.4
4	−31.4	−53.8	58	−49.9
5	7.3	2.2	−24.4	−1.7
6	8	38	−31.1	29.5
7	−6.4	29.7	−28.3	−13
8	−6.1	24.2	−40.6	−69

The minus sign (−) designates decrease in the percentage. No sign means an increase in the percentage.
Source: Ahmed, R.H. 2002. Effect of Gum Arabic on the Fecal Bacterial Mass in Healthy Human Volunteers. (MSc. Thesis), University of Khartoum, Khartoum, Sudan.

have had taken the same diet. Statistical analysis of the results by analysis of variance gave the insignificant increase. This is because of the great variations of the microbial load in the in the gut of the individuals, as they had not had the same history and did not live in the same conditions. To conclude with these results, we can say that GA dosing increases the fecal bacterial mass, which indicates an increase in the number of gut bacterial flora that ferments GA. The mechanism by which GA reduces urea nitrogen is attributed to the role of this flora. It is suggested that intestinal bacteria ferment dietary fiber, which provide them nitrogen incorporation, the net result of which is increased in the fecal bacterial mass, increase of fecal nitrogen excretion, and subsequently lower serum urea nitrogen.

For these reasons, supplementation of the diet with GA can be part of the treatment for uremic patients to decrease the workload on the kidneys. However, more detailed studies on the effects of GA and its precise action for the treatment of chronic renal failure patients are still lacking.

In conclusion, investigations needed to be made to correlate the fecal bacterial mass and the rate of increase and decrease of urea, creatinine, uric acid, and sodium and potassium ions.

References

Ahmed, R.H. 2002. Effect of Gum Arabic on the Fecal Bacterial Mass in Healthy Human Volunteers. (MSc. Thesis), University of Khartoum, Khartoum, Sudan.

Ali, B.H., Al-Qarawi, A.A., Haroun, E.M., Mousa, H.M., 2003. The effect of treatment with gum Arabic on gentamicin nephrotoxicity in rats: a preliminary study. Ren. Fail. 25 (1), 15–20.

Alkarib, S.Y, Saeed, A.M., Khalid, S.A., Groun, E.A., Ghalib, M.B., 2016. Gum Arabic role in high blood pressure among patients with stage III Chronic Kidney Disease. J. Pharmacol. Clin. Res. 1 (4), 555–569.

Al-Majed, A.A., Mostafa, A.M., Al-Rikabi, A.C., Al-Shabanah, O.A., 2002. Protective effects of oral Arabic gum administration on gentamicin-induced nephrotoxicity in rats. Pharmacol. Res. 46 (5), 445–451.

Babiker, R., Merghani, T.H., Elmusharaf, K., Badi, R.M., Lang, F., Saeed, A.M., 2012. Effects of gum Arabic ingestion on body mass index and body fat percentage in healthy adult females: two-arm randomized, placebo controlled, double-blind trial. Nutr. J. 11, 111.

Bliss, D.Z., Stein, T.P., Schleifer, C.R., Settle, R.G., 1996. Supplementation with gum Arabic fiber increases fecal nitrogen excretion and lowers serum urea nitrogen concentration in chronic renal failure patients consuming a low protein diet. Am. J. Clin. Nutr. 6, 392–398.

Calame, W., Weseler, A.R., Viebke, C., Flynn, C., Siemensma, A.D., 2008. Gum Arabic establishes prebiotic functionality in healthy human volunteers in a dose-dependent manner. Br. J. Nutr. 100 (6), 1269–1275.

Cherbut, C., Michel, C., Raison, V., Kravtchenko, T., Severin, M., 2003. Acacia gum is a bifidogenic dietary fiber with high digestive tolerance in healthy humans. Microb. Ecol. Health Dis. 15, 43–50.

Cresci, A., Orpianesi, C., Silvi, S., Mastrandrea, V., Dolara, P., 1999. The effect of sucrose or starch-based diet on short-chain fatty acids and faecal microflora in rats. J. Appl. Microbiol. 86 (2), 245–250.

Kuroda, M., Ohta, M., Okufuji, T., Takigami, C., Eguchi, M., Hayabuchi, H., Ikeda, M., 2010. Frequency of soup intake and amount of dietary fiber intake are inversely associated with plasma leptin concentrations in Japanese adults. Appetite 54 (3), 538–543.

Mohamed, S.S., El Sanousi, S.M., Sulieman, S.M., 2014. The effect of gum Arabic on colonic bacteria in healthy volunteers and patients with chronic renal failure on heamodialysis. J. Sci. Res. 6 (1), 170–178.

Murakami, K., Sasaki, S., Takahashi, Y., Uenishi, K., Yamasaki, M., Hayabuchi, H., Goda, T., Oka, J., Baba, K., Ohki, K., Watanabe, R., Sugiyama, Y., 2007. Nutrient and food intake in relation to serum leptin concentration among young Japanese women. Nutrition 23 (6), 461–468.

Nasir, O., Umbach, A.T., Rexhepaj, R., Ackermann, T.F., Bhandaru, M., Ebrahim, A., Artunc, F., Kempe, D.S., Puchchakayala, G., Siraskar, B., Föller, M., Saeed, A., Lang, F., 2012. Effects of gum Arabic (*Acacia senegal*) on renal function in diabetic mice. Kidney Blood Press. Res. 35 (5), 365–372.

Nasir, O., Wang, K., Föller, M., 2010. Downregulation of angiogenin transcript levels and inhibition of colonic carcinoma by gum Arabic (*Acacia senegal*). Nutr. Cancer 62 (6), 802–810.

Phillips, A.O., Phillips, G.O., 2011. Biofunctional behaviour and health benefits of a specific gum Arabic. Food Hydrocoll. 25 (2), 165–169.

Phillips, G.O., 1998. Acacia gum (gum Arabic): a nutritional fibre; metabolism and calorific value. Food Addit. Contam. 15 (3), 251–264.

Phillips, G.O., Ogasawara, T., Ushida, K., 2008. The regulatory and scientific approach to defining gum Arabic (Acacia senegal and Acacia seyal) as a dietary fiber. Food Hydrocoll. 22, 24–35.

Gastroprotective Activity of Gum Arabic: A Review

26

Suzy Munir Salama, Abdalbasit Adam Mariod**,†*
*University of Malaya, Kuala Lumpur, Malaysia; **University of Jeddah, Alkamil, Kingdom of Saudi Arabia; †Sudan University of Science & Technology, Khartoum, Sudan

Introduction

Gastric ulcer is part of the peptic ulcer disease that has threatened the global population in the past two eras, subjecting human life to high mortality (Malfertheiner et al., 2009). The prevalence of the disease ranges geographically and is linked to a contemporary style of living (Salama et al., 2016). However, the incidence of peptic ulcer started to decline seriously in the late 20th century, which may be due to the improvement of the therapeutic regimens that focus on suppressing gastric acid secretion and treatment of *Helicobacter pylori* infection (Ali, 2013; Malfertheiner et al., 2009).

The pathogenesis of gastric ulcer, as illustrated in Fig. 26.1, is a multifactor disease caused by a disproportion between the offensive factors destroying the gastric mucosa and the protective agents of the stomach wall (Sidahmed et al., 2013). Ulceration of the stomach starts with mucosal lesions (Qader et al., 2012) and ends with gastric cancer (Pope et al., 2017). Many factors can increase the incidence of gastric ulcer, including *H. pylori* infection, nonsteroidal antiinflammatory drugs (NSAIDs), psychological stress, and noxious agents such as ethanol, acidified ethanol, indomethacin, and acetic acid (Repetto and Llesuy, 2002). In the persistence of these factors, along with increased gastric acid secretion and suppressed mucous secretion, gastric injury takes place (Nordin et al., 2014).

Traditionally, the use of gum Arabic refers to the Early Dynastic Period (3400 BC). It is used by native Africans in preparing food as well as human and animal health. The different parts of acacia trees were used for many centuries in treating various health problems related to the alimentary canal, such as diarrhea, typhoid (Mahomoodally, 2013), and intestinal inflammation (Gamal El-din et al., 2003). In Iranian traditional medicine, gum Arabic is used as a gastric tonic in treating peptic ulcer (Farzaei, 2013).

In this review, we will display the outcomes of the earlier and recent studies conducted on gum Arabic, evaluating its gastroprotective activity and the trials of researchers to elucidate the mechanism of gum Arabic in protecting the stomach against ulceration.

Gum Arabic. http://dx.doi.org/10.1016/B978-0-12-812002-6.00026-9

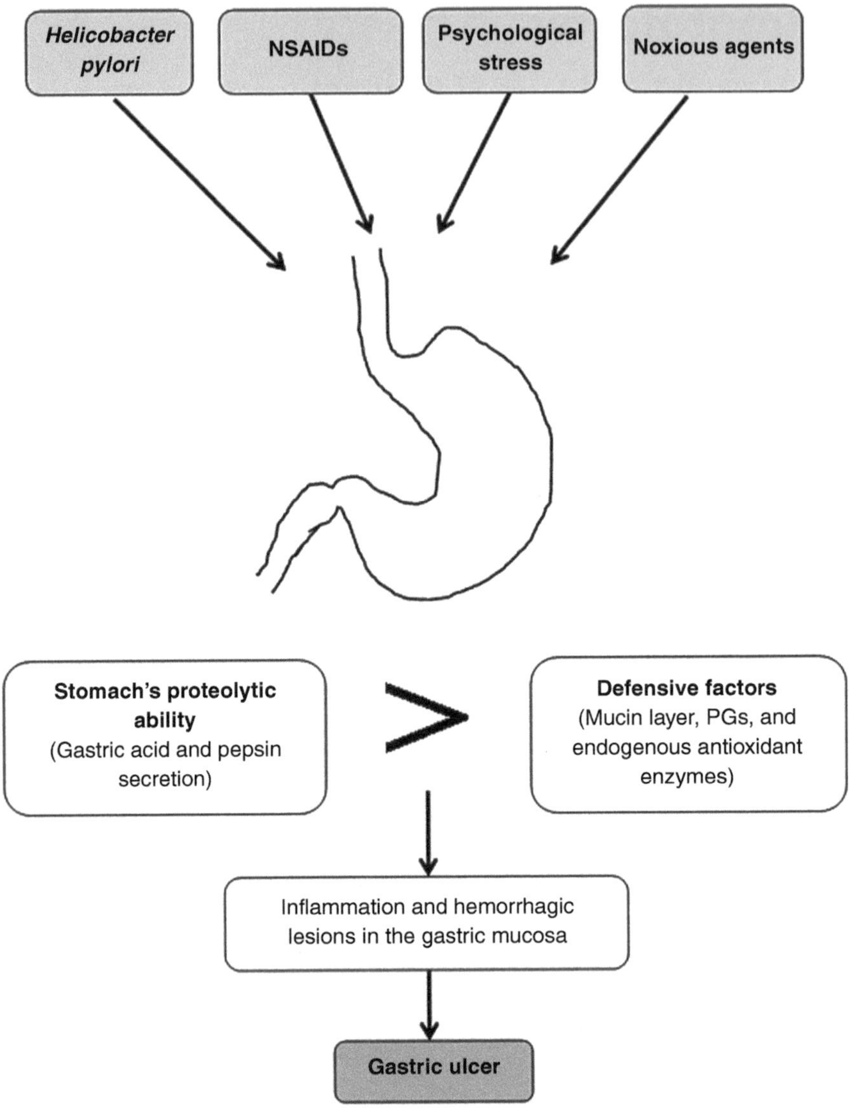

Figure 26.1 Pathogenesis of gastric ulcer. *NSAIDs*, Nonsteroidal antiinflammatory drugs; *PGs*, prostaglandins.

Gastroprotective activity of gum Arabic against insults

Ethanol-induced gastric ulcer

The mechanism of ethanol-induced gastric ulcer was explained by different hypotheses. Terano et al. (1989) clarified the ethanol-induced gastritis via oxidative stress, causing perforation of the mitochondria and gastric cellular necrosis (Terano et al., 1989). Brzozowski et al. (1998) related the gastric mucosal damage induced by ethanol to the increased permeability of gastric cell membranes to water and sodium following intragastric ingestion of ethanol (Brzozowski et al., 1998). Araki et al. (2000) attributed the lesions caused by ethanol to the constriction of mucosal blood vessels when ethanol comes in contact with the stomach walls, leading to inflammation and gastric injury (Araki et al., 2000).

Previous studies showed that oral gavage of 500 mg gum Arabic/kg to Wistar rats has reduced the ulcer index by 15% in acute gastric ulcer model (Rodrigues et al., 2009). Recent studies showed that oral administration of gum Arabic at doses of 500 mg/kg and 1000 mg/kg exhibit antiulcer activity in rats ulcerated by ethanol. Scientists explained the efficacy of gum Arabic against ethanol gastritis in terms of the cytoprotective potency of gum Arabic to gastric cells as well as its free radical scavenging capacity (AL-Yahya and Asad, 2016; Pal et al., 2012).

NSAID-induced gastric ulcer

Indomethacin and aspirin as NSAIDs are well known for their impact in suppressing inflammation due to arthritis and enhancing cardiovascular protection. On the other hand, they have been reported for their adverse effects in inducing lesions of the stomach walls if administered for the long term via their inhibitory impact to cyclogenases as well as consequential prostaglandin and mucus secretion (Nasif et al., 2011). Further, NSAIDs were claimed to inhibit mitochondrial oxidative phosphorylation, leading to the release of cytochrome C and initiating gastric cell death (Matsui et al., 2011). Indomethacin was found to increase the generation of leukotrienes and reduce mucosal blood circulation, leading to injury of the mucosal lining of the stomach (Gohar and Zaki, 2014). In vivo studies showed that gastric lesions are strongly linked to uncoupling of mitochondrial phosphorylation from adenosine triphosphate formation (Nasif et al., 2011; Robins, 1980). Another hypothesis strongly relates NSAID-induced gastric injury to the adherence of neutrophils to the microcirculation of stomach walls, generating reactive oxygen species (ROS), which result in blocking blood capillaries (Malfertheiner et al., 2009).

In animal models, it was reported that an intragastric dose of 20 mg/kg of gum Arabic to the animals revealed moderate protection against indomethacin-induced gastritis (Gohar and Zaki, 2014). In another recent study, two doses of gum Arabic (500 and 1000 mg/kg) were found to exhibit antiulcer potentiality against acute indomethacin-associated ulceration in rats (AL-Yahya and Asad, 2016). In the same approach, treatment of rats with aspirin followed by gum Arabic (1 g/kg/day) for 21 days revealed remarkable preservation of gastric mucosa (Nasif et al., 2011). The role of gum Arabic in preventing gastric injury may be explained in terms of the antioxidant capacity of its constituents in protecting the cells from the injurious effect of ROS (AL-Yahya and Asad, 2016).

Stress-induced gastric ulcer

Stress induces gastric ulcer via constriction of arteries, augmented contraction of gastric smooth muscles, and motility, resulting in ischemia. This status releases ROS, developing oxidative stress, which generates apoptotic mediators and death of gastric cells (Helal et al., 2011). Additionally, stress releases histamine, which induces gastric acid secretion, exposing the gastric mucosa to erosion and ulceration (AL-Yahya and Asad, 2016). Pylorus ligation as an example of stress was found to play crucial role in inducing gastritis through oversecretion of gastric acid in response to ligation (Håkanson et al., 1980).

Laboratory experiments manifested that oral administration of gum Arabic at a dose of 7.5 g/kg daily for 10 days before exposure of rats to cold stress has protected the animals' stomachs from injury. Researchers accredited the potentiality of gum Arabic in treating ischemic gastritis to the high constituents of polysaccharides, mainly arabinogalactan (Cipriani et al., 2009; Dror et al., 2006; Helal et al., 2011). Furthermore, intraduodenal supplementation of rats with 500 and 1000 mg gum Arabic in water/kg directly after pylorus ligation inhibited ulceration of the stomach via reduction of gastric acid production and elevation of mucin generation (AL-Yahya and Asad, 2016).

The role of gum Arabic in protecting the stomach from gastric lesions induced by experimental insults is summarized in Fig. 26.2.

Gum Arabic and gastric cancer

Angiogenin is a well-known protein that plays a pivotal role in tumor metastasis, stimulating the formation of new blood vessels via the process of angiogenesis in

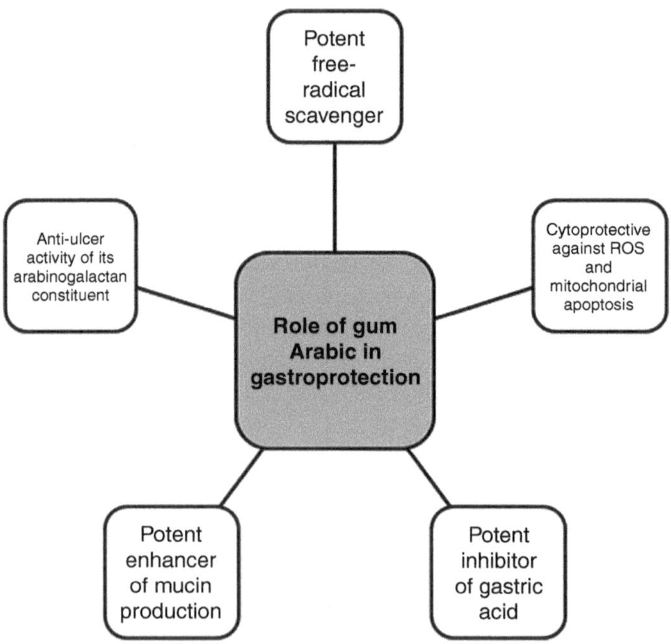

Figure 26.2 Role of gum Arabic in gastroprotection against gastric injury.

various cancerous diseases, including gastric cancer (Chatzileontiadou et al., 2016; Shimoyama and Kaminishi, 2000). Angiogenic inhibitors were recently reported for their therapeutic functionality in treating stomach tumor (Blank et al., 2015; Tu et al., 2003). Based on laboratory experimental animal models, it was claimed that oral administration of gum Arabic could significantly suppress the expression of angiogenin protein linked to chemical-induced colon carcinoma (Nasir et al., 2010).

Chemistry studies stated that gum Arabic contains mainly polysaccharides and a small percentage of protein–polysaccharide complex (Dror et al., 2006). Recently, in vitro, in vivo, and clinical researches noted that polysaccharides and protein–polysaccharide complexes isolated from some mushrooms exhibited immunomodulatory efficacy and were able to target tumors of the stomach (Cao et al., 2015; Ghosh, 2015; Li et al., 2014). In this context, more studies are required to focus on the antigastric tumor activity of gum Arabic.

Toxicity of gum Arabic

Earlier studies on the subchronic toxicity of gum Arabic revealed that the no-observed adverse effect level was tested to be 5.2 g/kg in male and 13.8 g/kg in female Wistar rats when supplemented daily to their diets (Anderson et al., 1982). Another subchronic designed toxicity rat model showed that the safe dose of the naturally processed SUPERGUM form of gum Arabic extracted from *Acacia senegal* tree was approximately 3000 mg/kg for both sexes of F344 rats (Doi et al., 2006). Additionally, administration of gum Arabic at doses of 5–40 g/day in drinking water of human volunteers for a month revealed safe doses, while 10 g/day was optimum. Moreover, gum Arabic at the mentioned doses offered a convenient environment to the bacterial flora examined (Calame et al., 2008).

Based on the Food and Agriculture Organization of the United Nations and World Health Organization, the use of gum Arabic was approved as a food additive (Kennedy et al., 2011). The US Food and Drug Administration (1973) had confirmed the safety of use of gum Arabic as a food supplement and generally regarded as safe. Moreover, the European Union (E414) and Codex Alimentarius Committee (INS414) have endorsed the addition of gum Arabic to food (Glover, 2012).

In conclusion, recent studies suggested gum Arabic as an effective natural antiulcer candidate against different insults including ethanol, aspirin, indomethacin, cold stress, and pylorus ligation. Gastroprotection of gum Arabic accounts for its antioxidant capacity, protecting the gastric cell from necrosis, inhibitory activity to the gastric acid secretion, and enhancement of mucous secretion. Additionally, the high polysaccharide constituents of gum Arabic experimentally improved its potency in protecting the stomach against injury. However, more studies are required to elucidate the detailed mechanism of gastroprotection of gum Arabic.

References

Ali, M.Z., 2013. Peptic ulcer disease. J. Khwaja Yunus Ali Med. Coll. 3, 270–272.

AL-Yahya, A.A., Asad, M., 2016. Antiulcer activity of gum arabic and its interaction with antiulcer effect of ranitidine in rats. Biomed. Res. 27, 1102–1106.

Anderson, D., Ashby, P., Busuttil, A., Eastwood, M., Hobson, B., Ross, A.M., Street, C., 1982. Subchronic effects of gum arabic (Acacia) in the rat. Toxicol. Lett. 14, 221–227.

Araki, H., Ukawa, H., Sugawa, Y., Yagi, K., Suzuki, K., Takeuchi, K., 2000. The roles of prostaglandin E receptor subtypes in the cytoprotective action of prostaglandin E2 in rat stomach. Aliment. Pharmacol. Ther. 14, 116–124.

Blank, S., Deck, C., Dreikhausen, L., Weichert, W., Giese, N., Falk, C., Schmidt, T., Ott, K., 2015. Angiogenic and growth factors in gastric cancer. J. Surg. Res. 194, 420–429.

Brzozowski, T., Konturek, P.C., Konturek, S.J., Kwiecien, S., Pajdo, R., Brzozowska, I., Hahn, E.G., 1998. Involvement of endogenous cholecystokinin and somatostatin in gastroprotection induced by intraduodenal fat. J. Clin. Gastroenterol. 27, S125–S137.

Calame, W., Weseler, A.R., Viebke, C., Flynn, C., Siemensma, A.D., 2008. Gum arabic establishes prebiotic functionality in healthy human volunteers in a dose-dependent manner. Br. J. Nutr. 100, 1269–1275.

Cao, X.Y., Liu, J.L., Yang, W., Hou, X., Li, Q.J., 2015. Antitumor activity of polysaccharide extracted from *Pleurotus ostreatus* mycelia against gastric cancer in vitro and in vivo. Mol. Med. Rep. 12, 2383–2389.

Chatzileontiadou, D.S., Tsirkone, V.G., Dossi, K., Kassouni, A.G., Liggri, P.G., Kantsadi, A.L., Stravodimos, G.A., Balatsos, N.A., Skamnaki, V.T., Leonidas, D.D., 2016. The ammonium sulfate inhibition of human angiogenin. FEBS Lett. 590, 3005–3018.

Cipriani, T.R., Mellinger, C.G., Bertolini, M.L., Baggio, C.H., Freitas, C.S., Marques, M.C.A., Gorin, P.A., Sassaki, G.L., Iacomini, M., 2009. Gastroprotective effect of a type I arabinogalactan from soybean meal. Food Chem. 115, 687–690.

Doi, Y., Ichihara, T., Hagiwara, A., Imai, N., Tamano, S., Orikoshi, H., Ogasawara, K., Sasaki, Y., Nakamura, M., Shirai, T., 2006. A ninety-day oral toxicity study of a new type of processed gum arabic, from Acacia tree (*Acacia senegal*) exudates, in F344 rats. Food Chem. Toxicol. 44, 560–566.

Dror, Y., Cohen, Y., Yerushalmi-Rozen, R., 2006. Structure of gum arabic in aqueous solution. J. Polym. Sci. B Polym. Phys. 44, 3265–3271.

Farzaei, M.H., 2013. Peptic Ulcer in Traditional Iranian Medicine Mohammad Hosein Farzaei, "Roja Rahimi," Zahra Abbasabadi, "Mohammad Abdollahi" Department of Traditional Pharmacy, Faculty of Traditional Medicine. Int. J. Pharmacol. 9, 108–124.

Gamal El-din, A.M., Mostafa, A.M., AL-shabanah, O.A., AL-Bekairi, A.M., Nagi, M.N., 2003. Protective effect of arabic gum against acetaminophen-induced hepatotoxicity in mice. Pharmacol. Res. 48, 631–635.

Ghosh, K., 2015. Review mushrooms: a source of immunomodulating and antitumor polysaccharides. J. Phys. Sci. 15, 239–252.

Glover, D.A., 2012. The Effects of Dietary Supplementation with GUM Arabic on Blood Pressure and Renal Function in Subjects With Type 2 Diabetes Mellitus. Cardiff University.

Gohar, A.A., Zaki, A.A., 2014. Assessment of some herbal drugs for prophylaxis of peptic ulcer. Iran. J. Pharm. Res. 13, 1081–1086.

Håkanson, R., Hedenbro, J., Liedberg, G., Sundler, F., Vallgren, S., 1980. Mechanisms of gastric acid secretion after pylorus and oesophagus ligation in the rat. J. Physiol. 305, 139–149.

Helal, O.K., Yousef, M.M., ELnaa, M., 2011. Possible protective effect of gum arabic on experimentally induced gastric ulcer in adult male albino rats: a histological and immunohistochemical study. Egypt. J. Histol. 34, 546–553.

Kennedy, J.F., Phillips, G.O., Williams, P.A., 2011. Gum Arabic. Royal Society of Chemistry.

Li, C., Tian, Z.-N., Cai, J.-P., Chen, K.-X., Zhang, B., Feng, M.-Y., Shi, Q.-T., Li, R., Qin, Y., Geng, J.-S., 2014. *Panax ginseng* polysaccharide induces apoptosis by targeting Twist/AKR1C2/NF-1 pathway in human gastric cancer. Carbohydr. Polym. 102, 103–109.

Mahomoodally, M.F., 2013. Traditional medicines in Africa: an appraisal of ten potent African medicinal plants. Evid. Based Complement Altern. Med. 2013, 14.

Malfertheiner, P., Chan, F.K., Mccoll, K.E., 2009. Peptic ulcer disease. Lancet 374, 1449–1461.

Matsui, H., Shimokawa, O., Kaneko, T., Nagano, Y., Rai, K., Hyodo, I., 2011. The pathophysiology of non-steroidal anti-inflammatory drug (NSAID)-induced mucosal injuries in stomach and small intestine. J. Clin. Biochem. Nutr. 48, 107–111.

Nasif, W., Lotfy, M., Mahmoud, M., 2011. Protective effect of gum acacia against the aspirin induced intestinal and pancreatic alterations. Eur. Rev. Med. Pharmacol. Sci. 15, 285–292.

Nasir, O., Wang, K., Föller, M., Bhandaru, M., Sandulache, D., Artunc, F., Ackermann, T.F., Ebrahim, A., Palmada, M., Klingel, K., 2010. Downregulation of angiogenin transcript levels and inhibition of colonic carcinoma by Gum Arabic (*Acacia senegal*). Nutr. Cancer 62, 802–810.

Nordin, N., Salama, S.M., Golbabapour, S., Hajrezaie, M., Hassandarvish, P., Kamalidehghan, B., Majid, N.A., Hashim, N.M., Omar, H., Fadaienasab, M., 2014. Anti-ulcerogenic effect of methanolic extracts from *Enicosanthellum pulchrum* (King) Heusden against ethanol-induced acute gastric lesion in animal models. PLoS One 9, e111925.

Pal, R., Hooda, M., Bhandari, A., Singh, J., 2012. Antioxidant potential and free radicals scavenging activity by pod extracts of *Acacia senegal* willd. Int. J. Pharm. Chem. Biol. Sci. 2, 500–506.

Pope, J.L., Tomkovich, S., Yang, Y., Jobin, C., 2017. Microbiota as a mediator of cancer progression and therapy. Transl. Res. 179, 139–154.

Qader, S.W., Abdulla, M.A., Chua, L.S., Sirat, H.M., Hamdan, S., 2012. Pharmacological mechanisms underlying gastroprotective activities of the fractions obtained from *Polygonum minus* in sprague dawley rats. Int. J. Mol. Sci. 13, 1481–1496.

Repetto, M., Llesuy, S., 2002. Antioxidant properties of natural compounds used in popular medicine for gastric ulcers. Braz. J. Med. Biol. Res. 35, 523–534.

Robins, P., 1980. Ultrastructural observations on the pathogenesis of aspirin-induced gastric erosions. Br. J. Exp. Pathol. 61, 497–504.

Rodrigues, R., Jorge, M., Sousa, I.D.O., Lima, J., Ruiz, A., Carvalho, J., Figueira, G., Foglio, M., 2009. Pharmacological activity from Arrabidaea Chica Verlot bioencapsulated with arabic gum. XVIIth International Conference on Bioencapsulation, Groningen, Netherlandspp. 24–26.

Salama, S.M., Gwaram, N.S., Alrashdi, A.S., Khalifa, S.A., Abdulla, M.A., Ali, H.M., EL-seedi, H.R., 2016. A zinc morpholine complex prevents HCl/ethanol-induced gastric ulcers in a rat model. Sci. Rep. 6doi: 10.1038/srep29646.

Shimoyama, S., Kaminishi, M., 2000. Increased angiogenin expression in gastric cancer correlated with cancer progression. J. Cancer Res. Clin. Oncol. 126, 468–474.

Sidahmed, H., Abdelwahab, S.I., Mohan, S., Abdulla, M.A., Taha, M.M.E., Hashim, N.M., Hadi, A.H.A., Vadivelu, J., Loke Fai, M., Rahmani, M., 2013. α-Mangostin from Cratoxylum arborescens (Vahl) Blume demonstrates anti-ulcerogenic property: a mechanistic study. Evid. Based Complement. Altern. Med. 2013doi: 10.1155/2013/450840.

Terano, A., Hiraishi, H., Ota, S.-I., Shiga, J., Sugimoto, T., 1989. Role of superoxide and hydroxyl radicals in rat gastric mucosal injury induced by ethanol. Gastroenterol. Jpn. 24, 488–493.

Tu, S.P., Jiang, X.H., Lin, M.C., Cui, J.T., Yang, Y., Lum, C.T., Zou, B., Zhu, Y.B., Jiang, S.H., Wong, W.M., 2003. Suppression of survivin expression inhibits in vivo tumorigenicity and angiogenesis in gastric cancer. Cancer Res. 63, 7724–7732.

Index

H

I

U

V

W

X

FSC
www.fsc.org
MIX
Papier | Fördert
gute Waldnutzung
FSC® C083411

Kolibri 360 GmbH
Ferdinand-Jühlke-Straße 7
99095 Erfurt, Deutschland
produktsicherheit@kolibri360.de